P9-DBL-333

THE SEVENTY GREAT MYSTERIES OF THE NATURAL WORLD

THE SEVENTY GREAT MYSTERIES OF THE NATURAL WORLD

Edited by Michael J. Benton

with 368 illustrations, 338 in color

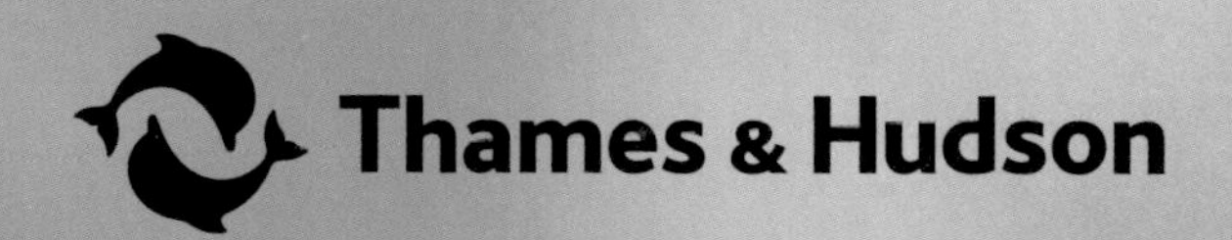

Contents

First published in 2008 in hardcover in the United States of America by Thames & Hudson Inc., 500 Fifth Avenue, New York, New York 10110

thamesandhudsonusa.com

Library of Congress Catalog Card Number 2008901129

ISBN 978-0-500-25143-0

Designed by Christopher Perkins

Printed and bound in China by Toppan Printing

Half-title *The 'Blue Marble', a NASA image of the Earth from space.*

Title page *Parson's chameleon* (Chamaeleo parsonii) *poised on a branch.*

How planets form: an artist's reconstruction of a protoplanetary disc.

Origins

The Earth

Close-up of a fly's compound eye.

Lava streams into the sea at Poupou, Hawaii.

Evolution

Biogeography & Environments

Saguaro cactus plants in the Sonoran Desert, Arizona.

Biodiversity: a golden tortoise beetle from the rainforest in Ecuador.

Plants & Animals

Animal Behaviour

Two red deer stags lock antlers.

Industrial pollution, Norilsk, Russia.

Global Warming & the Future

Understanding the Earth

Now there is one outstandingly important fact regarding Spaceship Earth, and that is that no instruction book came with it.
R. BUCKMINSTER-FULLER, 1963

From the earliest times people have been fascinated by the Earth and by Nature. Around 30,000 years ago humans began to paint images of the prey they hunted and the animals they feared, and presumably looked to wise men and women for advice on where the best herds might be found. The wonder of the returning seasons has also always encouraged endless speculation and concern. Subsistence farmers, then as now, worry whether the rains will come again, and have constructed elaborate religious rituals to ensure the success of their crops. In modern society, Nature is perhaps not so mysterious, and yet the reverence is still there. No one could fail to marvel at a mountain or a sunset, a giant redwood, a whale, a dinosaur fossil.

The fitful growth of science – first with the philosophers of Classical Greece, then, after a long hiatus, from the Renaissance onwards – has furnished us with a coherent framework within which to understand the natural world. We know that summer will return, and why. We know many

Right *This fossil of an ammonite – an extinct marine creature – found in Antarctica is evidence that there has not always been ice at the poles.*

Opposite *The 'dumbo' octopus, so called because of the large ear-like projections, is one of the many new species being disovered in the deep oceans: three new species were named in 2003.*

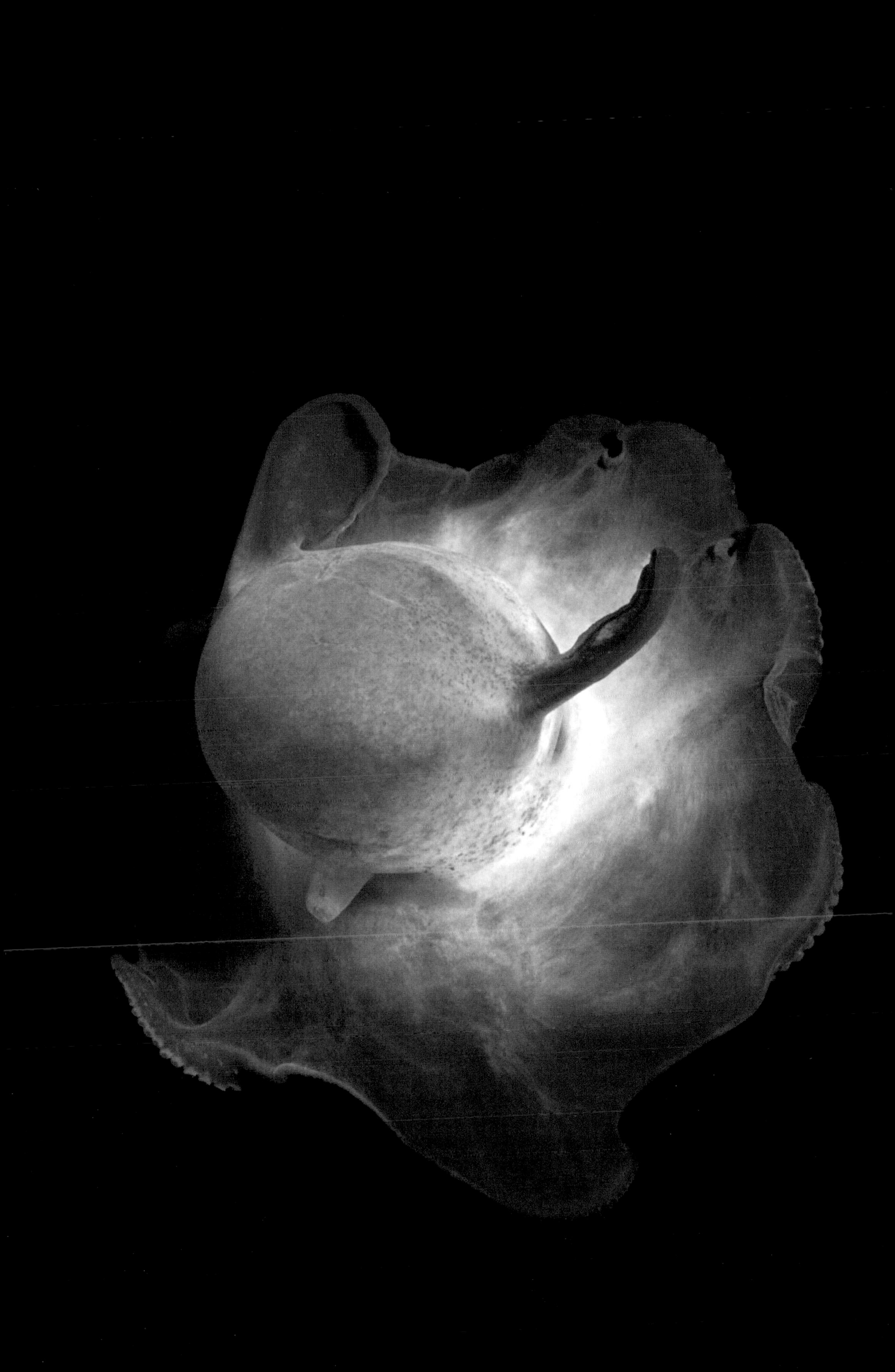

facts about dinosaurs – how long ago they lived and even something about why they died out. The growth of a modern understanding of the Earth and of Nature, really only in the past two centuries, has been one of the most astonishing achievements of humanity. In this book, we explore all the big questions – those that we now have a better knowledge of, and also some of the huge number that are still to be properly understood. Here, these vital themes are explored by over 60 of the finest science writers of today; all are experts from universities and government laboratories around the world.

In the first section we look at the origins of the Earth itself, of the life on it, and of complex animals and humans. The search for origins has occupied peoples of all cultures for millennia, and, despite intense study, many mysteries have still not been resolved. Next, 'The Earth' covers our world and some of the key questions in geology, exploring the internal structure of our planet, and why we have oceans and landmasses. Sudden events, such as volcanic eruptions, earthquakes, tsunamis and meteorite impacts are now partially understood, but much research remains to be done. Conjectures about the height and shape of mountains relate back to plate tectonics.

'Evolution' is an extraordinary topic in the natural sciences: so important and yet so misunderstood. We look at Darwin's theories, and how his views permeate all of modern biology. Yet it is also important to try to understand why the subject has proved to be so controversial. Genes are at the heart of evolution – they are fundamental to explaining why each species of plant and animal looks the way it does, and to explaining why there are now so many species on Earth – and yet Darwin knew nothing about them.

The distribution of plants and animals, which we examine in the section on 'Biogeography & Environments', is one of the marvels of our Earth. We look at why Australia has such different

An Apatosaurus *skeleton at the Carnegie Museum of Natural History, in Pittsburgh, measuring over 23 m (77 ft) long. This photograph emphasizes the huge size of the dinosaurs, and highlights key biological questions about how such enormous animals survived and were so successful.*

Every year millions of monarch butterflies fly from North America to Mexico – but many have never made the journey before, so how do they know where to go?

wildlife from the rest of the world, how the poles and deserts have changed over time, the extreme conditions in which life may be found, and how geography has been fundamental in evolution.

The sheer diversity of 'Plants & Animals' and their extraordinary modes of life are a constant source of wonder and delight. There are so many species on the Earth, especially of insects, and organisms range so enormously in size, that it seems anything might be possible. And yet basic engineering principles control what plants and animals can – and cannot – do.

In the section 'Animal Behaviour', we explore the distinction between instinct and learning, and how these styles of behaviour are manifested in the simplest to the most complex animals. Behaviour depends on the senses, but also on evolution, and there are remarkable parallels between the behaviour of certain animals and that of humans.

The final section, 'Global Warming & the Future', puts scientific flesh on the bones of current debates about global warming, human overpopulation, wildlife conservation, diminishing resources and sustainability. These themes are widely discussed, and yet often misrepresented and misunderstood. Scientists debate many of these key issues, and certain points at least are agreed by all. It is important for the citizens of the world to consider their future on the basis of adequate scientific knowledge, rather than the sometimes ill-informed statements of politicians and those with particular viewpoints.

Origins

The focus of much human enquiry from ancient times has been a search to understand origins. Where did humans come from, or flowers, trees or animals? How were the Earth, the Sun, the stars created? We cannot address issues of meaning here, but the scientific search for the origins of life and the Universe are of profound interest and importance.

Many scientific and philosophical ideas about origins that were current until around 200 years ago might seem rather fanciful to us today. Since then great advances have been made in physics, chemistry, biology and earth sciences. We can now date the age of the Earth; we have examples of some of the oldest and simplest fossils close to the origin of life; and palaeontological fieldwork has revealed extraordinary ancient fossils of plants and animals, dinosaurs, and humans. However, many aspects of these topics are still heavily debated.

Much is known about the deep structure of the Earth. We have moved a long way from the best speculations of late Victorian times, when Lord Kelvin saw the Earth as a great ball of iron, molten at the core and cooling rapidly. Modern geological methods that sample the chemistry of lavas derived from deep in the Earth's mantle and use seismic waves to reconstruct the physical structure of the interior of the Earth have refined our views substantially. This knowledge allows geologists to reconstruct even the core of the Earth. Combined with studies of neighbouring planets, such research can at least direct our thinking about the origin of the Earth and the solar system, even if in numerous ways they are still mysterious.

A view of the first life? Stromatolites, here at Shark Bay, Australia, are combinations of cyanobacteria and sediment; examples have been found in some of the oldest rocks on Earth.

Bonobos are our closest living relatives. Early evolutionists such as Charles Darwin and Thomas Henry Huxley saw the anatomical evidence; molecular studies of DNA now show we share most of our genetic code with these pygmy chimpanzees.

The other great mystery is, of course, the origin of life. We all have images of the fabled primeval soup – some kind of organic broth struck by lightning and generating strange, semi-living microscopic life forms. Laboratory experiments seek to mimic these earliest stages in the transition from non-living chemicals to life, and scientists sometimes claim that they have created life, or are on the verge of doing so. Life has not yet been created in a test tube, and perhaps never will be. The complexity and difficulty of crossing the divide from non-living to living thus still preserves some of the mystery of life.

The various stages in the history of life from the simplest virus-like organisms all mark major transitions. For most of its history, life consisted of single-celled microscopic organisms. The move to more complex cells with a nucleus, as seen in eukaryotes, and then to organisms consisting of multiple cells, and even larger forms that are visible to the naked eye, can now be traced, if obscurely, and the evidence is much debated. A few rare localities in the vast spans of Precambrian time give us snapshots of what might have been going on. But these fossils can only be preserved in unusual conditions, and because of their minute size and simplicity, it is easy to miss or mistake them. Precambrian palaeontologists are very careful to prove the reality of their fossils, because so much of our understanding hinges on such inconspicuous material.

It is hard to penetrate the mists of deep time to imagine the truly ancient, Precambrian world. It teemed with life, but mainly life we could not see. The burst of diversification that happened at the beginning of the Cambrian period, some 540 million years ago, is a revelation. There, suddenly, in the fossil record, we find large fossils with skeletons, shells and other tissues. They can be held in the palm of the hand – vaguely familiar-looking shellfish, arthropods, sponges. The Cambrian 'explosion' is still much debated, and the debates raise the constant anxiety of all geologists and palaeontologists – just how trustworthy is the ancient rock record? Does it give us the true picture of what was going on, or is it so patchy and incomplete that the limited windows on ancient life are really quite misleading?

There are many mysteries in the later history of life, through the past 500 million years. Dinosaurs always fascinate us, and while every new discovery answers a few questions, it opens up more: how did they become so huge, how did they function, and, of course, why did they die out? We look at two mass extinctions that bracket the age of dinosaurs, the end-Permian event at the start, and the KT event at the end. The rise of our own group, the mammals, is an endless source of fascination. When the mammals took over the world after the demise of the dinosaurs, how did they diversify into forms as disparate as bats and whales, horses and humans? And the focus of so much attention and research – the hunt for human origins – is as lively a topic, and as heated a source of dispute, as it ever was. What do we currently know about where we came from?

BERNARD WOOD

How did the Earth form?

1

I assume that, at the beginning, all of the planets and other bodies in the solar system were dispersed as small particles throughout the space which they now occupy.
IMMANUEL KANT, 1755

In this way, Immanuel Kant, the great German philosopher, attempted to explain the origin of the Universe, and of the Earth, at a time when he could do little more than speculate. The existence of clouds of interstellar dust and gas (nebulae) had been known to astronomers at least since the 17th century, and Kant proposed that our solar system developed from a rotating nebula that had undergone gravitational collapse so that the material was concentrated into the Sun at the centre with the planets orbiting around it.

His theory still forms the basis of ideas about how the Earth and other planets formed. Recent observations of many young stars (those less than 10 million years old) indicate that they are orbited by discs of dust and gas and that as they age these dusty discs disappear. The obvious conclusion – that the material of the discs accumulates slowly into planets – is supported by the expanding number of extrasolar planets which have been detected since 1991, when the first ones were discovered, as well as by study of extraterrestrial materials from our own solar system.

Meteorites and the age of the Earth

Almost all of the thousands of meteorites in museum collections around the world originally came from small (100 km/62 miles in size) asteroids in the belt between Mars and Jupiter. These meteorites are the oldest rocks available to us and they record events at, and even before, the beginning of the solar system. Radioactive dating using the decay of uranium to lead indicates that the most primitive examples are 4,567 million years old. Applying the same method to the Earth implies a similar, if slightly younger (by 30–100 million years), age for our planet.

The oldest meteorites contain grains of silicon carbide and other compounds; analysis of the ratios of two different isotopes of carbon (mass 12 (^{12}C) to mass 13 (^{13}C)) in them show that they were produced in an old star quite different from

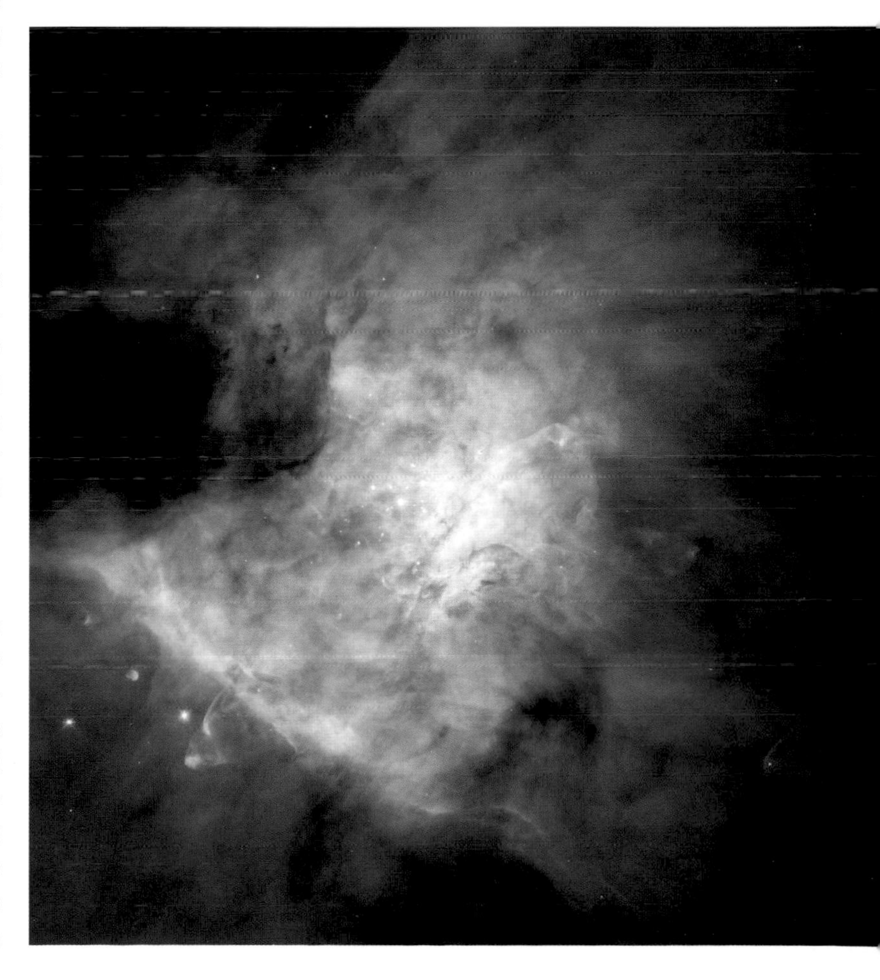

The centre of the Orion nebula, taken with NASA's Hubble Space Telescope: this view contains at least 153 protoplanetary discs ('proplyds').

Artist's impression of a brown dwarf encircled by a swirling cloud of planet-building dust – a protoplanetary disc.

the Sun. They also contain traces of short-lived radioactive isotopes produced in a giant star and scattered into 'our' pre-solar nebula by a supernova explosion. The most primitive meteorites (CI carbonaceous chondrites) have chemical compositions which, except for depletions in hydrogen and helium gas, closely mimic that of the Sun. This situation is exactly what would be expected if they were part of the left-over material from the collapsing solar nebula. These meteorites also have striking chemical affinities with the Earth, Mars and the Moon, indicating that these planets and asteroids were all made of very similar material, and, given their similar ages, at about the same time.

The conclusion from chemical study of meteorites (including a small number from Mars), the Earth and the Moon is that the dust (and to some extent gas) of the solar nebula accumulated first into clumps and then larger and larger bodies as the grains stuck together. The processes of accretion are not well understood until the bodies reach about 1 km (0.6 miles) in size; after this point the forces of gravitational attraction and collision dominated. The process of growth into small asteroids and eventually planets can, it is believed, be computer-modelled, and the models tested and refined using isotopic and chemical information.

Structure and chemical composition of Earth

The Earth has a metallic core which, from seismic data, is known to be very dense and predominantly liquid. This core is surrounded by the rocky mantle and crust, collectively known as the 'silicate Earth'. Silicate Earth is very depleted in 'sidero-

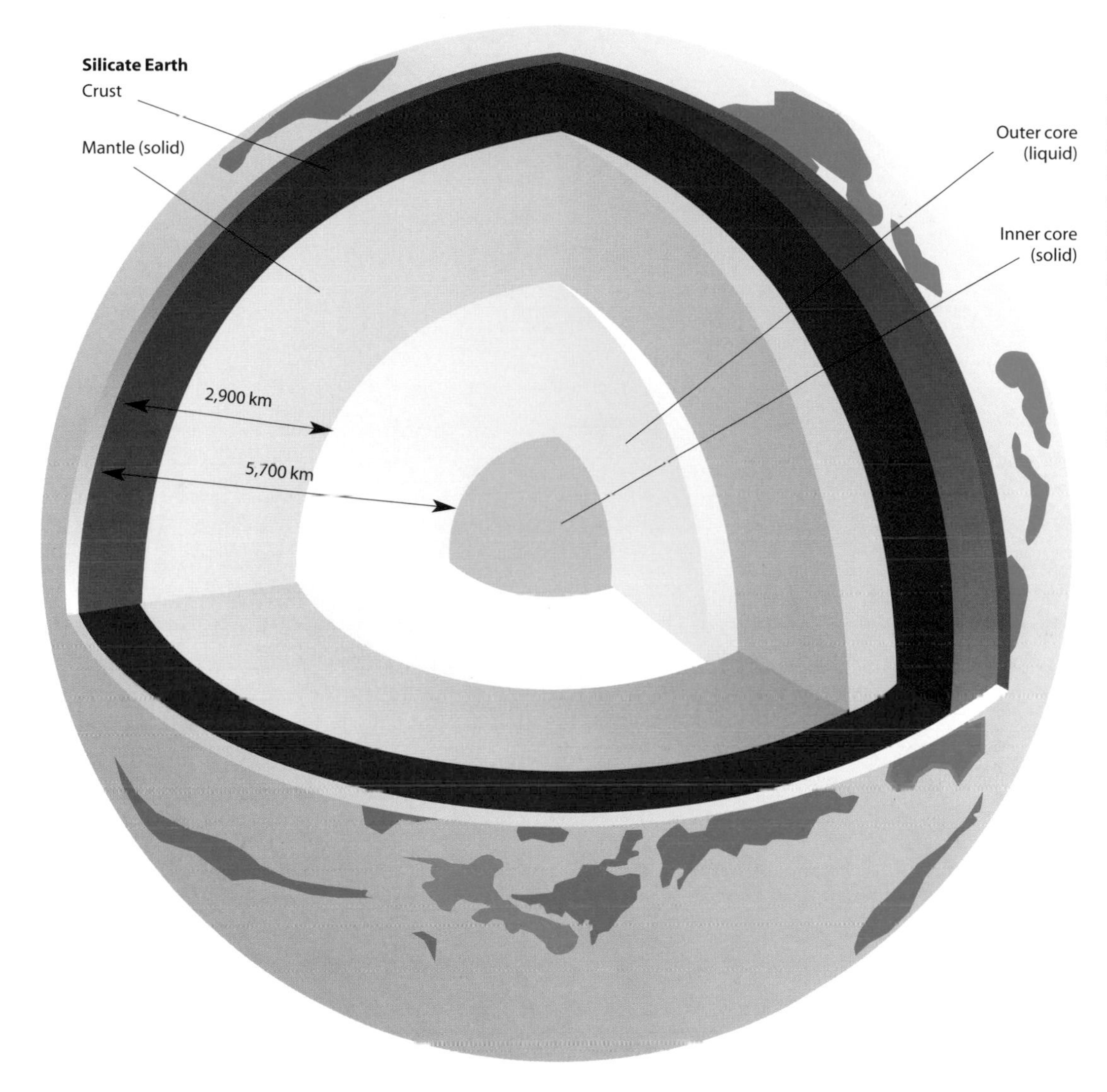

A diagram of the structure of the Earth. The core, which makes up about 32.5% of the whole, is mostly liquid, with a small solid centre. Together, the mantle (67%) and crust (less than 1%) comprise what is known as silicate Earth.

phile' elements – that is elements such as iron, gold and platinum which make very stable metals – compared to the primitive CI chondritic meteorites, indicating that these elements have been strongly extracted into Earth's core. In contrast, 'lithophile' elements (which do not enter planetary cores, e.g. calcium and titanium) are in the same ratios one to another in silicate Earth as in CI chondrites.

From the depletions of siderophile elements in the silicate Earth we can calculate that the Earth's core is about 85% iron and 5% nickel, with a large number of other elements making up the balance. Interestingly, the gold content of the core is estimated to be sufficient to coat the land surface of the Earth to a depth of half a metre. It is apparent that many asteroids also have cores, since a large number of metallic meteorites rich in iron and nickel have fallen on Earth. These 'iron meteorites' also have high concentrations of siderophile elements.

Separation of the core

One of the short-lived radioactive isotopes present at the beginning of the solar system was ^{182}Hf (hafnium-182) which decays with a half-life of 9 million years to stable ^{182}W (tungsten-182). After five half-lives (about 45 million years after the beginning of the solar system) all the ^{182}Hf had decayed. So for 45 million years the ratio of ^{182}W to ^{184}W (which was present from the beginning) increased, due to ^{182}Hf decay, but remained constant thereafter. Tungsten is siderophile while hafnium is lithophile, and therefore as cores

Above *The Moon formed through an impact between Earth and a Mars-sized body. Here separate images taken by the Galileo probe are combined.*

Right *The dating of core formation from the ratio of tungsten (W)-182 and -184 isotopes, compared with that of a primitive meteorite (carbonaceous chondrite). Early in the formation of the Earth the ratio increased to the point at which the core formed.*

formed on asteroids and planets, most tungsten, but no hafnium was extracted to the core, leaving a core with Hf/W of zero and a silicate part with very high Hf/W.

If core separation took place early enough, the silicate part will have a very high ratio of ^{182}W to ^{184}W resulting from the decay of ^{182}Hf, while the tungsten isotope ratio was fixed at the time of core separation. If core separation took place later, the isotope difference compared with carbonaceous chondrites will be smaller. By measuring the isotope ratios in meteorites we find that many asteroids formed cores very rapidly – within 1 million years of the origin of the solar system. On Mars, core formation took place over a few million years and on Earth over a 30-million-year period from the beginning of the solar system. Thus isotopic dating produces a picture of rapid growth of asteroids from dust (a million years) and more protracted growth of planets, presumably by collisions between fewer and fewer but larger and larger bodies.

Impacts and the formation of the Moon

By the time the Earth was about 10% of its current size, energies from collisions with smaller accreting bodies would have been sufficient to cause its rocks to undergo melting. As the Earth grew and collisions became infrequent and more energetic, it would have been periodically covered by a deep layer of molten rock – a 'magma ocean'.

Several lines of evidence indicate that the final and most energetic impact on the Earth led to the formation of the Moon. The Moon has very strong chemical and isotopic similarities to the Earth, indicating a close genetic connection. Silicate Moon, for example, has similar depletions in siderophile elements to silicate Earth, but the Moon has almost no core into which they could have been extracted.

The motion (angular momentum) of the Earth-Moon system is consistent with formation of the Moon from an impact between Earth and a Mars-sized body. Modelling of such an impact indicates that a small amount of mixed silicate material from both impactor and Earth would have been ejected to form the Moon, while the core of the impactor could have been left behind in the Earth. This explains the siderophile depletions on the Moon and its lack of a core. Isotopic measurements of Moon rocks indicate that the Moon formed around 40–50 million years after the origin of the solar system, placing the timing of the giant impact at the end of accretion of the Earth.

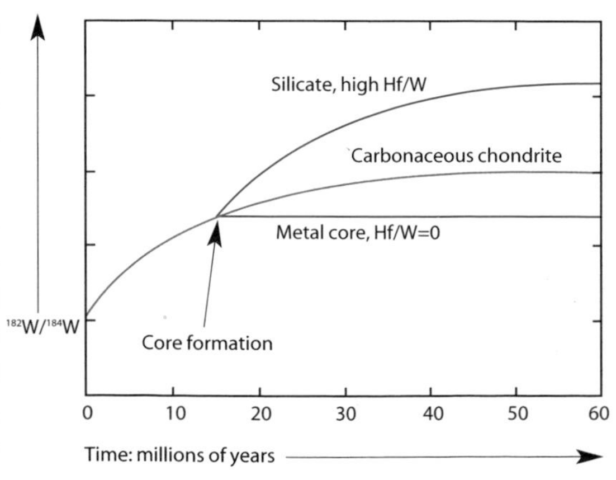

MARTIN BRASIER

The origins of life

Would it be too bold to imagine, that in the great length of time since the earth began to exist, perhaps millions of ages before the commencement of the history of mankind, would it be too bold to imagine, that all warm-blooded animals had arisen from one living filament?

ERASMUS DARWIN, 1794

How did life begin? This greatest of all puzzles brings natural science into direct confrontation with older ways of thinking. One age-old solution has been what we may call 'Top-Down' thinking, which invokes miraculous intervention from an unseen power. Such a view is traceable as far back as the earliest literate cultures of Mesopotamia in the 3rd millennium BC, and still holds sway in some conservative societies. Arguably much more ancient is the belief that life has always existed somewhere, and that the Universe is both eternal and perfectly fit for life. Such thinking can be detected in writings from Plato down to Isaac Newton. Indeed, some modern cosmologists have returned to it, with the concept of DNA drifting throughout the Universe.

Our classic scientific solution to this question – how did life begin? – is that of 'Bottom-Up' thinking. Starting with the Atomist philosopher Thales in Greece in about 600 BC, and certainly since

Hot-water-loving bacteria predominated at the surface of early Earth. These modern, colourful cyanobacteria live in a hot spring at 46°C (115°F) in Yellowstone National Park.

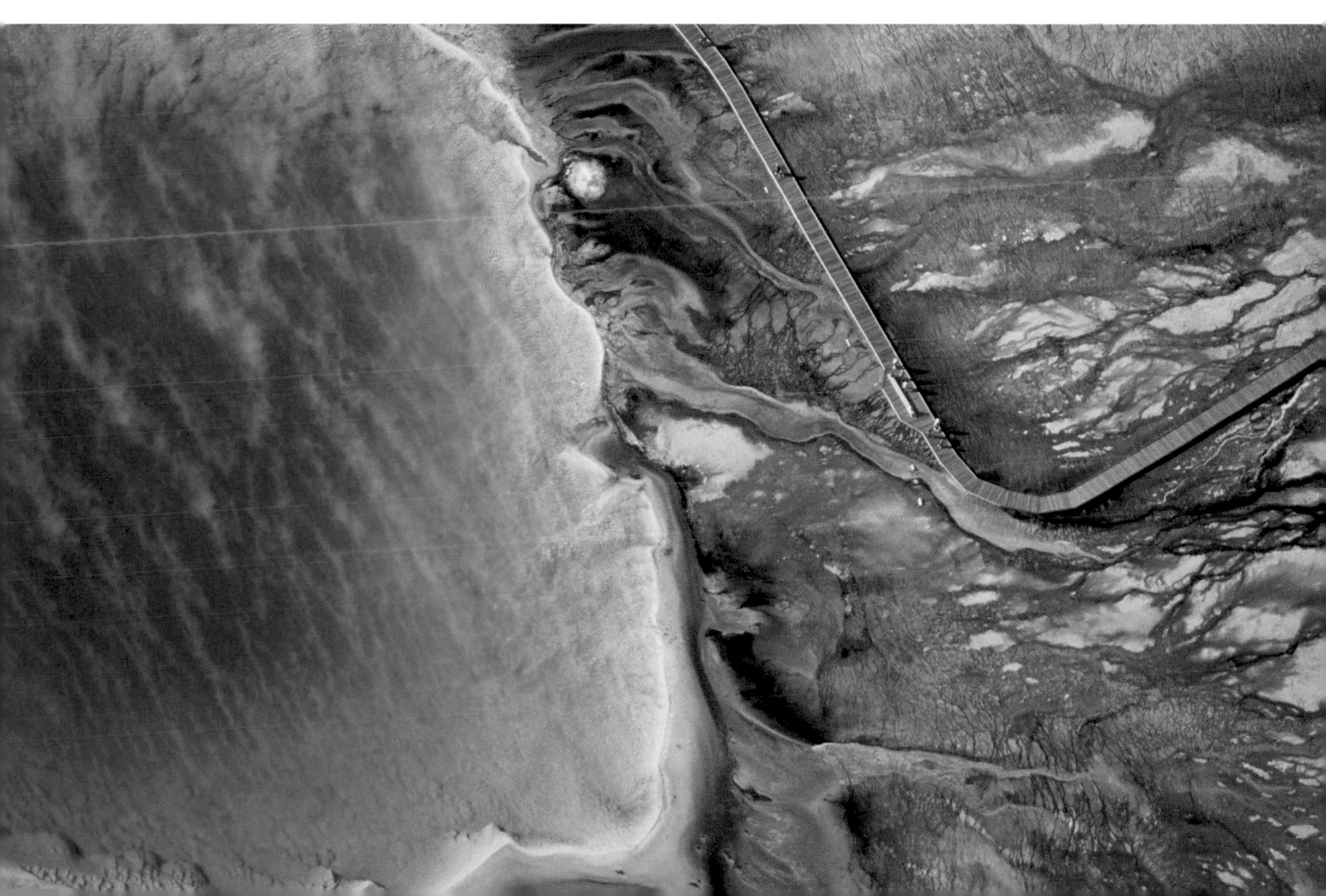

Charles Darwin in 1859, science has discovered that the universe is organized from the bottom upwards, from little things to big things – from atoms to animals as it were – according to beautifully simple rules for self-organization.

Stanley Miller reconstructs the famous experiment he and Harold Urey conducted in 1953 in their research into the origins of life.

What is life?

So what is 'life'? This conundrum is almost as hard to answer as the first question. For Top-Down thinkers, the answer once involved something called the 'vital spark'. But, like ghosts and fairies, no such spark has ever been detected by science. For Bottom-Up scientists, all answers are always open to doubt. But one key feature of life has emerged: it is a form of matter that shows a unique capacity for evolution by means of natural selection. It adapts to changing conditions. Since evolution by natural selection best explains everything we see in the living world, then life itself must surely have emerged as a result of natural selection acting upon inorganic matter under the conditions that pertained at the surface of the young Earth.

What, then, were those conditions? We know from centuries of chemical research that life forms are mainly made from about 16 natural elements. Some of the most important ones (hydrogen and oxygen) are freely available in water, which is vital because of its remarkable properties as a solvent and chemical buffer. That is why the search for early or remote life is also, typically, a search for water. Indeed, life has been called 'animated water'. Other major components of life can occur naturally as atmospheric gases (compounds of nitrogen, oxygen, hydrogen, sulphur and carbon), in the air or as dissolved components in water.

These gases are all essential components of amino acids and proteins in living matter. It may be no coincidence, therefore, that they appear to have been the predominant ones in the atmosphere surrounding the early Earth. But life also needs a whole range of other materials which are largely confined within minerals of the solid Earth. The earliest forms of life must, therefore, have had easy access to such vital minerals and have started out as a series of simple mineral reactions – as a branch of mineralogy.

With the exception of viruses, modern life forms are typically cellular and reliant upon three major components: the cell wall, whose role is to define a reaction chamber for chemical reactions; the cell contents, whose role is to undertake these reactions by means of proteins – what we call metabolism; and the genetic code, held within RNA or DNA, whose job is to act as an information store and to allow natural selection to operate on the products of reproduction.

For some, the challenge has been to understand how such seemingly complex and sophisticated components could have arisen together. For others, however, it now seems self-evident that each of these components must have built up slowly, and perhaps independently, via small steps that began with simple precursors found abundantly at the surface of the early Earth. Two great thinkers of the 20th century, Aleksandr Oparin and J. B. S. Haldane, came independently to the conclusion that, since the primitive atmosphere of Earth was without oxygen, the other gases would tend to promote the synthesis of the complex organic building blocks (amino acids), especially when stimulated by ultraviolet light or lightning. This was famously confirmed by the

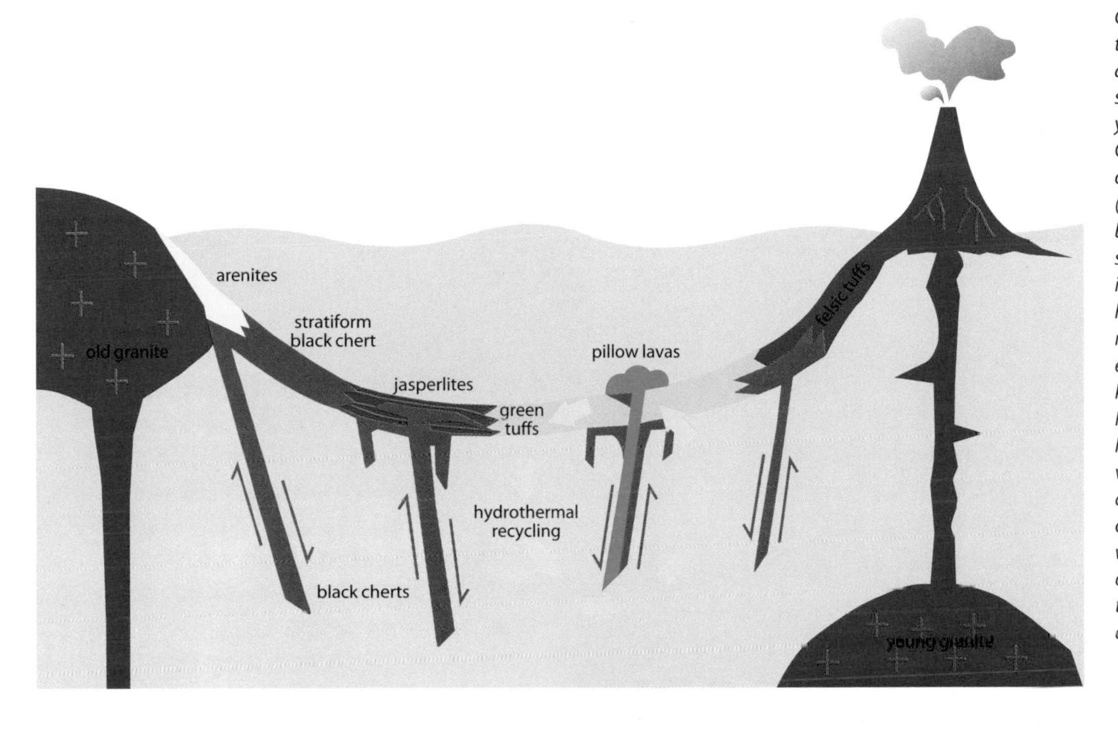

Cross-section through the crust and ocean of Earth some 3,500 million years ago. Carbonaceous organic matter (black) is found both within marine sediments, but also in pillow lavas and hydrothermal mineral veins that extended several kilometres deep. Pillow lavas and hydrothermal veins are potential cradles for the origins of life, which may have evolved several times, and in different forms.

experiments of Harold Urey and Stanley Miller in the 1950s, leading towards a model which imagined localized water bodies on the early Earth becoming filled with the scum of protolife. Intriguingly, we have since discovered that amino acids are synthesized within comets and meteorites by means of solar radiation. The building blocks for life were, and perhaps still are, almost everywhere.

Making the first living thing

A major question therefore remains: what were the steps that led towards the first nucleic acid replicators, such as RNA? It is here that a study of the early Earth again provides us with a helpful perspective. Geologists have found that the early Earth abounded not only with warm oceans and hot springs, but also with waters bearing a rich harvest of metals that now provides nearly all our gold, iron and nickel reserves. Chemists have found that fool's gold (pyrite, FeS_2) and iron and nickel compounds have impressive properties. They were probably able to act upon simple organic compounds, such as methane, leading towards the synthesis of more complex organic molecules, such as fats and sugars, and aiding the

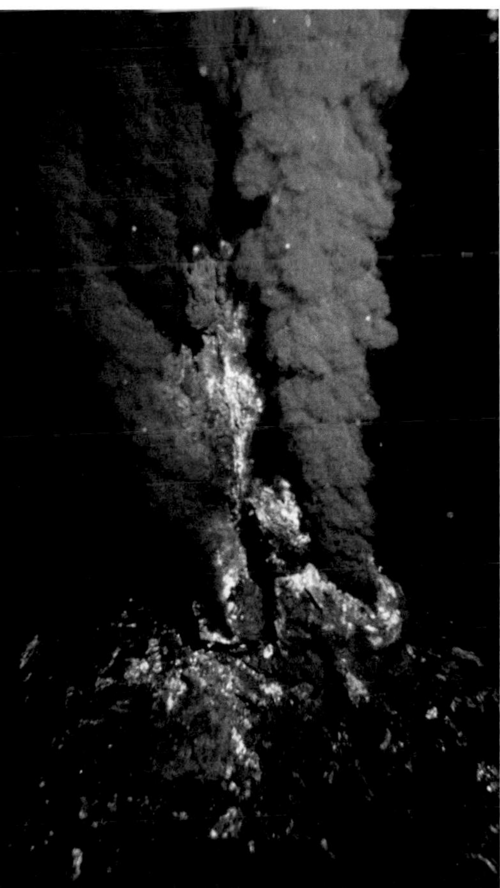

Hydrothermal vents on the deep seafloor – often known as black smokers because of the clouds of iron sulphide that surround them – have been found to support a huge range of microbes (see p. 167). It seems that such heat-loving, oxygen-hating and metal-processing microbes were among the earliest forms of life on the planet.

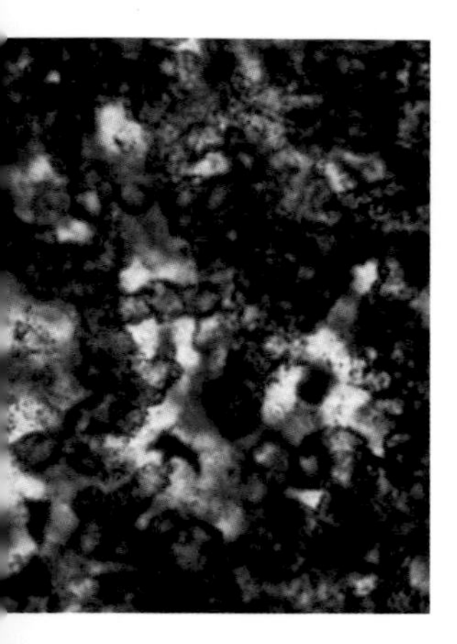

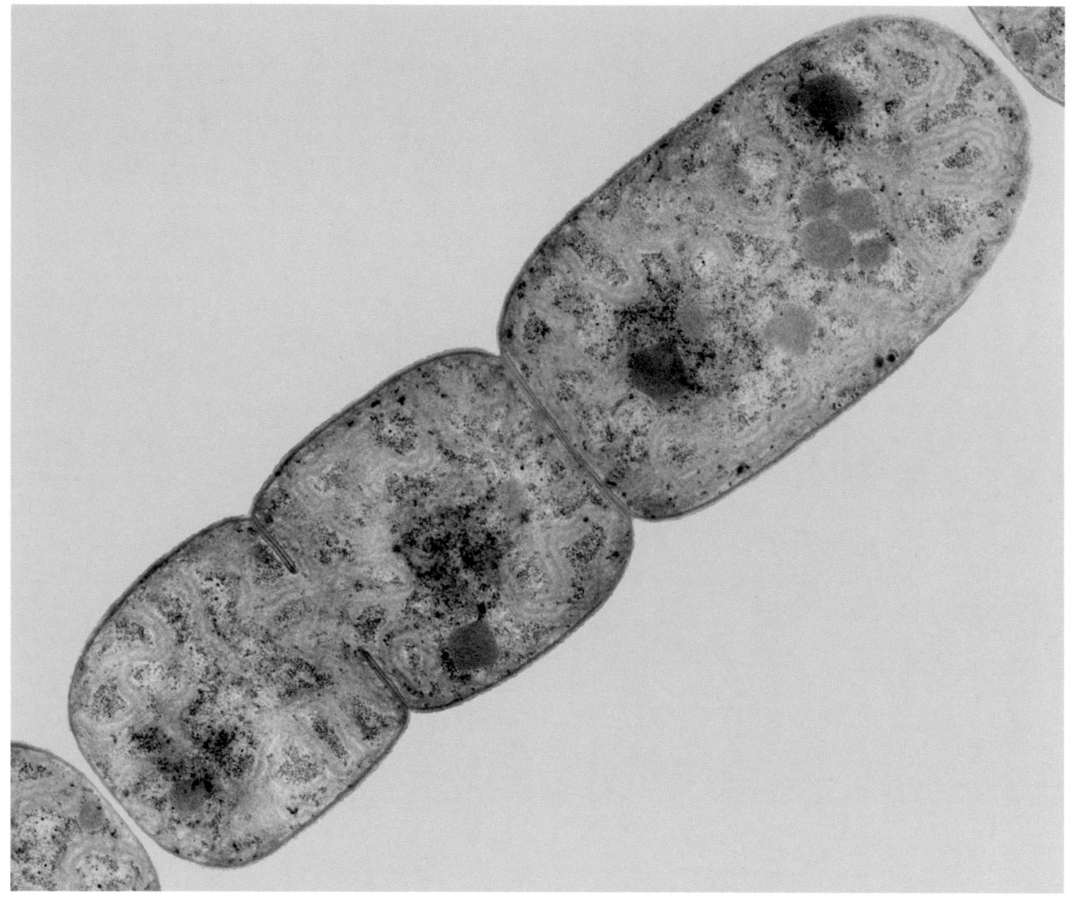

Above *Cell-like structures preserved in silica, recently discovered by the author in the Strelley Pool chert of Western Australia, some 3,430 million years old. Early microbes likely flourished in conditions resembling a modern toxic waste site.*

Right *Cells of living cyanobacteria arranged in a chain; the central cell is undergoing binary division. The green photosynthetic pigments stand out clearly. The first evidence for such microbes does not appear until after about 2.7 billion years ago, almost half way through Earth history.*

stabilization of large and complex organic products on metallic surfaces.

It also seems possible that clay minerals could have acted as a 'template' for the build up of RNA and DNA. Happily, it seems that clay minerals were also widely available at the surface of the early Earth. There may, indeed, have been many different kinds of primitive life forms at this time – each using different kinds of compartment (such as emulsions or minerals) and different kinds of replicator (like analogue or digital code; RNA or DNA). As far as we know, however, only one kind of life survived to the present time.

So when did life appear in the fossil record? Although this is a highly contentious issue, there is reason to suspect that bacteria-like microbes were around by about 3,400 million years ago. Those early life forms would appear extremely alien to us today because of their affinity for volcanic vents, hot springs and toxic metals, and their avoidance of oxygen. In other words, life probably included the kinds of microbes that we now find in the waters of abandoned zinc mines or toxic waste sites, such as the purple-, sulphur- and methanogenic bacteria (p. 171). Studies of molecular evolution have lent support to this view. Such hot-water-loving (hyperthermophile) bacteria predominated at the surface of the early Earth until the widespread appearance of oxygenic photosynthesis in marine cyanobacteria, some time after about 2.7 billion years ago (see also p. 69).

From that time onwards, the earliest denizens of the planet must have been increasingly obliged to retreat from the growing levels of atmospheric oxygen. Life's oldest relatives have now moved to the deeper layers of the sediment or, indeed, taken refuge inside the cell walls of eukaryotes (all organisms whose cells contain a nucleus carrying genetic material) – such as ourselves.

NICHOLAS J. BUTTERFIELD

The origins of multicellular life

Now and again there is a sudden rapid passage to a totally new and more comprehensive type of order or organization, with quite new emergent properties, and involving quite new methods of further evolution.
JULIAN HUXLEY, 1947

Unicellular organisms are the fundamental form of life on Earth, but it takes a highly differentiated multicellular organism – us – to appreciate this fact, and to contemplate the revolutionary consequences of integrated intercellular cooperation. Multicellular organisms constitute the majority of the planet's biomass, the majority of its recorded diversity, and impart an overarching control on the modern biosphere. Second only to the origin of life itself (p. 23), it is the evolution of multicellularity that makes Earth-like planets interesting.

The transition to a multicellular grade of organization entailed a fundamental shift in individuality, from free-living unicells to composite organisms with their own, emergent levels of heritable variation. The original advantage of such an association presumably lay in the physiological and ecological buffering afforded by increased organism size and, at some minimum cell population, the efficiencies of specialization. But there were also opposing interests. Why would an independent (and inherently selfish) organism abandon its own individuality to serve a homogenizing confederation? And even if it did, what prevents it from cheating the system by reverting to a unicellular existence?

The key to multicellular integration lies in genetic similarity. If all the constituent parts carry identical genes, then the effort is entirely cooperative and bears no altruistic or Hamiltonian cost (see p. 227). Conflict is still likely to arise as individual cells and cell lineages are exposed to differential mutation, but there is an elegantly simple, universal solution: at some stage in their lifecycle all multicellular organisms revert to a single cell, ensuring that each generation begins with a single, fully cooperative genome. Among eukaryotic organisms (those whose cells contain nuclei), this step is typically accomplished by the single-celled gametes and zygote associated with sexual reproduction.

Specialization and individuality

At one level, multicellularity is nothing more than simple mitotic cell division followed by the incomplete separation or secondary aggregation of daughter cells to yield a colony. True individuality, however, is only reached with specialization, such

Sporangia of the slime mould Lamproderma, *representing one of many independent evolutionary experiments in simple, protistan-grade multicellularity.*

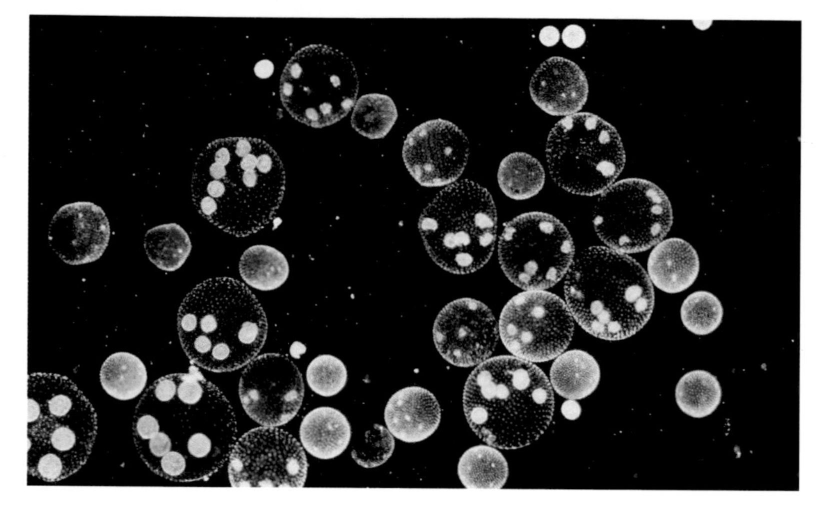

A population of the multicellular green alga Volvox, *showing various stages of its life cycle.* Volvox *has served as a model organism for investigating the mechanics of inter-cellular cooperation.*

that individual cells are no longer able to survive or reproduce on their own. This loss of unicellular independence is a natural byproduct of multicellularity, as the trade-offs between survival and reproduction are met by a division of labour within the organism. Under suitable circumstances, the fitness of the whole can be enhanced by differentiating a regenerative, essentially immortal lineage of reproductive germ cells, and outsourcing the business of organism function to populations of sterile, so-called somatic cells. It is this mutual germ-soma inter-dependence that suppresses the individuality of the constituent cells, and secures an emergent multicellular existence.

Further divisions of labour are possible with the differentiation of multiple different somatic cell-types, which in turn may associate to yield the higher-level properties of tissues, organs and organ-systems – and yet further degrees of individuality. Indeed, the number of differentiated cell-types offers a useful measure of evolutionary 'grade of organization': 2 or 3 types in complex bacteria and simple nucleated organisms (eukaryotes); 7 in the most complex seaweeds and fungi; up to 11 in sponges; 30 or so in vascular land-plants and simple bilaterally symmetrical animals; and as many as 200 for the most complex vertebrates.

When did multicellularity occur?

It is clear that multicellularity has evolved independently on numerous occasions – not only in the three multicellular kingdoms (plants, fungi and animals), but repeatedly among the bacteria and simple, sub-tissue grade eukaryotes collectively known as protists. Of the latter, the volvocalean green algae, especially *Volvox*, and the cellular slime moulds, especially *Dictyostelium*, have served as model organisms for investigating the mechanics of inter-cellular cooperation. The principal insight from these simple systems is that there are many fascinating, but ultimately limited routes to multicellularity.

Multicellular organisms can be tracked in the early fossil record, with mat-forming filamentous cyanobacteria known from at least the late Archaean (2,500 million years ago) and relatively larger multicellular eukaryotes from at least the early Mesoproterozoic (1,600 million years ago). Eukaryotic forms include early representatives of the red, green and 'brown' seaweeds, which are identifiable on the basis of their unique, independently evolved patterns of multicellularity. Evidence for tissue-grade animals, however, is not detected until the Ediacaran Period (635–545 million years ago), within shouting distance of the infamous Cambrian explosion of around 530 million years ago (p. 30).

A conspicuous component of the Precambrian record is represented by multicellular and/or macroscopic organisms with no living counterparts. Curiously, none of these failed experiments in multicellularity appears to have managed anything more than a simple protistan grade of organization, despite a billion-year window of opportunity. The delay has been widely attributed to limiting levels of atmospheric oxygen, but this fails to account for the belated appearance of tissue/organ-grade land-plants, or indeed any other lineage with modest oxygen requirements, such as most marine invertebrates.

Higher-level multicellularity

The simpler explanation is that the higher-order evolution of tissues, organs and organ systems was difficult. So difficult, in fact, that it has been managed only twice on this planet: once in animals and once (some 200 million years later) in plants. Such singular occurrences of tissue-grade multicellularity point to a fundamental, internal,

Far left *A 1,200-million-year-old fossilized red alga from Arctic Canada.*

Left *An accumulation of fossil Pteridinium from the terminal Proterozoic (Ediacaran) of Namibia – it probably represents a failed experiment in simple multicellularity.*

control on the evolution of development. All multicellular organisms employ a variety of primitive inter-cellular signalling and adhesion molecules, but a fundamentally higher level of coordination was required for the development of tissues, organs and organ systems. In plants and animals this leap forward has been achieved not so much by the evolution of novel genes, but rather the controls over gene expression, especially DNA-binding transcription factors, accompanied by truly enormous increases in the complexity of developmental programmes. This was the technological hurdle that so thoroughly frustrated the early evolution of tissue-grade multicellularity.

The evolution of tissue/organ-grade multicellularity revolutionized the biosphere by tapping into the emergent, but increasingly inter-dependent properties of proliferating cell lineages. In the case of animals, the discovery of a differentiated gut and nervous system swept aside 3 billion years of microbial domination, imposing a brave new world of hierarchically constructed food webs, ecological escalation and evolutionary dynamism. Together with the co-evolutionary assistance of multicellular plants and fungi, these remarkable consortia of cell lineages have continued to experiment with divisions of labour – everything from worker bees to the mass production of motor cars.

An example of the complex, co-evolutionary morphology and behaviour that arises from tissue/organ-grade multicellularity, and that has structured the biosphere for the past 500 to 600 million years. Unlike simple, protistan-grade multicellularity, tissue/organ-grade organisms have evolved only twice.

RICHARD FORTEY

The Cambrian evolutionary 'explosion'

The Cambrian explosion is the key event in the history of multicellular animal life.
STEPHEN J. GOULD, 1995

The apparently sudden appearance, or 'explosion', of diverse animals at the beginning of the Cambrian Period is considered to be hugely significant. This event was sufficiently profound to mark off the Phanerozoic Eon from the underlying Precambrian. And yet recent research has suggested to some that the Cambrian explosion never in fact happened.

The Cambrian Period began some 542 million years ago, and typical fossils include trilobites (relatives of modern crabs and insects), sponges, brachiopods and odd-looking sea urchins, as well as others whose relationships are far from obvious. These different animals all have a hard shell – or external armour – of some kind, composed of calcium carbonate, calcium phosphate or silica. They also all appeared within a short period of time close to the base of the Cambrian. Evidence from studies of molecules, however, suggests to some that these animal groups originated much earlier, deep within the Precambrian.

Paradoxides bohemicus *is the largest known trilobite from the Cambrian, being almost 1 m (3.25 ft) in length; it lived in shelf muds.*

It was clear that some event triggered the appearance of shell-bearing organisms, for diligent search of the late Precambrian strata beneath discovered no shelly fossils. However, the search *did* reveal that the shelly fossils were preceded by large, soft-bodied fossils of strange design, which are generally referred to as the Ediacara fauna, after a famous Australian locality.

The plot thickened with the description of rare cases of preserved Cambrian soft-bodied fossils, the most famous being the Burgess Shale in Canada. Here was a vast selection of strange animals, different from both those of older, Ediacaran times, but equally from their shell-bearing contemporaries. Many were arthropods, yet without any obvious relationship to the trilobites, which were often assumed to be the most primitive of the phylum. Yet others were so odd that they were routinely described as 'weird wonders' – they did not fit anywhere in the usual evolutionary scheme of things.

These discoveries led to a view of the Cambrian as a time of evolutionary experiment, when dozens – maybe hundreds – of 'hopeful monsters'

briefly crawled and swam about the sea floor. The evolutionary 'explosion' gave rise both to the familiar animal phyla and also to doomed designs that left no progeny.

This scenario has proven both attractive and controversial. When the diverse animal species appeared in the Cambrian they had already differentiated into many separate body plans. If they had indeed descended from a progressively ancient series of common ancestors, it follows that such animals must have been present in earlier strata. The Ediacaran soft-bodied animals (whatever they were) did not seem to fit the bill as 'ancestors'. So where were the ancestors of modern faunas hiding? One popular idea was that the organisms in question must have been very small. Perhaps the Cambrian 'explosion' was partly an increase in size, and an increase in burrowing activity, with the subsequent acquisition of shells. This scenario built on an older idea that the 'explosion' was driven by the appearance of predators. Once predation became important, so would protection, cryptic behaviour, burrowing, and so on, on the part of the prey, in a kind of biological arms race. Every new predatory innovation led to new responses in the preyed upon.

Did the explosion happen?

The notion of a Cambrian explosion was challenged in the mid-1990s by the first 'molecular clock' analyses. Comparison of the DNA of modern groups seemed to show that the major animal groups had diverged long before the Cambrian, deep within Precambrian time, lending credence to the idea of a slow-burning fuse of evolution lasting as much as a billion years predating the main 'explosion'. This molecular challenge sent shock waves through the palaeontological community. However, after much debate, a refinement of statistical methods in clock estimates in recent years has shortened the fuse once again, so that the latest estimates have the deep divergence of animal groups of the order of 50 million years before the Cambrian.

But how could such diverse body plans appear so rapidly? In fact, genes have now been discovered which trigger cascades of effects after only a relatively small mutation. Maybe those 'hopeful monsters' are not so improbable, after all.

There is no question that interaction between the newly evolved organisms after the initial 'explosion' prompted and encouraged further adaptation. By the Middle Cambrian, for example, *Anomalocaris* was a predator approaching 1 m (3 ft) long, which must have stimulated a variety of defensive responses among other organisms. The greatest innovative phase in the history of life still has secrets to reveal, which will derive from new discoveries in the field, refinements of analytical techniques and advances in genetic theory.

Above left *An example of a Burgess Shale fossil:* Hallucigenia, *armoured with dorsal spines.*

Above right *A reconstruction of some of the vast array of strange creatures represented in the Burgess Shale fossils.*

MICHAEL J. BENTON

The biggest mass extinction of all time

Killing over 90% of the species in the oceans ... is remarkably difficult ...
DOUG ERWIN, 1993

Below left
Lystrosaurus, *one of the few survivors of the end-Permian mass extinction.*

Below right
Before (top) and after (below) the crisis, in the shallow South China Sea. The scale of the loss of species is dramatic.

The largest-ever extinction event happened at the end of the Permian period, some 251 million years ago. While much more has been said and written about the extinction of the dinosaurs, and many other groups, at the end of the Cretaceous (p. 42), the end-Permian event was more severe. In fact, it is estimated that up to 95% of all species died out in a relatively short span of time – the nearest life on Earth has ever come to annihilation. And yet, until quite recently, no one had any idea what had caused this profound crisis.

What we do know is that most of the dominant animal groups in the sea disappeared, or were much reduced in diversity: corals, articulate brachiopods, bryozoans, trilobites and ammonoids. There were also dramatic changes on land, with widespread extinctions among plants, insects, amphibians and reptiles, which led in all cases to dramatic long-term changes in ecosystems.

What happened?

Around 1990, palaeontologists and geologists were unsure whether the extinctions lasted for 10

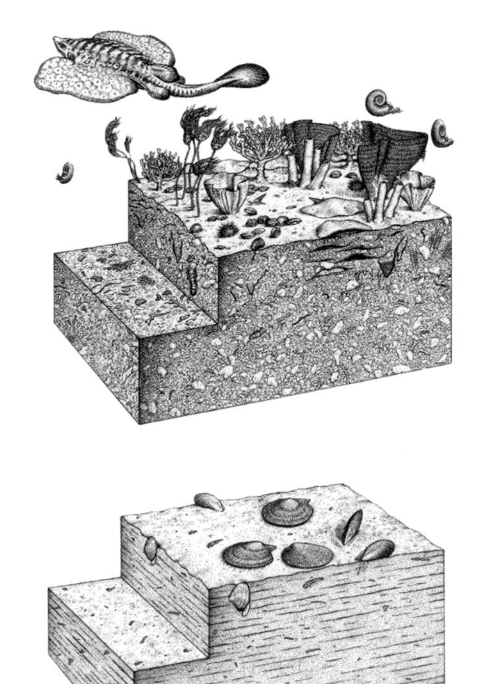

CLOSE-UP OF A MASS EXTINCTION

In 2000, Jin Yugan and colleagues published the detailed anatomy of the Permo-Triassic crisis, based on their studies of the Meishan section in southern China. They had collected thousands of fossils through 90 m (295 ft) of rocks spanning the Permo-Triassic boundary, and identified 333 species belonging to 15 marine fossil groups. During the 4 million years before the end of the Permian, 161 species became extinct, with extinction rates in particular beds amounting to 33% or less.

Then, just below the Permo-Triassic boundary, at the contact of beds 24 and 25, most of the remaining species disappeared, giving a rate of loss of 94% at that level. Species had been going extinct from time to time below this crisis level, but at a steady, or 'normal' rate. Following the crisis level, there followed some 500,000 years of time when new species arose fast and then died out fast; these are so-called 'disaster species'. At the end of this time of instability, extinction rates returned to normal and species began to live for longer.

A close-up of the Permo-Triassic boundary at the junction of the black and white clays in south China.

million years or happened overnight, or whether the main killing agent was global warming, sea level change, volcanic eruption, meteorite strike or anoxia (loss of oxygen). We are a little clearer on most of these points now.

When palaeontologists reconstruct the scene before and after the event, the differences are startling. Before, reef ecosystems teemed with dozens of creatures: corals and other reef-builders filter-feeding, and arthropods and molluscs crawling about the seabed, as well as fishes and ammonoids swimming above. Afterwards, the reefs had gone, and only two or three species of paper pectens and other lowly organisms were to be seen. These detailed pictures are based on the well-studied rock sections at Meishan in south China (see Box), but similar observations have been made elsewhere, from equator to pole.

Causes

The suddenness and magnitude of the mass extinction suggest a dramatic cause, perhaps asteroid impact or volcanism. Impact is the likely cause of the later extinction of the dinosaurs, but evidence for impact at the Permo-Triassic boundary is weak. It seems, alarmingly, that this greatest of all mass extinctions was caused by global warming driven by volcanic eruption.

Most attention has focused on the Siberian basalts – a huge volume of lava erupted in pulses over 600,000 years around the time of the Permo-Triassic boundary. Volcanic eruptions produce lava of course, but also gases such as carbon dioxide, sulphur dioxide and others that mix with water in the atmosphere to produce acid rain. Carbon dioxide is a well-known greenhouse gas (see p. 257), and would have warmed the atmosphere. But is there any geological evidence for these phenomena?

Studies of the sediments across the Permo-Triassic boundary show a change from oxygenated sediments with abundant life, to black mudstones, with very few fossils. These confirm that there was a dramatic phase of anoxia – loss of oxygen – one of

Palaeontologists climb up Butterloch Canyon in Italy to collect samples at the Permo-Triassic boundary, visible as a dark line of sediment running from upper left to right across the cliff face.

the consequences of global warming. As air and water temperatures rise, the amount of dissolved oxygen in the oceans decreases. On land too there is evidence that soils were stripped off, presumably when the larger plants were killed by acid rain, leading to erosion.

Geochemistry has provided additional clues. At the Permo-Triassic boundary, there is a dramatic shift in oxygen isotope values: a decrease in the value of the ratio of oxygen-18 to oxygen-16 of about 6 parts per 1,000, corresponding to a global temperature rise of around 6ºC (11ºF). Carbon isotopes also show a big increase in the light carbon isotope (^{12}C) – but the shift is too large to have been caused by the eruptions or by the burial of dead plants and animals. Something else is required.

The *deus ex machina*

That something else might be gas hydrates. These are formed from the remains of marine plankton that sink to the seabed and become buried. Over millions of years, huge amounts of carbon are transported to the deep oceans around continental margins and the carbon may be trapped as methane in a frozen ice lattice. If the deposits are disturbed, or if the seawater above them warms, the gas hydrates may be dislodged and methane is released, rushing to the surface.

The assumption is that initial global warming at the end of the Permian, triggered by the huge Siberian eruptions, melted frozen circumpolar gas hydrate bodies, and massive volumes of methane (rich in ^{12}C) rose to the surface of the oceans in great bubbles. This caused more warming, possibly melting further gas hydrate reservoirs. So the process continued in a positive feedback spiral that has been termed a 'runaway greenhouse'. The term 'greenhouse' refers to the fact that methane is a well-known greenhouse gas, causing global warming. Perhaps, at the end of the Permian, some sort of threshold was reached, beyond which the natural systems that normally reduce greenhouse gas levels could not operate. The system spiralled out of control, leading to the biggest crash in the history of life.

ANUSUYA CHINSAMY-TURAN

Were the dinosaurs warm-blooded?

Barring the invention of a time machine, no one will ever measure the body temperature or respiration rate of a living dinosaur.
JAMES O. FARLOW, 1992

The closest living relatives of dinosaurs are crocodiles and birds – so were dinosaurs cold-blooded, like the former, or warm-blooded, like the latter? The debate has raged for over 150 years. It might seem impossible ever to know, yet intriguing new evidence from bone structure might be providing us with some of the answers.

Ectothermy and endothermy

The term 'warm-bloodedness' – or more accurately 'endothermy' – refers to an animal's ability to keep its body temperature at an optimum level irrespective of the outside temperature. In contrast, the body temperature of many reptiles, such as snakes, lizards and crocodiles, fluctuates according to the environmental temperature. They are said to be 'cold-blooded', or ectothermic. This makes a difference. Endotherms are able to regulate their body temperatures at optimal levels to function effectively, and have high rates of metabolism (that is, they are able to maintain a sustained rate of energy production); they therefore also have sustained activity levels and faster growth rates.

Since the body temperature of ectotherms is dependent on the environmental temperature, their activity levels vary depending on whether they are warm or not. Many ectotherms adopt special behavioural patterns, such as basking in the sun to help keep their bodies warm. Temperature variations during the day and between the seasons clearly affect their lifestyles – for example, during the cold winter months ectotherms are barely active and hardly eat.

Dinosaurs were highly specialized reptiles that evolved from the archosaurs – these also gave rise to alligators and crocodiles, which are ectotherms. However, the descendants of dinosaurs are birds, and they are endotherms. This raises the compelling question as to whether dinosaurs had a physiology similar to their archosaurian relatives, or whether they were more like their descendants. Some scientists have suggested that non-avian dinosaurs were endothermic and that birds inherited this characteristic from them, while others have proposed instead that endothermy only evolved later in the evolutionary history of birds.

Whether dinosaurs were endotherms or ectotherms affects our understanding of them in terms of their activity levels and overall biology. But in general, all that survives for us to study are their bones and teeth. Unfortunately, none of the physiological indicators of endothermy (such as the amount of mitochondria present in cells, the capacity of their lungs, or their metabolic rates) are preserved in the fossil record. So the question

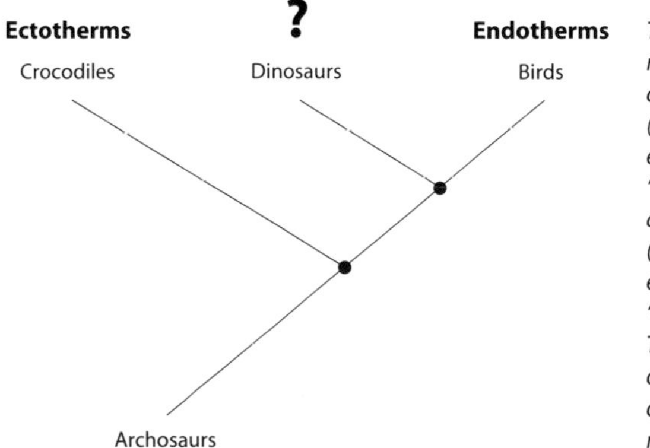

The closest relatives of dinosaurs are birds (which are endothermic, or 'warm-blooded') and crocodilians (which are ectothermic, or 'cold-blooded'). Thus the question of the physiology of dinosaurs still remains a mystery.

of dinosaur physiology and thermal biology is rife with speculation.

Originally, dinosaurs were considered to be sluggish 'over-grown lizards', and artists' reconstructions often reflected this. However, in the 1980s, perceptions of dinosaurs changed dramatically, and they began to be portrayed as highly active, agile animals that were more than likely endothermic.

Several hypotheses were put forward to support the 'hot-blooded' dinosaur argument, most of which were based on indirect, anatomical evidence, such as erect posture and brain size as deduced from cranial capacity. At the same time, the high concentration of blood vascular channels and the extensive development of resorption and redeposition of bone in dinosaur bones were also cited as indicative of endothermy. So what is the microscopic structure of dinosaur bone, and what could it mean in terms of physiology and overall biology?

Evidence from bone

Soon after an animal dies, all organic matter decomposes, including that in bone (such as blood and collagen). However, even though the soft tissues decompose rapidly, because the inorganic part of bone is so closely linked to the organic components, the hard tissues retain their structural organization. Thus, the microstructure of dinosaur bone remains virtually intact even after millions of years of fossilization, and this can be compared with the bones of living animals. Bone is a living tissue and various biological signals are recorded in its microstructure. For example, we know that in living crocodilians and lizards, the cycles of activity and inactivity during summer and winter respectively are clearly reflected as alternating wide and narrow growth rings (similar to those seen in trees). The wider rings reflect the period of fast growth during the favourable growing season, while the narrow rings indicate periods when growth has slowed or sometimes even stopped during the cold, unfavourable season. Thus, by counting the number of growth rings in the bone, an estimate of the age of the individual can be deduced – a study termed skeletochronology.

Endothermic animals like birds and mammals, generally do not form growth rings. They have rapid sustained growth without any alternating cycles of activity and inactivity (hibernating/aestivating animals being an exception).

Most theropod dinosaurs, the flesh-eaters (p. 37), show a cyclical pattern of bone microstructure. This alternating fast and slow growth pattern is like the growth rings present in the bones of modern 'ectotherms'. However, the bone relating to the fast-growing season is of a type that is formed at rapid rates, and often shows a large amount of secondary reconstruction, both features that are commonly found in modern 'endotherms'. Thus, these dinosaurs have bone microstructure characteristics that are intermediate between those seen commonly in living endotherms and ectotherms.

To complicate matters even further, basal (early) birds (such as *Rahonavis*, *Patagopteryx*, enantiornithines), also have cyclical patterns of bone formation, and hence cyclical growth rates.

Below *Three examples of bone microstructure: an alligator (ectotherm), with distinct cycles of growth (left); a mammal (a rabbit; endotherm) showing a richly vascularized bone tissue without any growth rings (centre); and* Tyrannosaurus rex *(right), showing growth rings and well-developed secondary osteons in the zonal regions of the bone.*

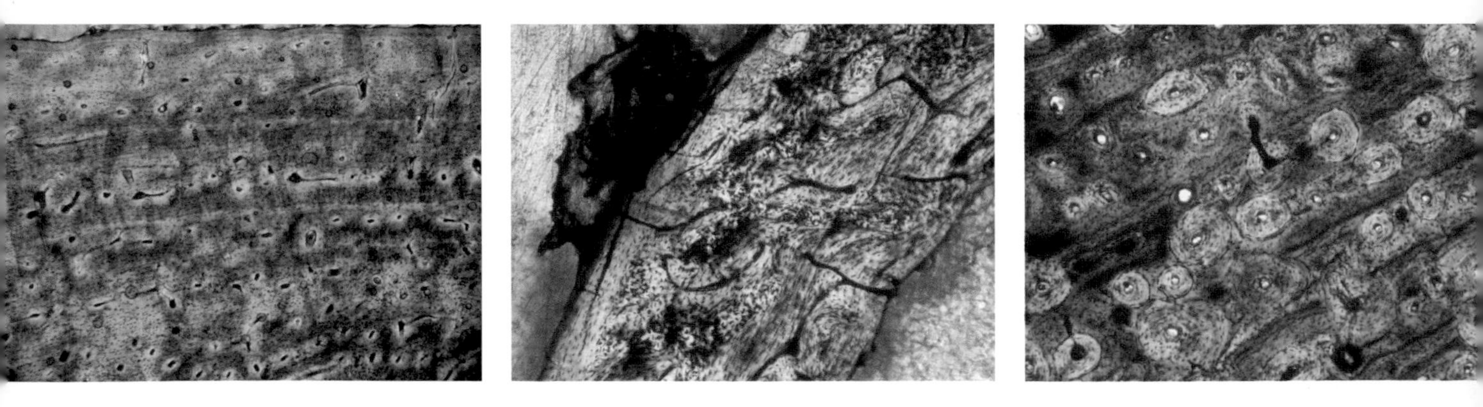

A reconstruction of Sinornithosaurus, *a feathered dinosaur from the Early Cretaceous of China. The actual fossils show that it was covered with downy fibres and that the arms and legs had long feathers.*

The fact that these Mesozoic birds have a pattern of bone deposition suggesting they were affected by environmental conditions raises the possibility that they could not have been classical 'endotherms' like modern birds. It is only in later fossil birds of the Late Cretaceous, about 70 million years ago (such as *Hesperornis* and *Ichthyornis*), that we find rapidly formed, uninterrupted bone patterns as in modern birds. So the ability to grow at sustained rapid rates, and perhaps also endothermy, may only have developed later in the evolution of birds, and was not inherited from their dinosaurian ancestors.

Evidence from feathers

Since 1990, exceptionally preserved specimens from China have shown that many theropod dinosaurs had feathers. Researchers have taken this as proof that they were full-scale endotherms, because feathers could be used for insulation and so maintain body temperature. However, even basal birds (including ones with well-developed flight mechanisms such as the enantiornithines) have cyclical bone patterns, and therefore experienced different growth strategies from modern birds. This implies that feathers need not necessarily be directly linked to the acquisition of endothermy. They could have first evolved for other functions, for instance for gliding and flight, or perhaps for sexual selection, and were only later used for insulation.

Even though the microscopic structure of bone provides fascinating insights into the growth patterns of dinosaurs, it is difficult to extrapolate from bone tissue structure to metabolic rates, thermal biology and physiology. Perhaps future research will shed light on this question, but for now, the verdict is still out as to whether dinosaurs were warm-blooded or cold-blooded or had a thermal physiology somewhere in between.

P. MARTIN SANDER

7

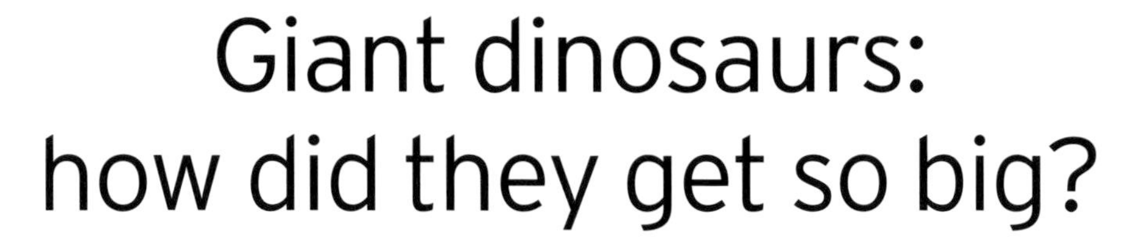

Giant dinosaurs: how did they get so big?

The never-since-surpassed size
of the largest dinosaurs remains unexplained.
G. P. BURNESS, JARED DIAMOND & T. FLANNERY, 2001

Opposite *The world's tallest dinosaur skeleton, 13.27 m (44 ft) high, belongs to the sauropod* Brachiosaurus, *and is newly remounted at the Natural History Museum of the Humboldt University, Berlin. Partial skeletons and isolated bones indicate that there were much larger sauropods than this, however.*

Below *The size of one of the largest theropods,* Spinosaurus, *compared to two large Sauropods and a human.*

Dinosaurs were the largest animals ever to inhabit the land. Only modern-day whales attain greater dimensions and body weight. The key mystery is why dinosaurs were so huge, and yet today there are, for example, no 50-ton elephants. So which were the all-time giants among the dinosaurs, and how did they achieve their extraordinary body size? And is bigger always better? Edward Drinker Cope, the 19th-century palaeontologist, showed that animals generally increase in body size from small ancestors to larger descendants. This has been termed 'Cope's Rule', and if it is valid, the question is not so much what drives body size, but what limits it.

Giant dinosaurs evolved many times, but they all belong to two lineages, the Sauropoda and the Theropoda. Sauropods are distinguished by their enormously long necks carrying a diminutive head. The theropods include all carnivorous (meat-eating) dinosaurs and all were bipedal. While sauropods are always large to giant, theropods evolved an amazing variety of body shapes and sizes, including the only dinosaurs surviving today, the birds. Topping out at 14 tons, giant theropods were not nearly as large as sauropods, but they were still 10 times as large as the largest carnivorous mammals today.

Gigantic sauropods and theropods

Dinosaurs appeared in the Late Triassic, about 230 million years ago, and the first large sauropods are known soon after. The oldest truly giant sauropods are from the Late Jurassic of Africa and the USA, around 150 million years ago. They include *Brachiosaurus*, which had a body mass of 50 tons and held its neck in giraffe fashion, with its head towering 13 m (43 ft) above the ground, as well as giant relatives of the familiar *Diplodocus*, such as *Seismosaurus* and *Supersaurus*. Largest of all is perhaps *Sauroposeidon* from the Lower Cretaceous of Oklahoma, USA (145 million years ago): some truly enormous neck vertebrae indicate that this animal was about 30% larger than even *Brachiosaurus*. Equally monstrous may have been the titanosaur sauropod *Argentinosaurus* from the Upper Cretaceous of Argentina, perhaps weighing around 70 tons. And the discovery rate of new giants in the 50-ton class seems

Brachiosaurus
Spinosaurus
Sauroposeidon

DIE WELT IM OBEREN JURA
Brachiosaurus
Dicraeosaurus
Elaphrosaurus
Allosaurus
Dysalotosaurus

even to be increasing, with *Puertasaurus* in 2005, *Turiasaurus* in 2006 and *Futalognkosaurus* in 2007.

Body-size evolution in theropods appears to have been more gradual than in sauropods, with the giants only found in Cretaceous rocks. Best known is *Tyrannosaurus* from the end of the Cretaceous period (65 million years ago) in North America. The largest *T. rex* skeleton, 'Sue' at the Field Museum in Chicago, had an estimated life weight of 9 tons. However, *T. rex* was probably not the largest of the theropod giants.

Spinosaurus (13 m/43 ft) from the mid-Cretaceous of North Africa, with its distinctive sail-back, was at 14 tons possibly the largest and heaviest meat-eater of all times. *Giganotosaurus* (10 tons) from the Upper Cretaceous of Argentina and *Charcharodontosaurus* (12 m/40 ft, 6 tons), also from the mid-Cretaceous of North Africa, may have been larger than *T. rex* as well. All of these giant theropods appear to have been close to the biomechanical limit of a strictly bipedal animal.

Environment or design?

Are the upper limits to body size set by the environment or by the design of the animals? Important environmental parameters that are known to have varied through Earth's history are sea level and the configuration of the continents, as well as atmospheric composition, both in terms of oxygen and carbon dioxide content. However, no correlation between body-size evolution in dinosaurs and these variables has been demonstrated, suggesting that the design and ecology of the animals themselves are the key to their gigantism.

Such factors are still at work in modern animals, because there is a strict relationship between land area inhabited and maximal body size. If individual body size increases, a given landmass will support fewer individuals, which will lead to lower population densities and a higher risk of chance extinction. The message here is clear: species that grow too large will risk becoming extinct.

Giant dinosaurs could adopt various strategies to try to avoid this, which may help explain why they could grow to such a remarkable size. They could make better use of existing resources (and so take up more energy). Alternatively, they could use less of these resources (and save energy). Finally, they could reduce the risk of chance extinction by ensuring increased rates of recovery after a population crash.

Tyrannosaurus rex *does battle with* Spinosaurus, *both meat-eating theropod dinosaurs, in a scene from the film* Jurassic Park III.

The largest Tyrannosaurus rex *skeleton ever found, nicknamed 'Sue', is 12.8 m (42 ft) long and on display at the Field Museum, Chicago.*

Reproduction by egg-laying instead of giving birth to live young may have boosted giant dinosaur population recovery rates, allowing much lower population densities than were possible in similar-sized mammals. And in order to save energy, sauropods and theropods may have used less energy in locomotion than similar-sized mammals. Increased energy uptake could have been facilitated by more nutritious food or more oxygen per breath.

Evidence that some of these advantages may have existed comes from the hollow spaces in the backbone and ribs that are found in theropods and sauropods. These same spaces exist also in birds, which suggests that the lungs of saurischian dinosaurs were bird-like. Bird lungs are twice as efficient as mammal lungs in taking up oxygen. They also include large airsacs, meaning that the density of the body of the giant dinosaurs would have been lower than most animals. Transportation cost would therefore have been considerably less for an elephant-sized sauropod than for an elephant. The long necks of sauropods may also have made more efficient feeding possible.

Gigantism in the carnivorous theropods must have been closely tied to the evolution of giant sauropods, because they provided ample prey that helped theropods to take up more energy than is available in modern ecosystems. This is not the whole explanation, however, because at least one giant theropod, *Tyrannosaurus*, lived in an ecosystem where there was no sauropod prey.

A combination of factors

Were the largest dinosaurs as large as a land animal ever could be? Mathematical calculations have been based on a simple observation that the diameter of an animal's legs is proportional to body mass (p. 190). At a certain body mass, the legs would be so thick that the animal would seize up and be unable to walk. Calculations suggest that this would happen when the hypothetical über-monster reached a body mass of 150–200 tons – and no dinosaur was that big, as far as we know.

There was probably no single factor that made giant dinosaurs possible, but a combination of several biological features. Attempting to sort out this complex picture is currently a very active field of palaeontological research, involving not only palaeontologists, but also scientists from many other fields, such as biology, ecology, physics and even materials science. In this way, research on extinct animals such as dinosaurs has the potential to tell us much about the living world.

Why did the dinosaurs die out?

The great extinction that wiped out all of the dinosaurs, large and small, in all parts of the world... was one of the outstanding events in the history of life and in the history of the earth.... It was an event that has defied all attempts at a satisfactory explanation.
EDWIN H. COLBERT, 1965

Opposite *Artist's impression of a meteor impact such as might have caused the extinction of the dinosaurs. In reality the impact was much larger.*

Dinosaurs had been successfully living on Earth for 190 million years when their reign came to an abrupt end at the boundary of the Cretaceous and Tertiary (K/T) periods. For centuries their disappearance remained a complete mystery. But since the discovery of evidence for a giant impact that coincided with their demise, the story has become a little less mysterious. So what is the evidence and what does it tell us?

Finding the clues

To trace the final days of the dinosaurs, we need to go to the badlands of the western interior of North America. A large shallow sea once occupied the middle of the American continent; on its western shores were vast coastal plains, incised by meandering rivers draining the emerging Rocky Mountains, with numerous swamps in between. This is dinosaur country, and their bones fill the natural history museums of the world today. The peat fillings of the swamps have now turned into lignite and coal layers, the lowest of which has often been referred to as the Z-coal. This is a handy coal layer, because it documents the first coal layer above the last remains of dinosaurs, and was therefore used as a marker of the K/T boundary. The coals above also contain numerous ash layers of remote eruptions, now turned into what are known as bentonitic clays.

Within the Z-coal there is a unique clay layer 1–2 cm (less than 1-in) thick. The same layer can be traced from Alberta to New Mexico, over a distance of 1,800 km (1,118 miles), getting very slightly thinner from south to north. Such a thin

Shocked quartz from the K/T boundary. The visible lines or lamellae are caused by the passage of a shock-wave (sound), which melts the crystal, an effect only found in meteor craters and nuclear explosion tests.

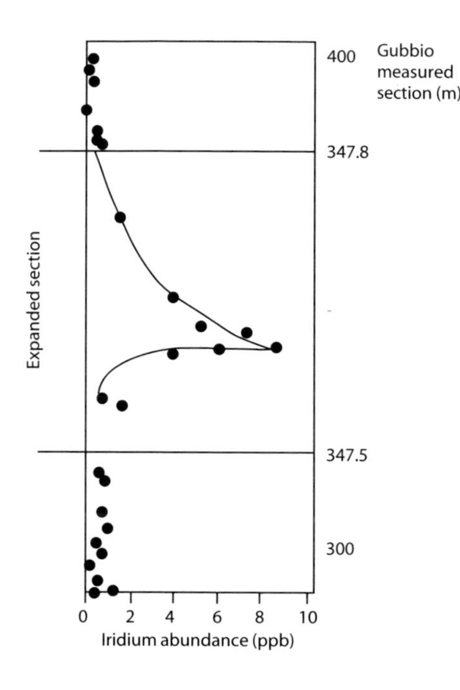

The iridium spike as first measured at the K/T boundary near the town of Gubbio, Italy.

clay layer could easily escape attention among the ash layers, but it was eventually identified because it marks a sudden change in the pollen assemblages. Plants are sensitive indicators of climate, thus their pollen data might provide insight into the changes. The pollen indicated that sub-tropical deciduous forests were destroyed, giving way to a carpet of ferns. After a few decades the pollen assemblages turned back to coniferous trees. Could this have been a deadly diet change for the dinosaurs?

Where did this clay layer come from? It resembled the volcanic ashes, and was likewise altered into clay. But a volcanic layer should contain crystals, which it did not. Moreover, an ash-layer extending over 1,800 km (1,118 miles) with only minor variation in thickness would be impossible to explain by a single volcanic eruption. Another clue came from the discovery of an iridium (Ir) anomaly precisely at the pollen change. Subsequent work in the badlands demonstrated that the iridium anomaly belonged to the clay layer.

Iridium is a remarkable element because it occurs in relatively large quantities in extraterrestrial material such as meteorites, but is quite rare in the crust of the Earth – about 500 ppb (parts per billion) to 0.02 ppb, respectively. Up to 25 ppb of iridium had been found precisely at the K/T boundary in numerous deep-sea localities, which led to the hypothesis of an extraterrestrial cause for the K/T mass-extinctions. We now know that this iridium peak is characteristic of a global clay-layer, 2–3 mm thick, which marks the extinction of many marine animal species such as ammonites, belemnites and all plankton with a calcareous skeleton. But is the clay-layer in the Z-coal the same as the one in the sea?

It must be, because the layer everywhere contains the same diagnostic products, such as small quartz and zircon minerals with unique shock-pressure features, and also the altered remains of

Below left
Concentric gravity anomalies of the Chicxulub crater – the initial indication that the crater was lurking below.

Below right
Map showing the ancient coastline of the Gulf of Mexico and the location of the Chicxulub crater.

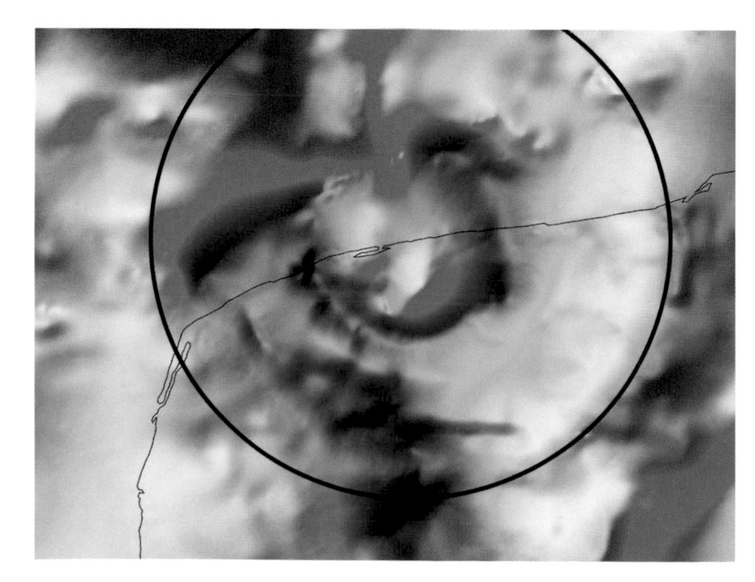

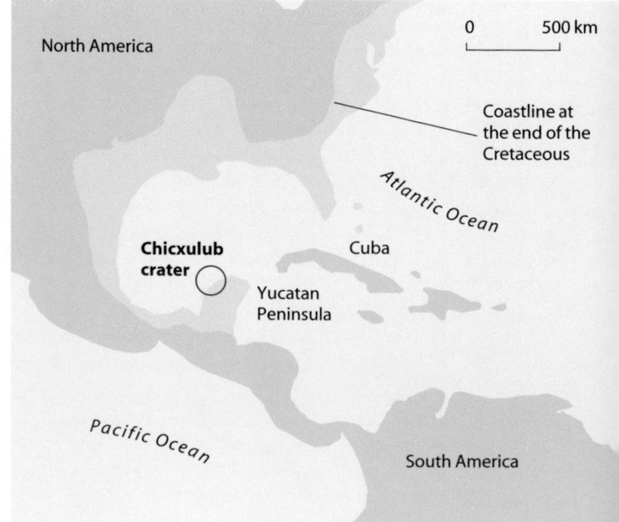

small glassy droplets and spherules that bear unique signatures of a distant meteorite impact. Such a thin layer may seem insignificant, but when you do the maths, spread over the entire world it translates into 800–1,000 cu. km of solid rock. It takes a lot of energy to emplace such a layer, and therefore a truly big impact must have happened somewhere on the Earth. On the basis of the amount of iridium, this meteorite is estimated to have had a diameter of about 10 km (6.2 miles). Arriving at a speed of 18 km (12 miles) per second it would have exploded with a force of 100 million megatons of energy and must have produced a crater 180–200 km (112–125 miles) wide.

Artist's impression of a giant impact at time zero + 0.1 seconds. The size of the impactor here is about 1,000 km (620 miles) – the Chicxulub impactor was about 10 km (6.2 miles).

Where was the crater?

Like a true detective story, the culprit was eventually tracked down by following a series of clues. No suitable crater was immediately visible from satellite observation, but it could have disappeared in the ocean, been eroded or completely filled with sediment. The amount of iridium did not vary significantly between different sites and thus provided no lead. However, the size of the shocked quartz increased clearly towards North America. Then in Texas and Mexico evidence was found for a major tsunami, which pointed to an impact in the Gulf of Mexico/Caribbean. At last, a series of circular geophysical anomalies, concentric with a ring of sinkholes in northern Yucatan, Mexico, betrayed the presence of a huge, 180-km (112-mile) wide crater, the Chicxulub crater, buried under a layer of sediments 3 km (almost 2 miles) thick. Subsequent analysis of the crater and worldwide ejecta at the K/T boundary demonstrated that this crater was indeed the source of the iridium anomaly and shocked minerals.

But why did this impact kill the dinosaurs? Such large impacts have happened before, but a connection with other major extinction episodes remains elusive. The key probably lies in the unfortunate combination of striking an unusual target at an unusual time. The impact target rocks include a 3-km (2-mile) thick layer of gypsum and limestone. The energy of the impact must have vaporized a large amount of this into dust, as well as aerosols of sulphur oxides and carbon dioxide, which spread ballistically over the entire globe at the top of the atmosphere. Sulphur oxide aerosols increase the albedo of the Earth and reflect sunlight back into space, thus cooling the Earth's surface. The dust blocked the Sun, and caused the collapse of photosynthetic plants and algae. All this happened on a timescale of months to years, and caused a huge mass-mortality, as witnessed by a sterile clay layer in the oceans.

The decaying corpses and trees released more carbon dioxide into the atmosphere, adding to the increased levels already caused by the impact. So after a few years of cooling, a greenhouse atmosphere warmed the Earth for several thousands of years. But is this enough to make all those animals extinct? This is where the unusual time comes into play. The global climate at the end of the Cretaceous was extremely mild and equable. No polar ice caps existed and dinosaur populations are known on all continents. The oceans were divided into (sub)tropical and temperate (boreal) realms. Organisms were likewise adapted to such mild environments, were specialized to a high degree and underwent relatively few evolutionary changes. How did the dinosaurs deal with all those sudden changes, then, if they were poorly adapted to climate variations? The answer is they couldn't, and that is why they went extinct.

JEAN-JACQUES JAEGER

Why do mammals rule the world?

I am fond of pigs. Dogs look up to us. Cats look down on us. Pigs treat us as equals.
WINSTON CHURCHILL

Mammals are warm-blooded animals that share a high metabolic rate and a body covered by hair. They feed their babies with milk from maternal mammary glands – the origin of their name. Humans are mammals of course, and so it is easy to say that 'mammals rule the world'. But do they, and if so, why?

Mammals are extremely diverse, in both size and food requirements, and are represented in all the major domains – terrestrial, aerial and even marine. Giants occur on land, such as the elephants, and in the sea – the whales. In addition, each continent and each large island houses its own specific endemic groups of mammals, whereas cosmopolitan species such as the mouse and rat have been spread only recently by humans. Such diversity is unprecedented in the fossil record and raises the interesting question of how this has come about

The first mammals

The first mammals arose about 230 million years ago. At that time, they already had the skeletal characters that allow palaeontologists to identify them as mammals, and they had probably also acquired warm-bloodedness and hair, features not preserved in the fossil record directly. These soft characters can, however, be identified by indirect evidence: microscopic bone structure indicates warm-bloodedness and pits in the bones of the snout suggest they had sensory whiskers, and so hair.

These earliest mammals were rare and tiny, and their biomass on the Earth's surface was probably negligible when compared to that of their contemporaries, the dinosaurs. The rise of the dinosaurs, and the limited success of mammals in the Mesozoic, may have been related to climatic conditions: climates were rather stable and the tropical zones extended further north and south than they do today, leaving only a small area of temperate climate – if any. Warm-blooded mammals have to use energy to fuel their muscles, and they also have to produce heat, which a cold-blooded reptile does not. Therefore, warm and stable climatic conditions can be considered as a bonus for cold-blooded reptiles and a disadvantage for warm-blooded mammals. Whatever the true reason, mammals during the Mesozoic remained small.

But the Mesozoic mammals continued to evolve. Some of the key changes occurred in their teeth, with many mammal groups developing more efficient chewing during the Cretaceous. Reptiles have numerous, simple teeth that are mostly used to pierce and hold their prey. Mesozoic mammals added two new functions to their teeth – cutting and grinding. Incisors and canines are used for piercing, premolars for cutting and slicing, and molars for grinding. This improved dentition meant the action of digestive enzymes was speeded up, allowing mammals to assimilate food much faster than reptiles.

In addition to this innovation, other changes can be observed in the fossil remains. Many other soft tissue and physiological characters evolved simultaneously, such as increasing brain complexity and intelligence, improved sensory organs, especially the senses of smell and vision, a double blood circulation (with oxygenated

A female mammal feeding its offspring. All placental mammals, such as this lioness, produce milk from their mammary glands, and the number of glands matches the typical numbers of young, four or five in this case.

blood from the lungs kept separate in the heart from the oxygen-poor blood returning from the body), increased blood pressure and gestation of young in the uterus rather than egg-laying.

The two-step diversification of modern mammals

The first diversification of modern mammals occurred during the Cretaceous, an event possibly connected with the evolution of flowers. About 100 million years ago, the angiosperms, or flowering plants, diversified rapidly worldwide. These plants commonly need insects to pollinate their flowers. So the rapid diversification of angiosperms led to an explosion of new insect groups, especially social insects such as bees, wasps and ants; insectivores such as mammals then also reaped the benefit. However, this radiation of mammals in the mid- and Late Cretaceous did not supplant the dinosaurs. Indeed, the new, Cretaceous mammals were still mostly small, with few of them larger than a cat. It seemed the dinosaurs were immovable.

The crucial event that triggered the rise of modern mammals was the extinction of the dinosaurs 65 million years ago (p. 42). Dinosaurs and pterosaurs disappeared from the land, leaving many empty niches, especially those of large herbivores and carnivores. Surprisingly, mammals did not immediately jump into these empty niches. A recent biodiversity data analysis, using the supertree method which combines molecular phylogenies and morpho-anatomical trees, concluded that the ancestors of modern groups did not increase significantly in diversity until more than 10 million years later.

This apparent delay in diversification is perhaps hard to understand, since the ancestors

Above *A small primitive mammal,* Bishops whitmorei *(foreground), from the Early Cretaceous of South Australia, some 100 million years ago. This mammal experienced freezing winters, and its basic warm-bloodedness and hair were essential for survival in these conditions.*

Right *Diagram of diversification of mammals through time, showing a current debate between molecular evidence, which suggests an ancient origin, and morphological and fossil evidence, which suggests a more recent origin, after the K/T boundary and the extinction of the dinosaurs.*

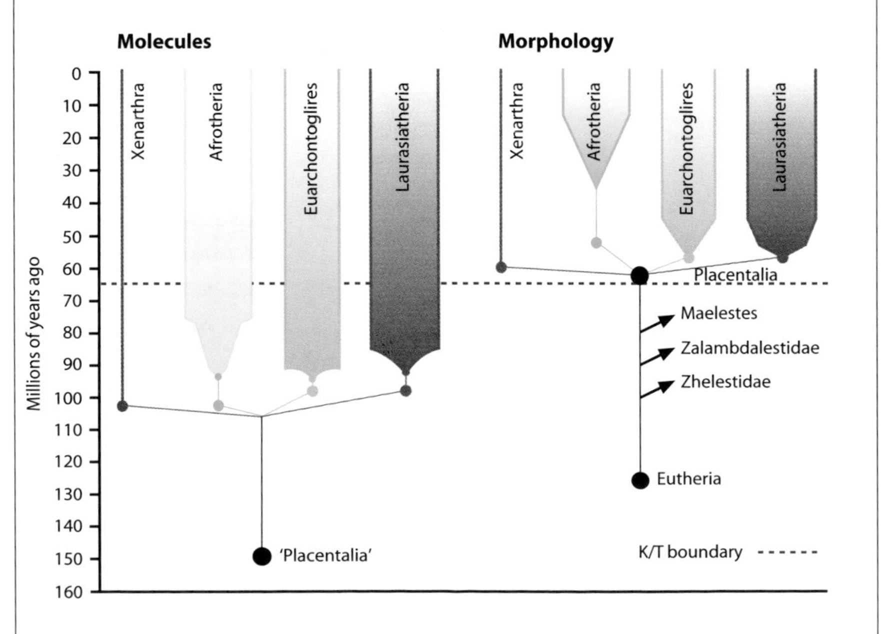

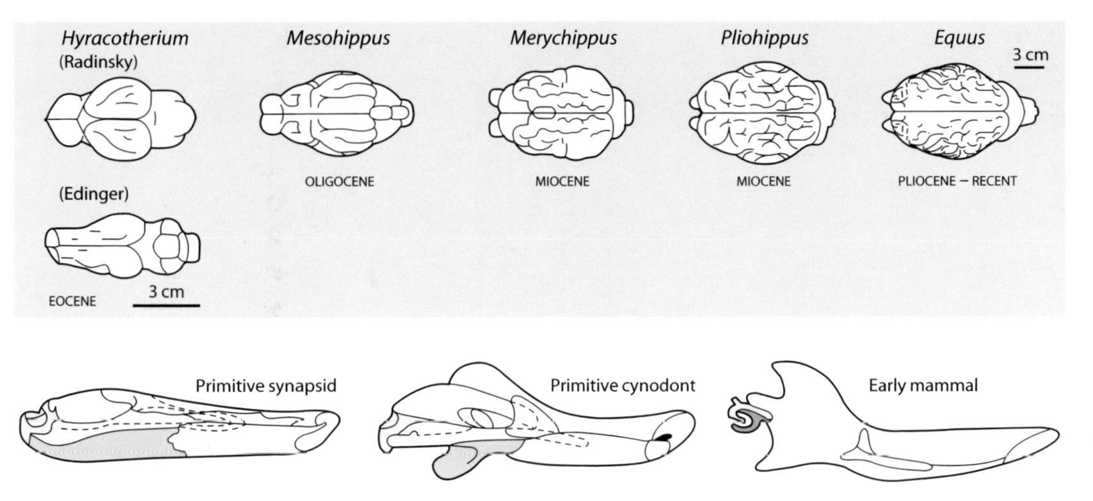

Above left *The evolution of the horse brain from* Hyracotherium *to* Equus *shows expansion of the neocortex, the 'thinking' part of the brain through time.*

Below left *Evolution from reptile to mammal: the jaws of a primitive synapsid, a primitive cynodont and an early mammal, showing how the five or six lower jaw bones become modified through time. The articular (yellow) and angular (blue) enter the ear of the mammal, and the mammalian jaw is composed solely of the dentary, the tooth-bearing element.*

of monkeys, cattle, elephants, bats and rats were already present before the demise of the dinosaurs. The delay in diversification may be because the loss of the dinosaurs, as well as other groups such as the flying pterosaurs, was so profound ecologically, that it took some time for ecosystems to stabilize. Indeed, during the 10-million-year post-extinction period, it was not clear which animals might finally replace the dinosaurs in all their niches: there were giant predatory birds that ate ancestral horse-like animals, some crocodilian groups became terrestrial flesh-eaters in some parts of the world, and there was a plethora of unusual mammal groups that subsequently went extinct.

Perhaps it was the cooling climates of the Cenozoic, a slow process that led to the Pleistocene ice age around 2 million years ago, that disadvantaged surviving reptile groups and allowed the warm-blooded mammals and birds to diversify.

Brain evolution and success of the mammals

We humans are rightly proud of our large brains, and indeed a large brain characterizes mammals in general. There is good evidence that both brain volume and brain complexity increased substantially during the Cenozoic – the past 65 million years. These changes especially affected the neocortex, which is classically associated with various aspects of intelligence. Development of the neocortex was particularly important in human evolution, but also in the evolution of most other modern mammal lineages during the Tertiary.

Intelligence is the ability to solve problems and to learn. Many animals behave largely instinctively (p. 217), and they cannot adapt readily to unexpected situations. The mammalian neocortex provides the power to adapt and to live a complex life. It is unclear why the neocortex expanded in so many mammalian lineages – perhaps if predators were becoming wilier, their prey were obliged to increase their intelligence at the same time or they would have been killed off.

The modern mammal groups prevailed in the end, and their success may relate to a number of anatomical and biological factors, such as their intelligence, excellent senses, parental care, warm-bloodedness and adaptability. Adaptability may be the key: mammalian teeth, for example, allow mammals to adopt a vast array of diets, from fruit-eating to mollusc-eating, and their warm-bloodedness gives them a wide geographic range, from the tropics to the poles. Mammals also, of course, include a wide range of body sizes, from tiny shrews that weigh a few grams to the 100-ton blue whale, and occupy habitats from the deep ocean to the tree tops.

So, the success of the mammals can be tied to at least five factors, two of them external and three internal: the extinction of the dinosaurs, the fact that the Cenozoic world was cooler than the Mesozoic, their warm-bloodedness, their adaptable teeth – and their increasing intelligence.

BERNARD A. WOOD

The hunt for the earliest human ancestor

In each great region of the world the living mammals are closely related to the extinct species of the same region. It is therefore probable that Africa was formerly inhabited by extinct apes closely allied to the gorilla and chimpanzee; and as these two species are now man's nearest allies, it is somewhat more probable that our early progenitors lived on the African continent than elsewhere.

CHARLES DARWIN, 1871

Opposite
A common chimpanzee; together with bonobos, chimpanzees are our closest living relatives.

Charles Darwin was aware of the evidence that links modern humans to the rest of the natural world. One of the achievements of modern biology is the confirmation of the close relationship between modern humans and the African apes. But which of the African apes is our closest living relative? Where should we look for the fossil remains of the creatures that lie at the base of the modern human twig in the tree of life (p. 116)? And which rocks, and of what age, are likely to yield those fossils?

Our closest living relatives

For more than 150 years scientists have known that the twigs closest to ours on the tree of life belong to the African great apes, namely, in alphabetical order, bonobos (or pygmy chimpanzees), chimpanzees (common chimpanzees) and gorillas (lowland and highland). Until relatively recently the conventional wisdom was that bonobos, chimpanzees and gorillas are more closely related to each other than any of them is to modern humans. Indeed, we were considered special enough to justify having a whole Linnaean family, called the Hominidae (hence our old name 'hominids'), to ourselves.

However, in the last half-century researchers have been able to compare modern humans with the great apes at the molecular level, for instance by studying blood proteins, such as albumin and haemoglobin, and now DNA. These comparisons have revealed a different pattern of relationships: the consistent message revealed by the molecules is that modern humans and bonobos/chimpanzees are more closely related to each other than either is to any other great ape. So, the first of our questions is answered. Our closest living relatives are bonobos and common chimpanzees. By convention we now call modern humans and all the extinct creatures on our twig of the tree of life, hominins, and bonobos and common chimpanzees and all the extinct creatures on their twig of the tree of life, panins.

Where should we look for fossil evidence?

Darwin suggested that because living bonobos and chimpanzees are confined to Africa, it makes sense to look within Africa for fossil evidence of the creatures at the base of the modern human twig. Because of what has been discovered about the close molecular ties between modern humans and living bonobos and chimpanzees, that logic is even sounder today than it was in Darwin's day. Fossil hunters thus long ago abandoned Asia as the possible location of our earliest ancestors and now focus their efforts on Africa.

Most mutations occurring in our DNA make no significant difference to us or to our lives. These so-called 'neutral' mutations occur at a constant rate, so the longer two adjacent twigs of the tree of life have been separate, the greater the number of differences that will have accumulated in their DNA, or in the proteins coded by that DNA. This system amounts to a 'molecular clock'. But how do we calibrate a molecular clock? The answer is to check it against some well-dated fossil evidence that relates to the splitting of two of the twigs on the tree of life. When researchers do this, the differences between equivalent molecules in modern humans and in bonobo/chimps

suggest their twigs have been separate for between 6 and 4 million years. So it would be unwise to look for fossil evidence from the base of the modern human twig of the tree of life in rocks that date to much before, or after, that time.

What have we found?

There are many differences between the skeletons of living modern humans and bonobo/ chimps, particularly in the brain case, face and base of the cranium, teeth, hand, pelvis, knee and the foot. However, scientists searching in 6–4 million-year-old sediments for fossil evidence of the earliest human ancestors must consider the possible differences between the first hominins and the first panins. These are likely to have been subtler and less obvious than the differences we see between contemporary hominins and panins.

Compared to panins, the earliest hominins would probably have had smaller canine teeth, larger chewing teeth and thicker lower jaws. There would also have been some changes in the skull and skeleton of the earliest hominins, linked both with the fact that they spent more time upright and with their greater dependence on the hind limbs for bipedal walking. These changes would have included a forward shift of the hole in the base of the cranium where the brain connects with the spinal cord, along with wider hips, straighter knees and a more stable foot.

The first person to claim they had discovered early – if not the earliest – evidence of our lineage in Africa was Raymond Dart. In 1924, while rummaging through a box of what others believed were monkey fossils from a cave called Taung, in southern Africa, he came across a skull that showed a mix of an ape-like small brain together with teeth that were more like those of a modern human. Dart called his new species *Australopithecus africanus* and boldly declared that it was an early ancestor of modern humans. We now know that at 2–3 million years old, *A. africanus* is too young and not primitive enough to be at the bottom of the hominin twig of the tree of life.

For the past 50 years the search for our earliest ancestors has focused on East Africa. In the 1970s Don Johanson and his colleagues found evidence of another early hominin called *Australopithecus afarensis* at a site called Hadar in Ethiopia. The most famous evidence of this hominin is the 3.2-

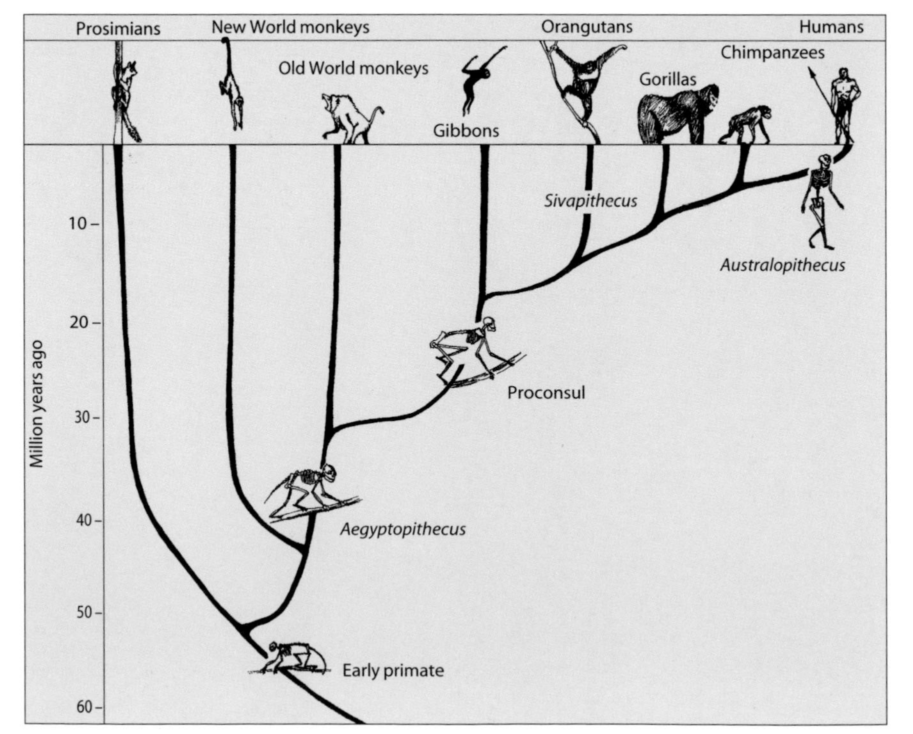

The closest twig to ours on the tree of life belongs to chimpanzees and bonobos.

million-year-old 'Lucy', as well as the 3.5-million-year-old set of footprints found by Mary Leakey at Laetoli in Tanzania. A recent discovery is Dikika child, an infant of the same species. But with hind sight *A. afarensis* was also neither sufficiently primitive nor old enough to be the first hominin.

Currently two early hominins from Africa, *Orrorin tugenensis* from Kenya and *Sahelanthropus tchadensis* from Chad, are vying for the coveted title of being 'the earliest hominin'. The bones of *O. tugenensis*, collected by Brigitte Senut and Martin Pickford, have been dated to around 6 million years old, so they are the right age to be our earliest ancestor, but the evidence thus far collected for *O. tugenensis* is not good enough to be sure that it is indeed a hominin and not a closely related ape.

The remains from Chad that make up the fossil record of *S. tchadensis* were collected by a team led by Michel Brunet and are believed to be 6 to 7 million years old. If the molecular biologists are correct, then this puts them just before the time when we would expect to find evidence of the first hominin. There is a better fossil record for *S. tchadensis* than there is for *O. tugenensis*, but it is still not good enough to be able to say for certain whether the former really is the common ancestor of all later hominins.

Researchers, encouraged by the discoveries in Chad, are now busy searching in countries such as Libya, Algeria, Niger and Mali for new sites that may yield good enough fossil evidence of the right type and age to be, if not the earliest hominin, then at least one of the earliest of our hominin ancestors.

Above *'Lucy' is the best known example of* Australopithecus afarensis. *At 3.2 million years old, however, this hominin is too young to be the earliest human ancestor.*

Above left *Zeresenay Alemseged holds a piece of the fossil skeleton of Dikika child, one of the most complete examples of* A. afarensis, *which was found in 2000 in Ethiopia.*

Left *The skull of* Sahelanthropus tchadensis, *possibly an early hominin ancestor, found by Michel Brunet in Chad.*

The Earth

The Earth is not as solid as people once thought. When Alfred Wegener first put forward his theory of continental drift almost a century ago, he was beset by critics. How could the solid Earth move, and where was the energy, the mechanism, that could shift huge segments of the Earth's crust around like flotsam floating on the surface of the sea?

A modern view of the Earth arose only in the 1950s and 1960s, when the mechanism of plate tectonics was formulated. The engine for movement was not some sort of outboard motor propelling each continental plate over a sea of molten magma. Rather, the oceans and continents were all in motion, and the motor was more passive – great convection cells driven by the vast heat in the core and lower mantle of the Earth. Huge, slow-moving gyres of heat rise from the depths, approach the surface, and diverge below the crust. Centimetre by centimetre, the diverging spirals pull the ocean floor apart along the mid-ocean ridges, and fresh basalt lava bubbles up, creating new crust. Elsewhere, at some ocean margins, ocean crust is subducted beneath continental plates. Where continental plates collide, such as in the Himalayan region, mountains may shift and rise, and measurements can show by how much, though ensuring the accuracy of the measurements is critical.

Despite our improved understanding of the Earth, mysteries do still remain. At the greatest depths, the core of the Earth is a subject of significant interest. The core is dense and vastly heavy because of its depth and its metallic composition – it largely consists of iron. Geophysics can reveal

Mount Etna erupting, with the city lights of Catania behind. Volcanic eruptions reveal the power and mystery of the forces that act deep within the Earth.

Mount Everest is the highest mountain on Earth. It sits above a massive continental margin collision zone, where the subcontinent of India is thrusting northwards beneath the main Asian plate. Geographers have long struggled to measure the exact height of Everest accurately.

the shape and physical properties of the core, and we know, for instance, that it is the motion in the liquid outer core that generates the Earth's magnetic field by a dynamo process.

Knowledge of the core, mantle and crust is essential also for understanding the present nature and make-up of the Earth's surface. There was a time when the surface was molten, and the history of differentiation of the solid and molten parts of the Earth many billions of years ago can be discerned from the study of ancient rocks at the Earth's surface, but also from experiments and chemical calculations.

Increased understanding of the Earth's depths can also have commercial advantages. Oil, and other hydrocarbons, are essential to our daily lives and fuel global economies, and a knowledge of where oil comes from is thus crucial. It may now seem obvious that oil was formed from dead plants and animals, but new studies have reopened the debate about this most useful of the Earth's treasures.

Two great mysteries are the source both of the oceans and of the atmosphere. It is well known that today most of the Earth's surface is covered by water, and this has been the case for a very long time. But it is only now beginning to be understood how the water separated from the molten rock of early Earth, and gathered at the surface.

The early atmosphere of the Earth is also hard to comprehend. Oxygen is crucial to us and to much of life today, and yet to begin with Earth had no atmosphere at all, and when an atmosphere began to develop, at first it contained next to no oxygen. Current thinking is that life itself generated the oxygen. The simplest life forms today are anaerobes, which respire without oxygen – indeed oxygen can be fatal to them. The first life on Earth was like this. Then, atmospheric oxygen levels rose in two or more steps during Precambrian time, and a great deal of research is beginning to show when and how this happened.

Volcanoes may seem familiar from the numerous images of eruptions seen in the press and on television. What is perhaps less well known is that there are many different kinds of volcanoes, and their positions on the Earth and their styles of eruption link directly to plate tectonics. Volcanically active parts of the world lie along the great plate boundaries. But it has been hard to study how volcanoes erupt, partly because they do so unpredictably (one aim of volcanology is to identify tools to predict eruptions), and also because they are dangerous places to be. The search continues, though, to understand how volcanoes erupt and how particularly large eruptions have affected humanity, as well as how certain interesting minerals, such as diamonds, form within particular kinds of volcanoes.

Volcanoes are one of the visible signs of the inner workings of the Earth, and are related to earthquakes, caused by movements at plate boundaries. Both can trigger tsunamis by displacing large masses of water in the oceans. These can travel at tremendous speeds and reach the shore as tidal waves, with devastating effects.

Disasters can also come from extraterrestrial sources. Until the 1960s, few geologists accepted that the Earth was subject to bombardment by meteorites just as much as the Moon or any of the other planets. The evidence is there in the rock record and on satellite photographs. Meteorite strikes are of all sizes, from the minute to the huge, which have had catastrophic consequences for life on Earth – and might again in future – making their study one of immense relevance.

DAVID GUBBINS

Why does the compass point north?

The mysterious course of the magnetic needle is equally affected by time and space, by the sun's course, and by changes of place on the Earth's surface.
ALEXANDER VON HUMBOLDT, 1849

In England today a compass points about 5° west of north. A hiker in the hills can afford to ignore this, a yachtsman navigating across the English Channel cannot. In New Zealand the error is over 20° and must be taken into account. This compass error, or *declination* to give it its scientific name, has changed a great deal in the past – by some 20° in England in the last couple of centuries. Early navigators knew declination changed from 'westing' to 'easting' as they crossed the Atlantic Ocean, and later found the line delineating the two, where the compass was true, to be gradually moving westwards. This gave rise to what for a long time was viewed as the main change in the magnetic field, the *westward drift*.

Much more dramatic changes have occurred in the distant past, however. A million years ago – not long in geological terms – the compass would have pointed south. These *polarity reversals* have occurred every few hundred thousand years throughout Earth's history. Just 40,000 years ago the poles flipped for a few centuries in what is known as an *excursion*. There have been 10 or more such excursions since the last reversal proper. Palaeomagnetism, the study of the magnetization of rocks, has uncovered much of the rich and long history of our magnetic field. Most spectacularly, new rocks formed at mid-ocean ridges as tectonic plates move apart (p. 61) are magnetized at formation to leave a pattern of magnetic 'stripes' on the ocean floor corresponding to the reversal record.

It was once thought that the compass was attracted to a large magnetized object in the northern sky, and then to a mountain on Earth. Variations in declination from place to place led

A world sea chart by Edmond Halley showing magnetic declination of the compass. Note the zero contour running down the central Atlantic. Early navigators knew that this line, where the compass points to true north, was drifting west. It now runs through the land mass of the Americas.

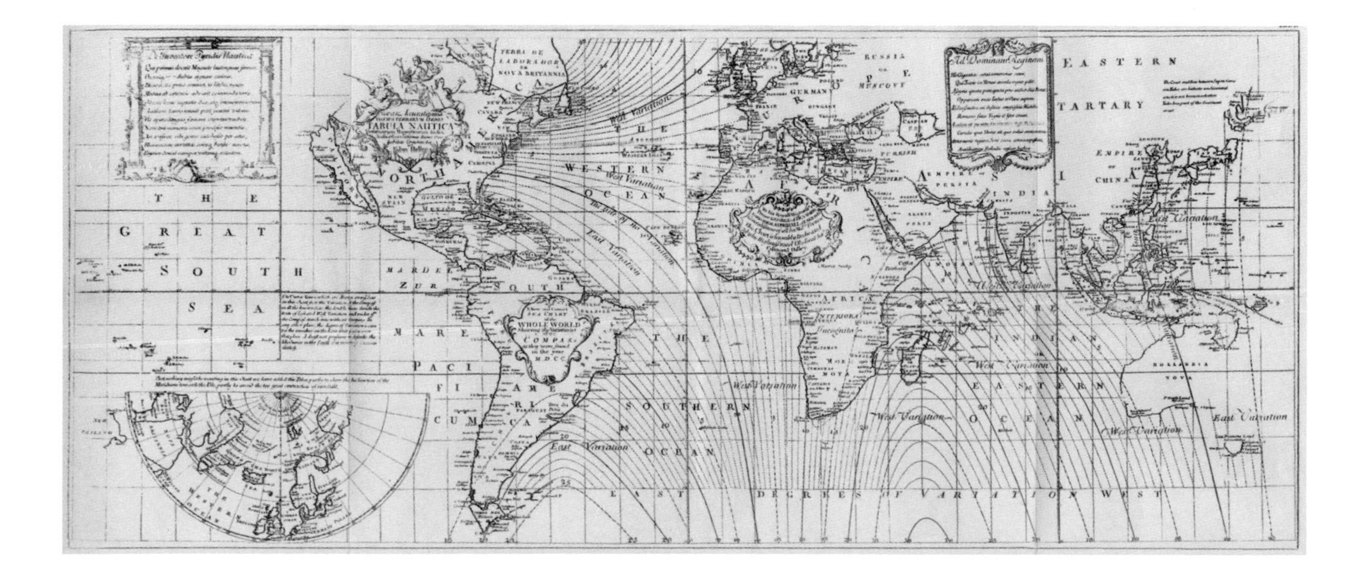

Magnetic stripes on the ocean floor. As new crust is formed at mid-ocean ridges it is magnetized by the ambient magnetic field. Magnetic reversals mean the crust is magnetized in alternate directions according to its age, leaving a detailed pattern of past plate motions on the ocean floor.

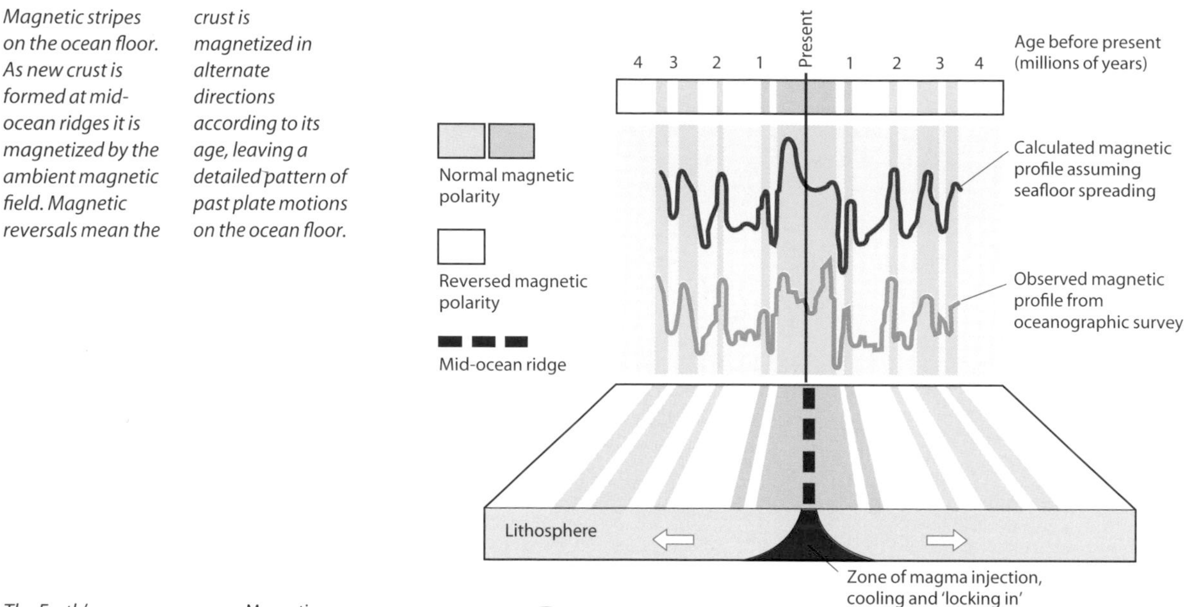

The Earth's magnetic field: the inclination of the lines of the field – the dip – changes as you move north or south. Compasses must be weighted according to their latitude in order to compensate for the downward magnetic force. The total force is horizontal near the equator – the line of zero dip is called the 'magnetic equator'.

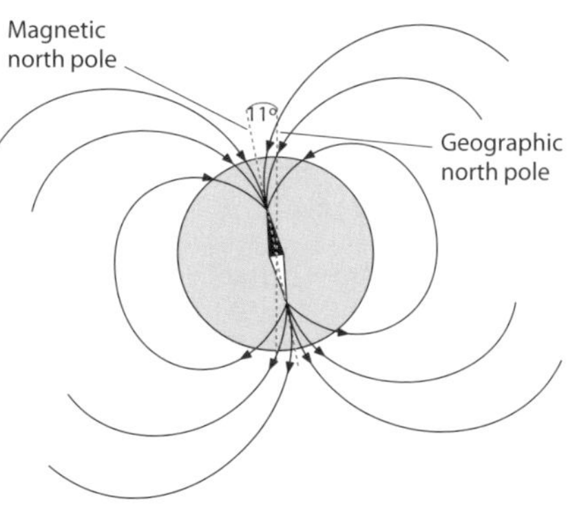

Edmond Halley to search for two north magnetic poles rather than one, and the changes with time led him to the notion of moving sources inside a hollow Earth – a remarkable prediction of the liquid core. We now know the Earth possesses an iron core, the outer part of which is liquid (p. 19). The Earth's interior is extremely hot, about 5,000°C (9,032°F), above the melting point of iron, but as pressure increases with depth, the melting point rises, causing the innermost 1,200 km (745 miles) to freeze. Fluid motion in the liquid outer core generates the magnetic field by a dynamo process, exactly the same as in a modern car alternator.

Dynamics

Liquid iron in the core moves under the action of three main forces: buoyancy, rotational and magnetic. Buoyancy forces are essential and provide the energy required to power the dynamo. They arise from density variations caused by differences in temperature and composition. Rotational (Coriolis) forces provide no energy but organize the flow and effectively align the magnetic field with the Earth's spin axis. Laboratory experiments show that convection driven by buoyancy in a rotating container takes the form of small rolls aligned with the spin axis. Magnetic forces counteract the rotational force to some extent, making the convection rolls larger and limiting the flow speed. The key process causing self-generation of the magnetic field is the most difficult to study, but it is now thought that the essential ingredient is the *helicity* of the flow, or the degree of 'push and twist' it possesses, which strengthens an existing field.

Energetics

Not much power is needed to drive the Earth's dynamo – a couple of power stations would do it – but it needs to be in the right form. The standard model now adopted invokes two sources of buoyancy from the slow cooling of the Earth, which both releases heat and freezes the liquid core from the bottom up. The solid inner core grows by accretion, releasing latent heat; it also consists of a more pure form of iron than the outer core, and impurities in the form of lighter elements are left behind in the outer liquid core to rise, to drive more fluid motion. The currently favoured light element is oxygen, but sulphur and silicon are also candidates. This energy source, coming ultimately from the Earth's gravitational energy, is more efficient at generating magnetic field than heat because it does not suffer from any thermodynamic inefficiency. Calculations suggest the cooling rate is a few hundred degrees throughout Earth's history and that the inner core started to form about a billion years ago.

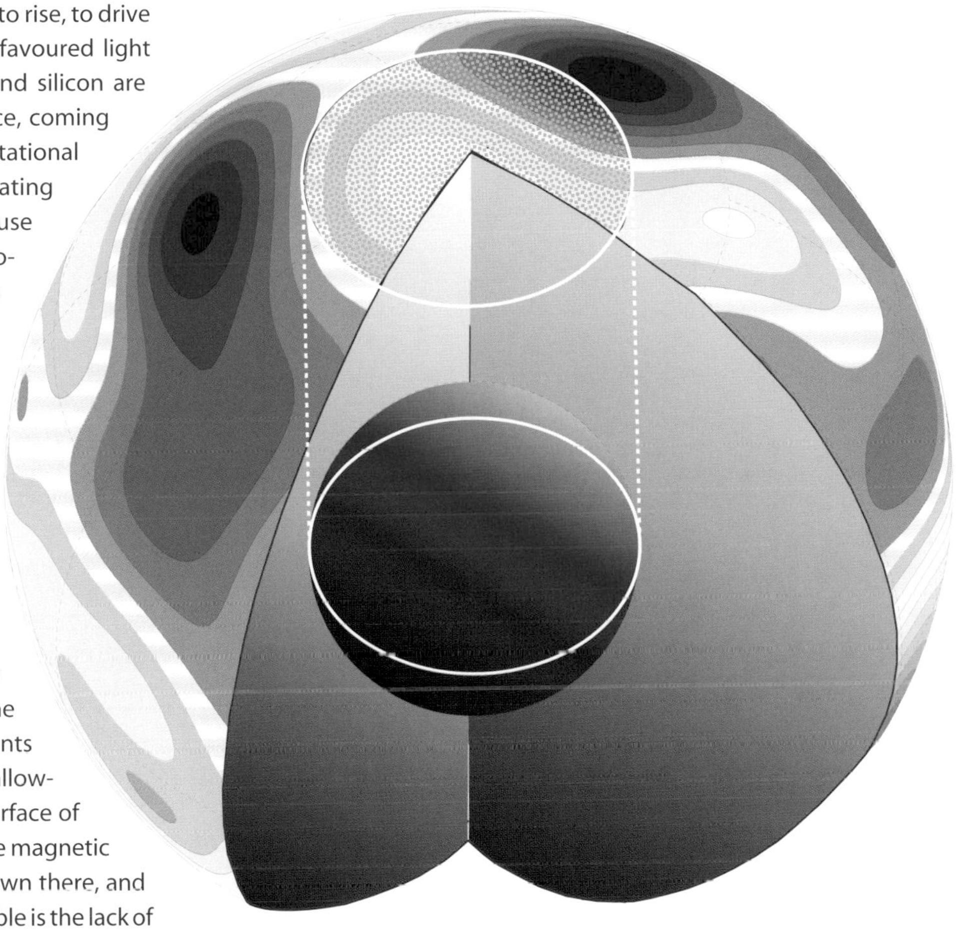

The core magnetic field

Nowadays the Earth's magnetic field can be mapped in great detail by satellites, aircraft and ground observatories. These maps may be continued down into the interior, provided no electric currents or magnetic sources lie in the way, allowing maps of vertical force at the surface of the Earth's core to be drawn up. The magnetic field is about 10 times stronger down there, and much more complex. Most noticeable is the lack of field near the poles; instead, it is mostly concentrated into four symmetrical main lobes centred around 60° N and S, and 120° E and W. There are numerous 'magnetic equators' and a large patch of field in the reversed direction beneath the south Atlantic: this patch has grown in historical times and if it continues to grow at the present rate it could presage another excursion or even reversal about a thousand years from now. The patches near the equator are drifting west and account for the westward drift seen at the Earth's surface; there is very little change in the Pacific.

The symmetry of the four main lobes across the equator clearly shows the influence of rotation. It is possible they represent the ends of aligned convection rolls, although this is almost certainly an over-simplified interpretation. Their displacement away from the poles demonstrates the influence of the solid inner core, which in the rotating sphere divides the fluid into two dynamically distinct regions, separated by the *tangent cylinder* which contains the inner core and has the same axis as the spin axis of the Earth. The magnetic field is generated preferentially just outside the tangent cylinder by shear of the fluid there. It is now possible to simulate the Earth's dynamo on computers, and these simulations reveal this effect clearly. Computer modelling is in its infancy and it is still impossible to reach the right parameters for the Earth's core – notably its rapid rotation rate – but they can already reproduce several important features of the Earth's field.

The core surface field showing the effect of the tangent cylinder, and that the magnetic field is concentrated outside the cylinder. Note the symmetrical four-fold pattern and the absence of a magnetic field over the poles.

Effects of the mantle

Some surprisingly long-term features can be found in the geomagnetic record. Most noticeable are long intervals when reversals ceased – the *superchrons*. The last was in the Cretaceous interval and corresponds to the ocean floor that now is

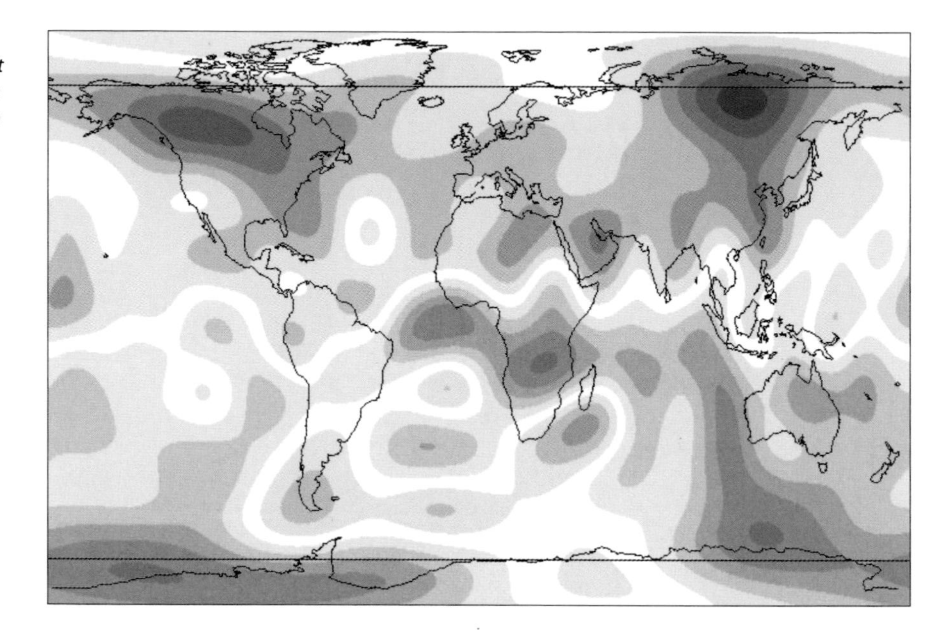

The vertical magnetic force at the surface of the liquid iron core in historical times. Note the symmetrical pattern of the four largest concentrations of flux in high latitudes.

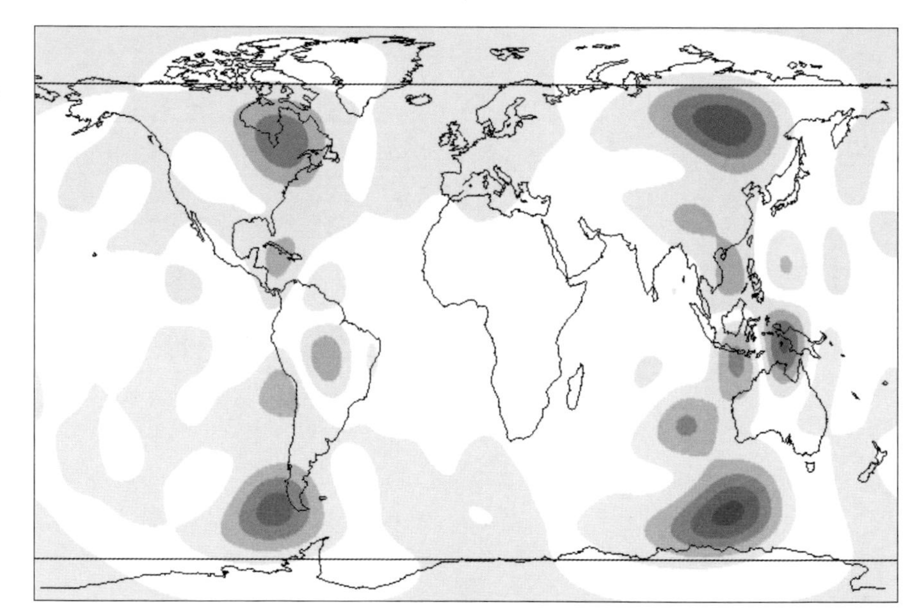

Computer simulation of the Earth's dynamo, in which heat is lost preferentially around the Pacific Rim, which is cooled by subduction of plates. The main lobes are in the same locations as the present field, showing the influence of the mantle on the Earth's dynamo.

found at the edges of the Atlantic Ocean and the western Pacific. Liquid iron in the core turns over every thousand years or so, and so it is hard to see how the dynamo can change on time scales as great as 100 million years. This time scale is right, however, for a turnover of the solid mantle. More observations suggest the influence of the mantle: the four main lobes appear to be a semi-permanent feature, as do weak time variations in the Pacific; the poles, when they reverse, seem to follow the Pacific Rim. These are controversial observations, but can be explained if more heat is drawn from the core around the Pacific Rim, where the mantle has been cooled by persistent subduction of plates over the last 100 million years.

Geomagnetism is one of the oldest sciences that continues to present intellectual challenges as well as practical applications – from one of the most difficult problems set for modern computing, through detailed information on past movement of continents, to new methods used in the search for oil and mineral deposits.

CHRIS HAWKESWORTH

How continents and oceans form

Besides I could not tell upon what geological theory to account for the existence of such an excavation. Had the cooling of the globe produced it? I knew of celebrated caverns from the descriptions of travellers, but had never heard of any of such dimensions as this.

JULES VERNE, 1864

At the present time, around 70% of the Earth's surface is covered by the oceans and around 30% by land, very largely in the continents. Key questions are, how does this division arise, and has it always been the same? Jules Verne was scientifically well read for his day, and his imagining a great ocean in a vast subterranean cavern was not so ludicrous then as it might seem now. But it is still puzzling to account for the volume of water on the Earth, and also how the relative proportions of ocean and continent are maintained.

The distribution of ocean and land reflects the different compositions and volumes of the rocks that underlie the oceans and comprise the continents, as well as the amount of water on Earth. An interesting feature of the Earth is the distribution of elevations above and below sea level: the average height of the land is 0.8 km (½ mile), and 21% of the Earth is between 0 and 1 km (0.6 mile) above sea level. However, 53% of the Earth's surface lies under 3–6 km (2–3.7 miles) of water, and the average depth of the oceans is 3.7 km (2.3 miles). The rocks of the Earth's crust in effect float

A satellite image of the Earth: today, 70% of the Earth's surface is covered by water, and only 30% by land.

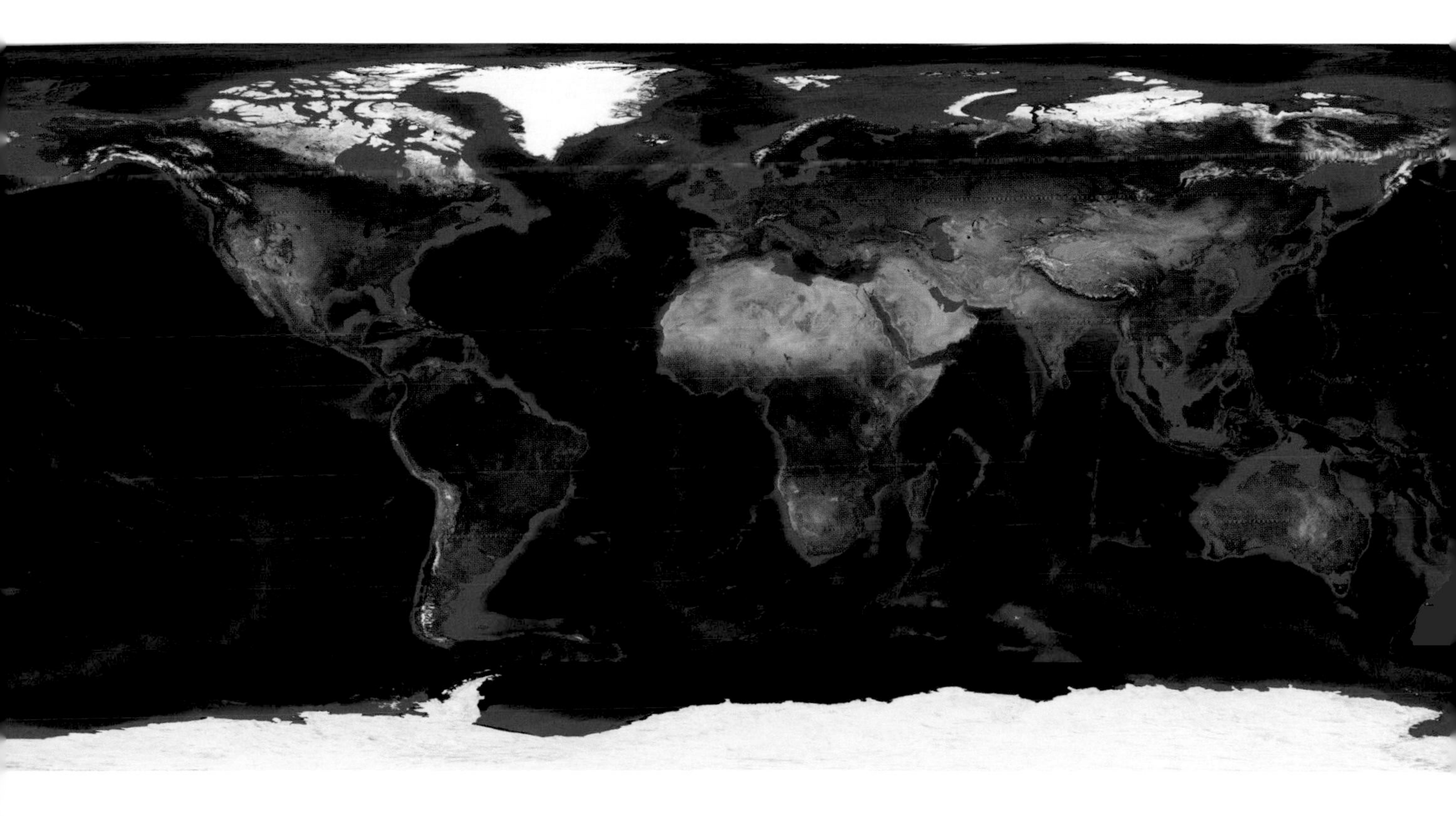

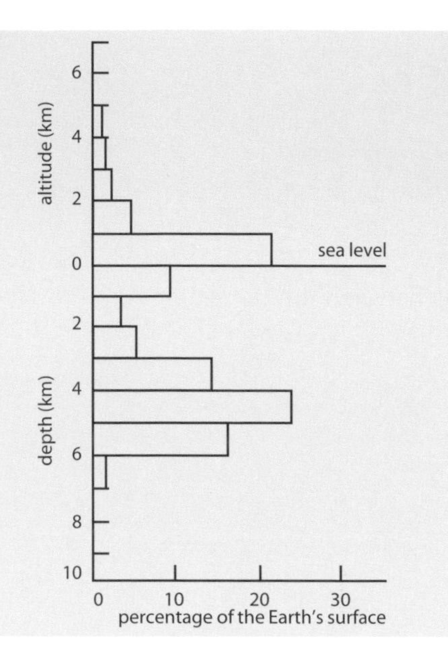

A diagram showing the distribution of elevations on the Earth relative to sea level. The clear difference in distribution indicates that the rocks of the ocean floor and the continents have markedly different compositions.

on the underlying mantle, and this clear bimodal distribution of elevations highlights the fact that there are two dominant compositions with different densities – one beneath the oceans and the other in the continents.

Rocks of the oceanic and continental crusts

The oceans are underlain by rocks of basaltic composition, with a density of around 3 g per cubic centimetre, whereas the average composition of the typical rocks that form the continents has higher quantities of the slightly lighter silica, silicon dioxide (the main constituent of sand grains), and so overall a lower density (of around 2.8 g/cc). The rocks of the oceanic crust are created at volcanic centres along the mid-ocean ridges, where plates pull apart and magma rises up from the mantle, and then return to the mantle where plate margins collide, at subduction zones.

It follows that the rocks under the oceans are not very old – in fact they all formed in the last 180 million years. In contrast, the more buoyant rocks of the continental crust are less readily subducted back into the mantle, and are on average much older. The rocks of the continents are therefore an archive of how conditions have changed throughout much of the Earth's history. The oldest mineral is 4.4 billion years old (a zircon from Western Australia), the oldest rock is 4 billion years old, and the average age of the continental crust is thought to be around 2 billion years.

The rocks are also formed in different ways. The basaltic rocks under the oceans are generated by partial melting of the upper mantle, whereas those of the continents are more evolved chemically. They are thought to have been generated in a two-stage process, such as by the remelting of basalt that was itself originally derived from the mantle.

The continental crust

Although there has been continental crust on the Earth for over 4 billion years, most of the rocks we actually see are relatively young: 80% of the rocks in the continents were formed in the last 1 billion years, and only 1% of the crust is older than 2.5 billion years – yet half of the continental crust had been generated by that time. This apparent paradox highlights both how difficult it is to study how most of the continental crust was generated,

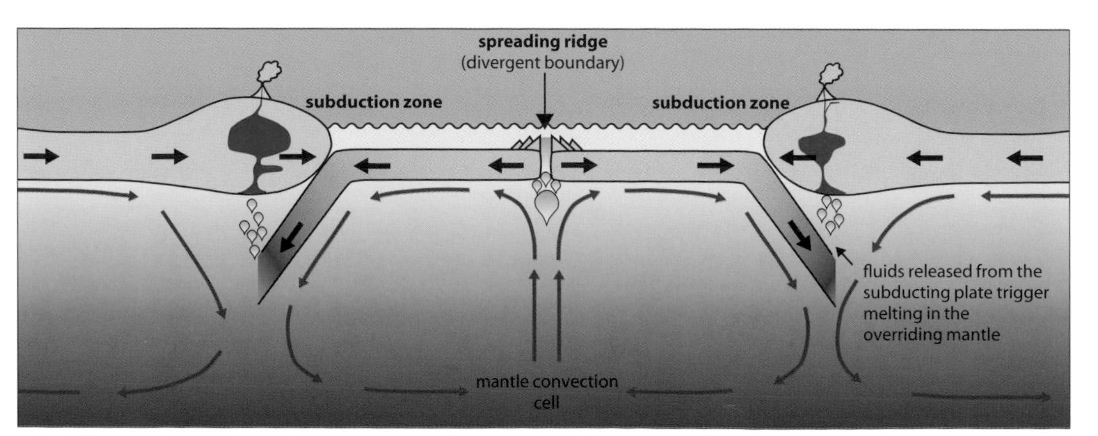

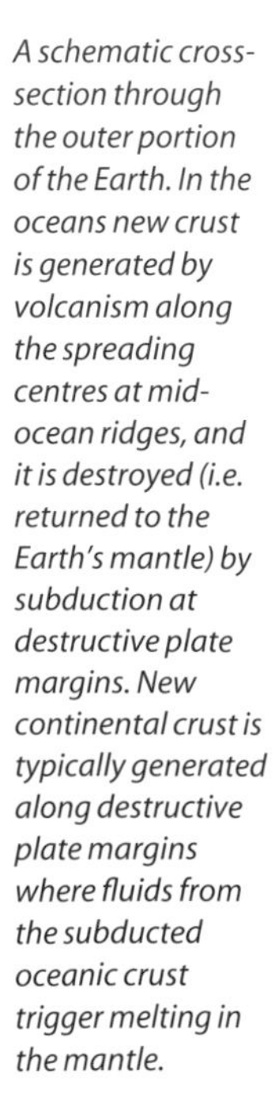
A schematic cross-section through the outer portion of the Earth. In the oceans new crust is generated by volcanism along the spreading centres at mid-ocean ridges, and it is destroyed (i.e. returned to the Earth's mantle) by subduction at destructive plate margins. New continental crust is typically generated along destructive plate margins where fluids from the subducted oceanic crust trigger melting in the mantle.

and also that the ages of the crust and the rocks of the continents are two different things.

The age of the crust is when it was generated from the underlying mantle, while the age of a rock is clearly when it formed – when an igneous rock crystallized or a sediment was deposited. However, many rocks contain material derived from other rocks that were already part of the continental crust. So 'crust formation ages' will be older than the ages of the rocks themselves, and will be different in different places.

More widely, the generation of the continental crust is a major feature of the differentiation of the silicate Earth (p. 19). Crustal rocks were then eroded and returned to the mantle via subduction zones. Chemical reactions involved in weathering of the crust typically require water and CO_2, via carbonic acid. Weathering of the crust therefore draws down CO_2 from the atmosphere, and so influences global climate; it may even have been the cause of the oldest known glaciation, at 2.9 billion years ago.

In studying the age distributions of rocks from the continental crust, it is striking that there are peaks representing times when relatively large volumes of new continental crust appear to have been generated. These peaks occur at 2.7, 1.9 and 1.2 billion years ago; none have been identified in the last billion years. Assuming that the peaks are representative of the history of the crust, and are not some artifact of preservation, the implication is that there has been a fundamental change in the processes involved in the generation of the continental crust through time.

The Jack Hills rock formation, Australia: the rocks here are over 3 billion years old and contain crystals of the mineral zircon which are up to 4.4 billion years old.

Continental crust is broadly generated in two different settings: at present most new continental crust forms along plate margins, while some may be generated within plates, away from the margins. Crust generation along plate margins is in response to shallow level processes: the tectonic movements between plates on the Earth's surface. In contrast, crust generated within plates may be linked to the upward movement of material from deep in the mantle, often termed mantle plumes. In this scenario, the generation of crust is in response to deep-seated thermal instabilities in the Earth's mantle.

One outcome of this is that rates at which crust was generated would have been very different at different times. Earlier in the history of the Earth, more energy was produced from the radioactive decay of isotopes of uranium, thorium and potassium, and it is easier to envisage that more crust was generated as a result of mantle plumes. One interpretation of the relative lack of peaks of crust formation after 1 billion years ago is that there was a shift in the balance of the way new crust was generated, with magmatism (the formation of igneous rocks) in response to plate tectonics becoming more dominant.

Volume of the oceans

The area of the Earth's surface covered by either continents or oceans depends on the volumes of water and continental crust. The amount of continental crust reflects the balance between the volumes generated from magma on the one hand and the amounts subsequently destroyed by erosion and being recycled back into the mantle on the other. We cannot estimate either with any confidence at present. The size of the oceans varies depending on how much water is locked up in ice sheets and hence the extent to which ocean water has retreated or encroached on to the land. Some 18,000 years ago, for example, towards the end of the last Ice Age, sea level was 130 m (427 ft) lower than it is today, and it was possible to walk from England to Holland on dry land, while a rise in sea level is one predicted result of global warming.

It is possible to estimate the total amount of water currently in the oceans and ice sheets, around 98% of which is in the hydrosphere and atmosphere. In principle, the total amount of water in the Earth can also be estimated from analyses of meteorites of material similar to that which formed the Earth, which suggests it initially contained around 2% by weight. We do not know how much water has since been lost, but it has been argued that there is sufficient water within the Earth below the ocean floor to fill between two and five ocean basins. Even Jules Verne could little have imagined perhaps that the water within the Earth may be locked into rocks to depths of 2,900 km (1,800 miles) – the base of the Earth's mantle.

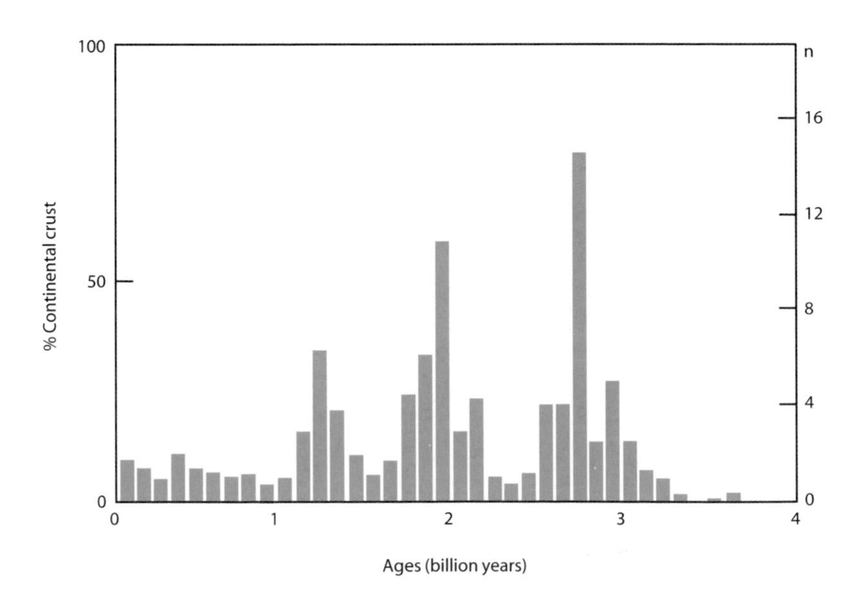

The distribution of the ages of rocks that represent new continental crust from 3.8 billion years ago to the present day. There are three peaks of crust generation at 2.7, 1.9 and 1.2 billion years ago; none have been identified in the last billion years.

Above *Lava flows on Tungurahua volcano, in the Andes of Ecuador. The Andes are on the edge of a subduction zone and are very active volcanically.*

Left *Lava drapery of Pu'u 'O'o-Kupaianaha, Hawaii. The drapery was formed when lava flowed over a sea cliff on to the new land forming at its base.*

GIORGIO PORETTI

Is Mount Everest getting higher?

All human works are subject to error, and it is only in the power of man to guard against its intrusion by care and attention.
George Everest

Opposite *The north face of Mount Everest at sunset.*

One day in the spring of 1852 at Dehra Dun, in the foothills of the Himalaya, the door of the office of the Director General of the Survey of India opened. Radanath Sikhdar, the chief of the team of human computers, entered. 'Sir I have discovered the highest mountain of the world ... it is Peak n. XV'. These words, reported by Colonel Francis Younghusband, have become a legend in the measurement of Mount Everest.

The height of a mountain is determined by three main factors: relative sea level; the accuracy of the estimates of the elevation of the points from which measurements are taken; and the thickness of snow on the summit. Optical measurements (using a theodolite) taken from a long distance away are heavily influenced both by the refraction of the atmosphere because of pressure and temperature differences between the points of observation in the valley and the summit, and by plumb-line deflections.

The first serious attempts to measure the heights of the great mountains of the Karakoram-Himalaya chain were made in the early 19th century. William Lambton and George Everest, founders of the Survey of India, were engaged to map the subcontinent. From seven stations along the India-Nepal border, measurements showed that Mount Everest (at that time called Peak XV) was the highest mountain in the world. In 1856 its height was determined to be 29,002 ft (8,840 m).

Measurements made between 1880 and 1904 gave a height of 8,882 m (29,140 ft), and a more accurate measurement by B. L. Gulatee in 1954, just after Edmund Hillary and Tensing Norgay had conquered the peak, gave a new height of 8,848 m (29,028 ft) above sea level, due to previous errors in measuring. So has Everest been rising, or not?

Taking measurements

The Earth is not a perfect sphere, nor is the level of the surface of the sea exactly the same distance from the centre of the Earth at all points. The shape of the ideal sea-level surface over the globe is known as the *Geoid*, and its shape is determined from gravity measurements. These are easy to determine on the oceans, but much harder under mountains.

The coordinates of points on the Earth are measured with reference to a geometrical surface called the *Ellipsoid*, and for every point the difference between the Geoid and the Ellipsoid, labelled N and called the *height anomaly*, must be known. On Earth the value of N can reach ±100 m (328 ft).

Over the past 70 years, improved gravity measurements have allowed geographers to refine their assessments of global and local geoids. The value of N in the Mount Everest area has been calculated and recalculated to a range between –25 and –35 m (82 and 115 ft). The current best measurement of N, made in 2005 by the Chinese survey, was –25.2 m (83 ft). This value must be subtracted from the Ellipsoidal heights to yield the true height above sea level.

In 1975 and 1992, Chinese and Italian expeditions took instruments to the top of Everest to measure the height of the mountain, both with and without its snow cover. In 1975 the depth of the snow was measured as 92 cm (36 in) above the rock, while in 1992 it was 2.55 m (8.4 ft). For an expedition that took place in 2004, a portable

An Italian climber operating the ground-penetrating radar on the slope of the Southwest Ridge of Mount Everest in May 2004.

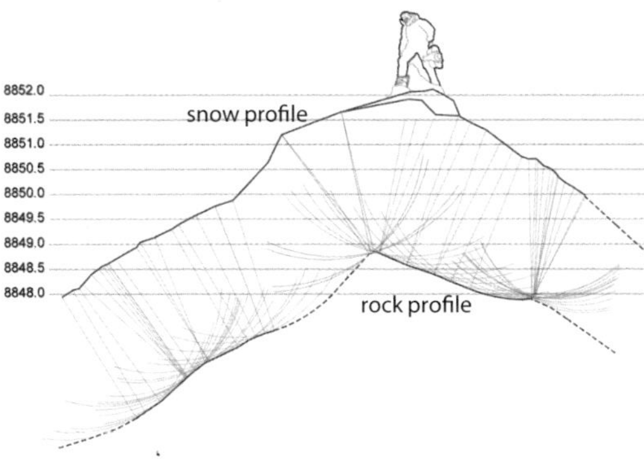

Cross-section along two profiles – the rock and the snow – at the summit of Mount Everest.

ground-penetrating radar (GPR) coupled with a global positioning system (GPS) was designed at the University of Trieste (Italy). The instrument (called a 'snow radar') was carried to the top and pulled across the summit along eight profiles, recording the positions of the points and the depth of the snow.

This allowed the scientists to draw one profile of the summit 'on the snow' and another 'on the rock'. The two summits (snow and rock) are about 1 m (3.3 ft) apart in the direction of the prevailing wind. The depth of the snow below the snow summit was 3.78 m (12.4 ft), while above the rock summit the snow was 3.04 m (10 ft) deep.

The coordinates of the snow summit were determined from the GPS recordings, and, taking into account the most recent value of N (–25.199 m/82.7 ft) from the Italian data, our measurements were:

	Latitude	**Longitude**	**Height**
Snow summit	27°59′16,963″	85°55′31,736″	8,848.36 m
Rock summit	27°59′16,998″	85°55′31,723″	8,845.32 m

A Chinese team led by Chen Junyong found a value of 3.5 m for the depth of the snow, 8,844.43 m (29,017 ft) for the elevation of the rock and 8,847.93 m (29,028 ft) for the snow surface.

Our conclusion is that Everest is not in fact rising at a meaningful rate – the apparent growth in height has more to do with improvements in techniques of measurement. The peak of Everest has moved, however, in the past century – by 8 m (26.2 ft) in a north-northeast direction, and is still moving at a rate of 4–5 cm (1.6–2 in) per year, showing that India is still a mobile tectonic plate, thrusting northwards into the main Asiatic plate, convincing evidence that plate tectonics are active.

Where did the oxygen in the atmosphere come from?

14

A breath that fleets beyond this iron world
ALFRED LORD TENNYSON, 1877

Today the air consists of 21% molecular oxygen (O_2), but this was not always so. Indeed, for almost the first half of Earth history, the air contained less than one part per million of oxygen. With so little oxygen, animals and multicellular plants could not evolve, and only single-celled organisms existed (p. 27).

The atmosphere first became oxygenated at 2.4–2.3 billion years ago. But at that time, oxygen concentrations probably only reached a few per cent of present levels. It was not until a second increase in oxygen, around 580 million years ago, that widespread animal respiration became possible. While oxygen appears to be essential for complex life, planets are constructed with chemicals that consume oxygen, so oxygen should not accumulate. Earth's oxygen-rich atmosphere is therefore rather mysterious. Fortunately, the comings and goings of oxygen on the modern Earth provide hints for solving this great puzzle.

The oxygen balance sheet: gains and losses

The principal source of oxygen is biological: it is a byproduct of a particular type of photosynthesis called oxygenic photosynthesis. This is a process whereby certain bacteria and green plants convert carbon dioxide and hydrogen extracted from water into organic matter such as carbohydrates, using the energy from sunlight. Oxygen is released from the process as a waste product:

Carbon dioxide + water + sunlight = organic matter + oxygen

$$CO_2 + H_2O + \text{sunlight} = CH_2O + O_2$$

The reverse process is aerobic respiration, whereby the organic matter is combined with oxygen to provide metabolic energy, releasing carbon dioxide and water.

Cyanobacteria are the most numerous oxygenic photosynthesizers – the oceans teem with them, and their ancestors were probably just as plentiful. DNA analysis shows that chloroplasts – bacteria-sized sites of photosynthesis within plants and algae – are descendants of cyanobacteria. Evidently, long ago, a cyanobacterium took up symbiotic residence inside another cell that was the ancestor of modern plants and algae.

At first sight, then, we might suppose that when cyanobacteria evolved, oxygen would have accumulated as a result of photosynthesis. But geochemical evidence suggests that cyanobacteria produced oxygen for several hundred million years or more before oxygen levels rose at 2.4 billion years ago. To understand what happened requires us to consider carefully the loss of oxygen, as well as its source.

Both respiration and the decay of organic matter consume oxygen, reversing the gains through photosynthesis, thus producing no net oxygen. But a tiny fraction (0.1–0.2%) of organic carbon escapes oxidation by being buried in sediments (mostly marine), leaving oxygen behind.

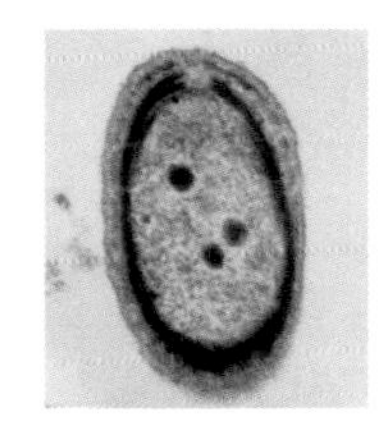

Prochlorococcus *is the most numerous cyanobacterium in the oceans. Ancestral marine cyanobacteria on the early Earth were the architects of oxygen in the atmosphere.*

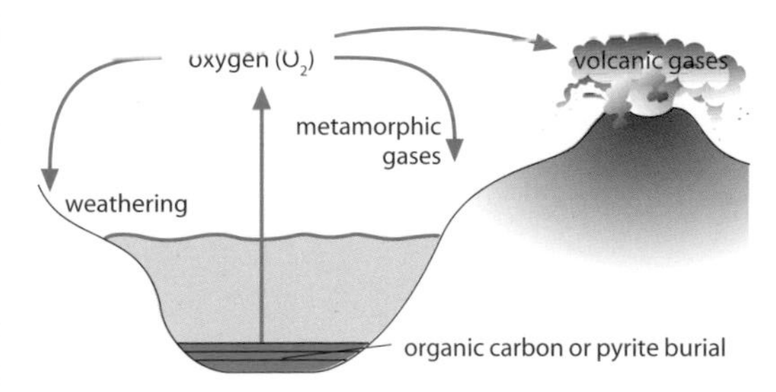

Oxygen is produced when organic carbon and pyrite are buried; but oxygen is consumed by 'reductants', substances that react chemically with oxygen, which are released from hot rocks that melt (volcanism) or metamorphic rocks that do not melt. Other reductants include minerals on land.

In the Grand Canyon of Arizona, thick red beds of Permian-age Esplanade Sandstone form the top of this tower of rock, O'Neill Butte. Red beds are found only in rocks younger than 2.4–2.3 billion years ago, when the atmosphere was oxygenated.

Thus, on geological timescales, the net gain of oxygen is actually equivalent to the burial rate of organic carbon.

Some organic matter is used by microbes to make pyrite (FeS_2) from seawater sulphate. Consequently, the burial of pyrite accounts for roughly 40% of net oxygen production today. The ancient oceans before 2.4 billion years ago were largely devoid of sulphate, so at that time organic carbon burial was the main cause of oxygen production.

Oxygen is easily lost. Geothermal activity and volcanoes release gases, such as hydrogen, which consume oxygen. In the ocean, oxygen reacts with dissolved minerals and gases from hot seafloor vents. Finally, oxygen dissolved in rainwater reacts with minerals in the process of weathering. For example, 'red beds' are riverbanks, deserts and floodplains with a reddish pigmentation arising from an iron oxide coating on mineral grains produced from the reaction of iron minerals with atmospheric oxygen.

When losses balance production, the amount of atmospheric oxygen remains steady. Today, about 80–90% of the oxygen produced from organic and pyrite burial is lost to oxidative weathering, while 10–20% reacts with reduced gases in the atmosphere and is also lost. 'Reduced' or 'reducing' gases are those that tend to react with oxygen and become oxidized, e.g., hydrogen, which oxidizes to form water vapour.

Anaerobic air and the advent of oxygenic photosynthesis

Before photosynthesis developed, the breakdown of water vapour (H_2O) in ultraviolet sunlight would have produced oxygen. This process is only a net source of oxygen when hydrogen subsequently escapes into space, so that the products of water decomposition cannot recombine. When water vapour condenses in the lower atmosphere, the hydrogen is prevented from escaping. Consequently, before life, volcanic gases would have scavenged oxygen down to less than one part per trillion.

Surprisingly, atmospheric oxygen did not accumulate as soon as photosynthesizers evolved. Biomarkers (diagnostic organic molecules) in 2.7 billion-year-old sedimentary rocks indicate the presence of cyanobacteria as well as eukaryotes (single-celled organisms with cell nuclei) that used local O_2 – some 300 million years before oxygen rose at 2.4 billion years ago.

A plausible explanation for why little atmospheric oxygen accumulated is that a glut of reductants – chemicals that consume oxygen – depleted it. Then, a time came when the oxygen produced from organic carbon burial exceeded the geological sources of reductants. At this tipping point, oxygen flooded the atmosphere until oxygen levels reached a plateau where oxygen production was balanced by losses to continental weathering.

But what caused this rise of oxygen? Either oxygen production increased or consumption decreased. Some favour the former idea, arguing that the growth of early continental shelves promoted organic carbon burial and thus oxygen production. Others question this hypothesis: organic matter extracts the light carbon isotope, carbon-12, from seawater, but seawater carbon, recorded in ancient limestones, does not show a steady decrease in carbon-12 content.

Instead, oxygen consumption by reducing gases could have abated. Proponents of this theory note that excess hydrogen-bearing reducing gases in the pre-oxygenated atmosphere would cause hydrogen to escape to space, which oxidizes the planet and diminishes further release

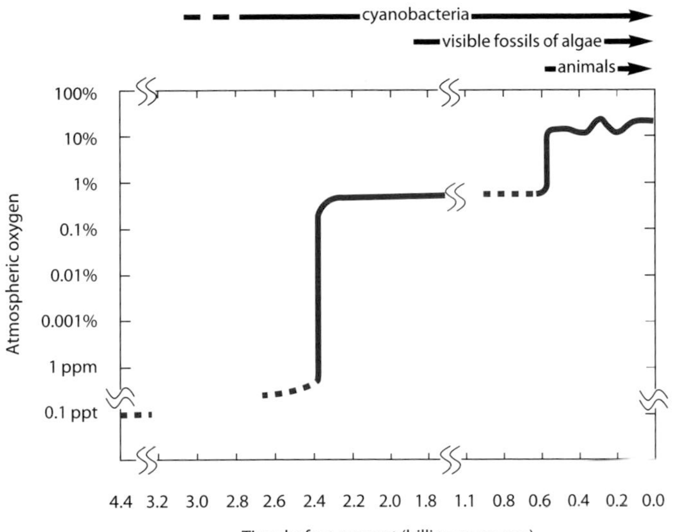

The red line shows estimated amounts of atmospheric oxygen over time, based on geological evidence and models. Before life existed, at 4.4 billion years ago, the oxygen concentration was less than 1 ppt (parts per trillion).

At 2.4 billion years ago, oxygen levels rose from less than 1 ppm (parts per million), up to 0.3–0.6%. Shortly after 0.58 billion years ago, oxygen rose to levels that supported animals. The history of fossils is shown above the graph.

of reductants. Perhaps methane was the key. In an anoxic atmosphere, methane reaches concentrations hundreds of times greater than today's 1.8 parts per million. Ultraviolet light decomposes methane in the upper atmosphere, causing hydrogen to escape and therefore Earth to oxidize. So high quantities of methane may have subtly encouraged oxygen to rise.

The 'Great Oxidation Event' and the 'boring billion'

The rise of oxygen 2.4–2.3 billion years ago is called the 'Great Oxidation Event'. Despite this name, oxygen levels remained limited and subsequent biological evolution progressed slowly. These years of stasis have been dubbed the 'boring billion'. During this time, much deep seawater may have remained anoxic, limiting biological evolution.

Oxygen finally rose a second time around 580 million years ago from a few per cent to greater than 15% of present levels. Afterwards, Ediacaran animal fossils (see p. 30) appear at 575 million years ago, and then Cambrian animals after 542 million years ago. The cause of this second rise of oxygen remains unsolved, but various different ideas have been suggested. Geological proposals include the enhanced production of clays that adsorbed and buried organic matter, or the construction of a supercontinent whose weathering flushed nutrients to the sea, encouraging more oxygen production. Biological proposals are that weathering was accelerated by lichen, or that faecal pellets from newly evolved zooplankton hastened organic burial. Alternatively, moderately high levels of biogenic methane throughout the 'boring billion' promoted hydrogen escape and oxidized the Earth.

In the Phanerozoic eon (542 million years ago to present), oxygen levels have probably varied between extremes of 10% and 30%. Thanks to high oxygen, the world has remained fit for animals, but exactly how oxygen rose from much lower Precambrian concentrations still remains somewhat enigmatic.

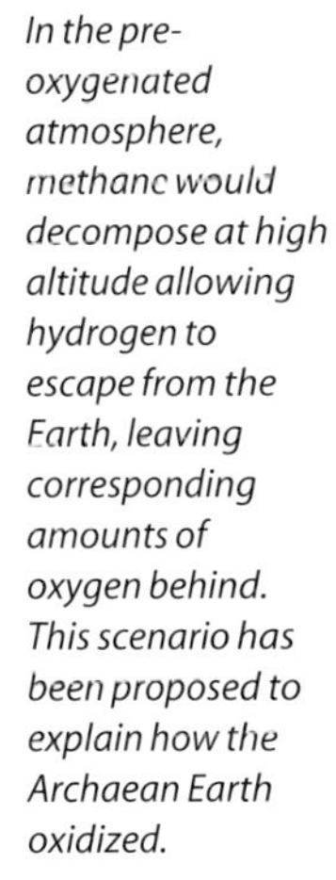
In the pre-oxygenated atmosphere, methane would decompose at high altitude allowing hydrogen to escape from the Earth, leaving corresponding amounts of oxygen behind. This scenario has been proposed to explain how the Archaean Earth oxidized.

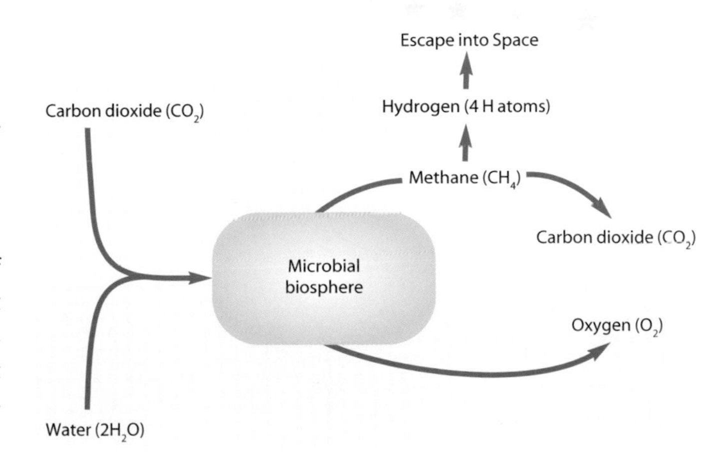

STEPHEN SPARKS

What makes volcanoes explode?

The sea seemed to roll back upon itself, and to be driven from its banks by the convulsive motion of the earth; it is certain at least the shore was considerably enlarged, and several sea animals were left upon it. On the other side, a black and dreadful cloud, broken with rapid, zigzag flashes, revealed behind it variously shaped masses of flame: these last were like sheet-lightning, but much larger.

PLINY THE YOUNGER, ON THE ERUPTION OF VESUVIUS THAT BURIED POMPEII, AD 79

Opposite *The eruption of Mount St Helens in 1980 sent ash and gases to heights of over 18 km (11 miles).*

Below *When two plates collide one is pushed back into the Earth's interior – the wet rocks release water, causing the hot interior of the Earth to melt and form explosive volcanoes.*

Volcanic explosions are one of the great spectacles of nature – their power can be awesome and highly destructive. The existence of volcanoes shows that we live on a very dynamic planet, and to understand why they explode we must first ask why there are volcanoes at all. Living at the Earth's surface, we are used to the idea that rocks are cold, solid and rigid. However, in the Earth's interior the heat generated by natural radioactivity has kept the rocks extremely hot for more than four and a half billion years. Only 100 km (62 miles) below the Earth's surface, temperatures increase to well over 1,000°C (1,832°F). The rocks are hot and plastic and can flow, even though they are crystalline solids. The critical point it that they are close to the temperatures where they start to melt. In most places, and for the majority of time, the hot rocks of the Earth's interior are not molten, but it doesn't take much extra energy to make them melt and form molten rock – the result is volcanic activity.

The great tectonic plates that form the outer skin of the Earth are pushed apart and collide as a consequence of the flow of the hot, but solid rocks of the Earth's interior. At the ocean ridges the plates pull apart and the hot rocks of the Earth's interior rise up to the surface and melt, forming volcanoes made of basalt lava. Where plates

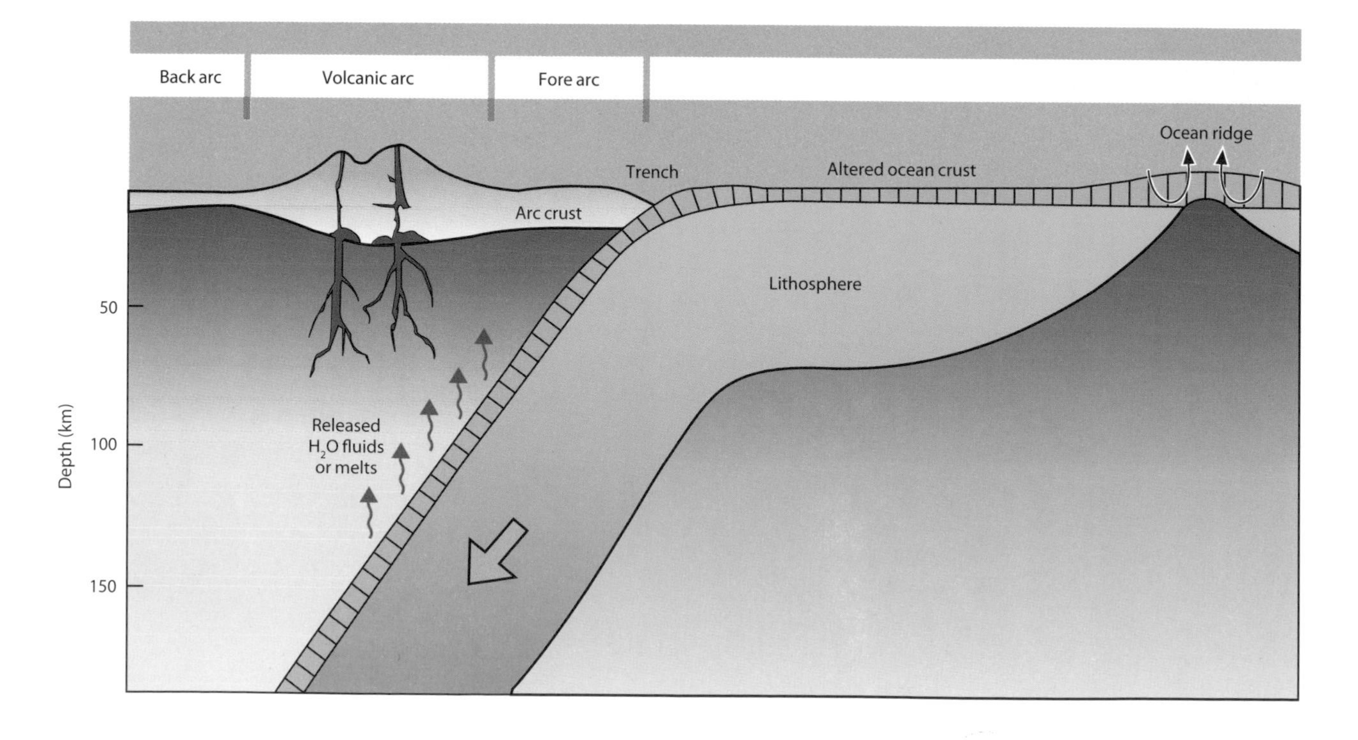

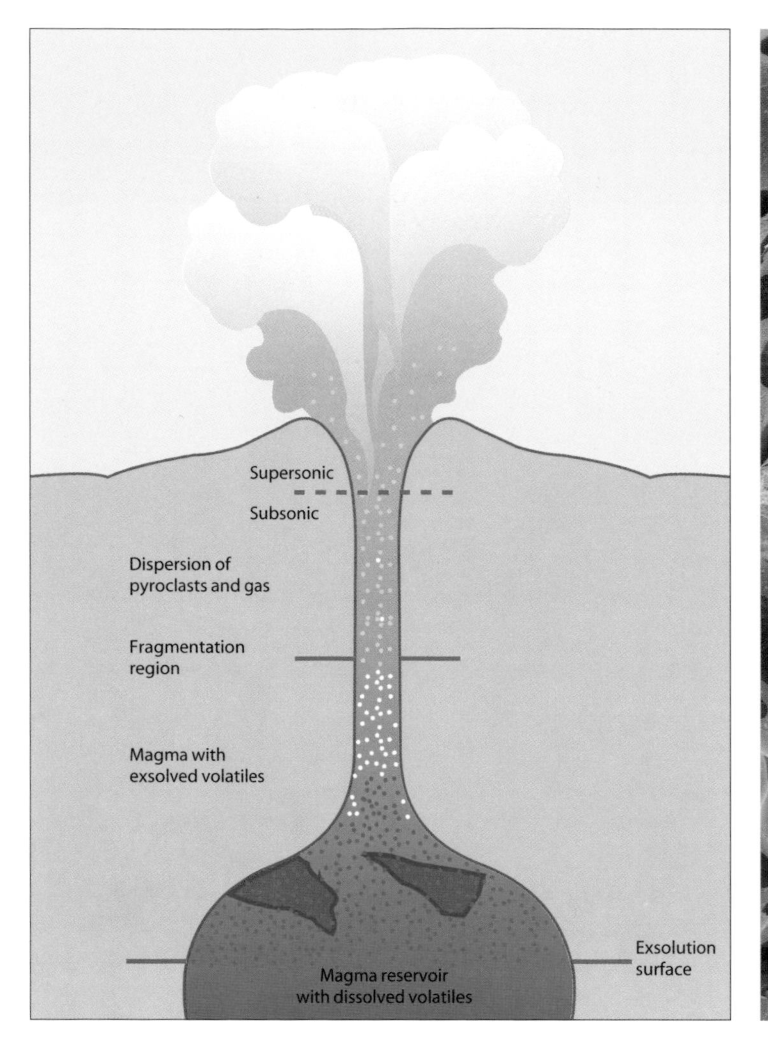

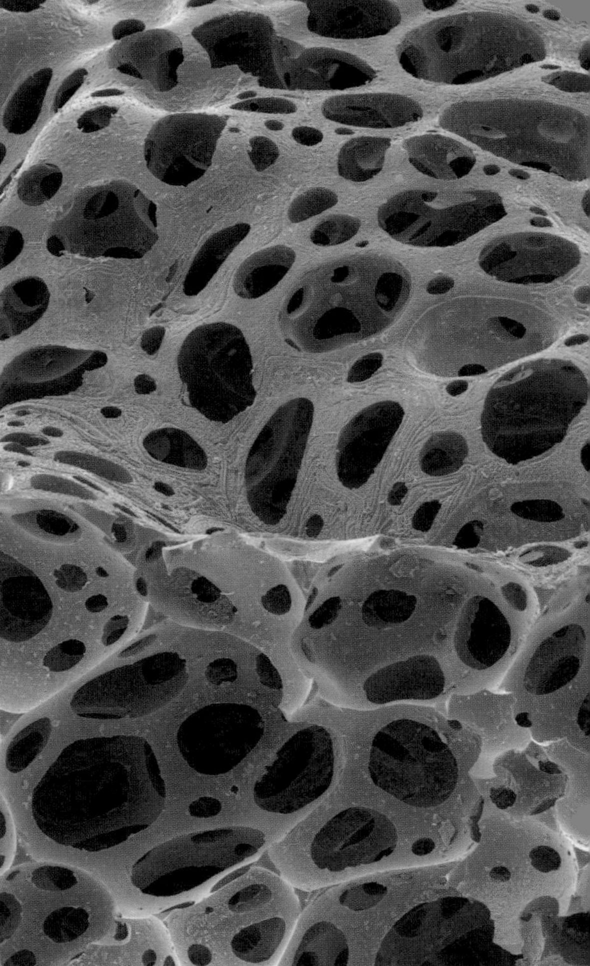

Above left *An explosive volcanic eruption: magma from beneath a volcano flows up to the surface. Expanding gases break the magma into gas and hot particles.*

Above right *This very lightweight basaltic pumice forms when lava containing gas cools rapidly.*

Opposite *Pu'u 'O'o fountain, Kilauea volcano, Hawaii.*

collide, one of the plates is pushed back into the Earth's interior. The sinking plate takes rocks down which have been soaked in sea water; this water acts like salt on ice, reducing the melting temperature of the hot mantle rocks and causing them to turn liquid. In this situation, volcanoes like Mount St Helens form.

Volcanic gases and bubbles

At the high pressures of the Earth's interior the molten rocks can dissolve lots of gases, such as water, carbon dioxide and sulphur dioxide. However, at the Earth's surface, where pressure is much lower, these gases are almost insoluble in rock melts. Therefore, as magma (a term for mixtures of molten liquid rock, solid crystals and gases) rises through cracks in the Earth to erupt, the pressure decreases and the dissolved gases separate from the melt. The process is exactly the same as when a bottle of fizzy drink is opened. Such drinks are manufactured by dissolving gas (usually carbon dioxide) under pressure. When the pressure is lowered by opening the can or bottle, the gas cannot be held dissolved in the liquid and myriad tiny bubbles of gas spontaneously form. The separation can be quite violent, as when a can is shaken and then opened, allowing the gas bubbles to form quickly. This is exactly what happens in many explosive eruptions where magma nears the Earth's surface and gas bubbles form rapidly.

So what exactly is happening when a gas bubble forms? When molecules of gas are dissolved in a liquid they are dispersed through the

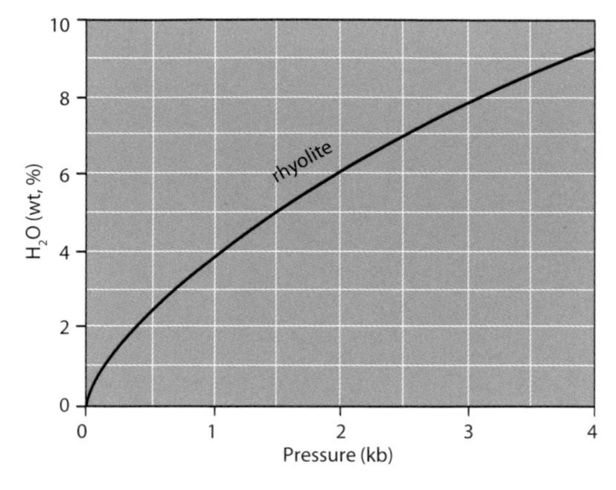

Water can dissolve in molten rocks at high pressure, but is not soluble at atmospheric pressure. Thus when magma with dissolved gas rises to the Earth's surface, bubbles rapidly from and cause explosions. The chart shows the amount of water that can be dissolved in magma with increasing pressure in kilobars. Note that 2 kilobars is about 6 km (3.7 miles) deep in the Earth's crust. Rhyolite is a volcanic rock that often erupts explosively because of its high viscosity.

much larger number of molecules of the liquid. However, at low pressures it is more energetically favourable for the gas molecules to come together than remain dispersed. If a small number of gas molecules are close to one another they start to collect to form a very tiny bubble; as other dissolved gas molecules travelling randomly through the liquid by diffusion accidentally come near the gas bubble, they will move from the liquid into the bubble, causing it to grow. This process of bubble formation is called *exsolution*.

Molecules of water (the most important volcanic gas) expand one thousand times when exsolving from magmas at surface pressures. However, if the liquid and the bubbles are not allowed to expand, the pressure inside the bubbles must increase. Huge pressures can result, and this is the essential reason why magmas with dissolved gases can explode.

Mechanisms of volcanic explosions

Not all volcanoes or eruptions are explosive – lava flows are cases where the gas bubbles out quietly. For a volcanic explosion to occur, either the gas bubbles have to form very fast or the mixture of melt and bubbles is prevented from expanding, causing the pressure to build up to very large values. In fact both these processes are usually involved.

Many of the explosive fire fountains of volcanoes on Hawaii result from rapid exsolution of gas caused by a flow of magma from deep inside the volcano where the pressure is high. The flow is sufficiently fast that the bubbles do not have time to separate quietly: here, the energy for the eruption comes about because of the huge expansion of exsolving gas.

The majority of volcanic explosions, however, are due to confinement of the magma that prevents the gas bubbles from expanding easily so that very high pressures result. Such confinement is primarily due to the fact that most magmas do not flow easily. The property that describes how easily a liquid flows is its *viscosity*. Magmas have viscosities that are typically many orders of magnitude greater than everyday liquids. For example, the magma of Mount St Helens is millions of times more viscous than honey. Why does this matter? Because when a bubble forms, the surrounding liquid has to move outwards; if the viscosity is very high the bubbles simply cannot expand fast and the pressure builds up inside them.

Another property of magma that explains volcanic explosions is that although liquid, magma can suddenly break like a solid if applied pressure is sufficiently high. Many materials exhibit this strange property, flowing under small external pressures and breaking when high pressures are applied. Thus in very viscous magmas like that of Mount St Helens, pressurized bubbles form as they rise towards the Earth's surface. The pressure in the gas bubbles is high compared to the external pressure in the atmosphere because the high viscosity of the magma prevents the gas bubbles expanding easily.

When the pressure difference is great enough the magma spontaneously fragments into a mixture of pumice, ash and gas. The released energy accelerates the gas and magma fragments into the atmosphere to reach incredible speeds of hundreds of kilometres per hour. As the mixture mixes with and heats the air, an eruption column can rise to tens of kilometres high in the atmosphere, and rocks with masses equivalent to a car can be thrown by the force of the explosion to distances of a few kilometres.

Finally, explosions can also occur by heating water rapidly in a lake or the sea. The heat from the magma boils the water and large pressures can result to cause another type of explosion.

RICHARD J. BROWN

The formation of diamonds

16

Those who contemplate the beauty of the earth find resources of strength that will endure as long as life lasts.
RACHEL CARSON, 1965

Diamonds possess a hypnotic allure that few other of the Earth's numerous and varied resources can muster. Many know the qualities of diamonds, their hardness, rarity and lucid beauty, but few consider the extraordinary circumstances under which they form or indeed the remarkable, violent journeys that bring them to the Earth's surface.

To understand how diamonds form, we must both comprehend what a diamond is, and also appreciate the nature of the deep parts of the Earth in which they grow. Diamond is a form of carbon, the fourth most abundant element in the solar system. To transform carbon into diamond requires high pressures, of 45 to 60 kilobars, and temperatures of 900–1,300°C (1,652–2,372°F), equating to depths of 150–200 km (90–125 miles) in the Earth, and in some cases up to 660 km (410 miles), at the mantle transition zone. Diamonds form in the upper mantle – the thick, plastic layer of the Earth beneath the solid outer crust.

Most diamonds are 3.3–0.9 billion years old and have been stored in the Earth for long periods before reaching the surface. Based on analyses of microscopic mineral inclusions trapped in diamonds as they grew, scientists have identified two main types: those created within eclogite (E-types) and those created within peridotite (P-types).

Eclogite is a coarse-textured metamorphic rock comprising garnet and pyroxene, which forms in a high-temperature, high-pressure environment in deep crustal metamorphic regions. It has a similar chemistry to basalt and is thought to derive from the subduction and metamorphism of basaltic oceanic crust. Peridotite is a term that encompasses a range of rock types (e.g. dunite, harzburgite and lherzolite) containing different proportions of the minerals olivine, clinopyroxene and orthopyroxene. It is the most common and abundant rock in the Earth's mantle.

The origin of carbon

The source of the carbon (C) in diamonds has been a subject of much heated debate for over a

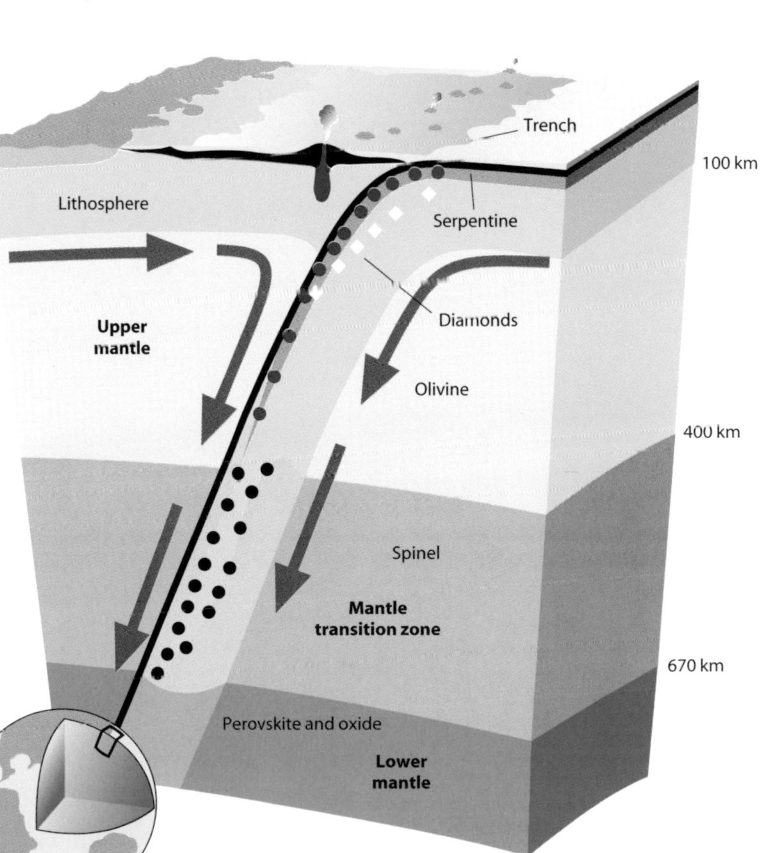

Cutaway of the outer 700 km (435 miles) of the Earth. Diamond formation is thought to occur in the upper mantle down to the transition zone.

Right *Peridotite, one of the diamond-hosting mantle rocks, comprising abundant green crystals of olivine and pyroxene crystals.*

Opposite *Diamond as it occurs at the Earth's surface in pyroclastic kimberlite. Jwaneng Mine, Botswana.*

century. Scientists have tried to understand where it comes from by analysing the ratio of stable carbon isotopes, reported as $\delta^{13}C$ values (the difference in measurement between a standard material and the sample in question). The two main diamond types have different isotope signatures: P-type diamonds are characterized by a $\delta^{13}C$ range of −2 to −9, while E-type diamonds have a much wider range of + 3 to −34. The narrow range of $\delta^{13}C$ values of P-type diamonds indicates a homogeneous source, perhaps from a convecting zone in the upper mantle. Originally the carbon may be derived from the primary components of the primitive Earth, which accumulated in the mantle about 4.5 billion years ago.

The carbon isotope signature of E-type diamonds is similar to that of marine carbonate minerals and organic hydrocarbons found in oceanic crust at the Earth's surface, and it has been proposed that the carbon in these diamonds derives from subducted oceanic crust, which has over many billions of years been transported to great depths in the mantle (over 660 km/410 miles). However, new analytical techniques have allowed other stable isotopes within diamonds (nitrogen, oxygen and sulphur) to be measured, and this has thrown doubt on the suggestion of a subduction origin for the wide $\delta^{13}C$ values in E-type diamonds. Recent studies suggest instead that the carbon may be introduced into eclogites and peridotites via a carbonate-bearing melt or fluid, a process known as metasomatism. The debate continues, but further study of stable isotopes should help pin down the source of the carbon in diamonds.

Journeys to the surface

Diamonds are not found evenly distributed at the surface of the planet, but are restricted to the stable interiors of the continents, known as cratons, for example in southern and south-central Africa, Canada and Russia. These cratonic regions are areas of abnormally thick continental crust, up to 200 km (125 miles) in depth. They are characterized by very long-term tectonic stability and have not been affected by plate movements or mountain building. This stability results in low geothermal gradients compared to oceanic parts

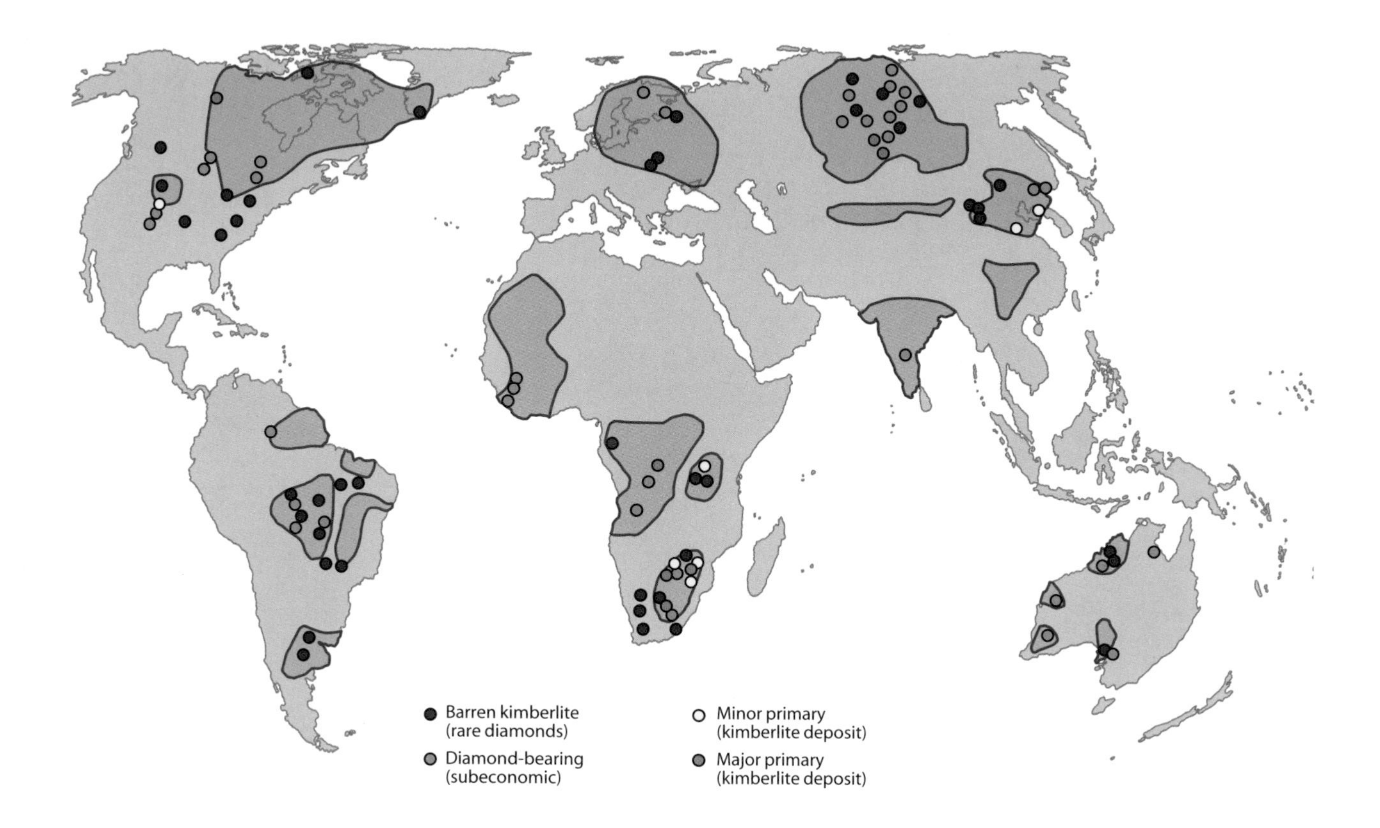

Map showing the distribution of kimberlites and diamond deposits worldwide. They are restricted to the stable cratonic regions of the continents (in blue).

of the world. Conditions here allow diamonds to be stored beneath cratons for very long periods.

Diamonds are transported to the Earth's surface by deep-sourced magmas (mainly kimberlite magma) derived from the partial melting of peridotite. Exactly where kimberlite magmas are generated has remained a topic of contention, with some scholars proposing formation at depths of 150–200 km (90–125 miles), while others suggest deeper sources, at 400–600 km (250–370 miles), and associate them with deep mantle upwellings, or plumes.

As the kimberlite magmas ascend through the upper mantle beneath cratons they can pick up chunks of diamondiferous peridotite and eclogite. During transport these chunks of mantle rock may break down and release their diamonds into the magma. For the diamonds to escape being corroded by the transporting magmas during ascent, the magmas must travel very fast through the upper mantle and crust – at speeds in the order of 10–30 km (6–18 miles) per hour. When the magmas reach the surface they erupt explosively, creating deep, carrot-shaped volcanic vents known as kimberlite pipes. As the eruption wanes, these vents become filled with pyroclasts (fragmented magma), disaggregated mantle material (eclogite and peridotite rocks and their constituent crystals), fragments of crustal rocks, and, in a very few cases, trace quantities of diamonds (only a few per cent of the approximately 3,000 known kimberlite bodies in the world contain diamonds).

The way in which kimberlite magmas erupt at the Earth's surface is not well understood, because none has erupted for the last *c.* 50 million years. Therefore, scientists must reconstruct the eruption dynamics by carefully studying the rocks preserved within the kimberlite pipes, and there is still much left to learn.

Diamonds are formed from carbon under extreme pressure-temperature conditions deep in the Earth and are transported upwards by unusual magmas and erupted explosively at the Earth's surface. Quite literally, diamonds would have fallen from the sky during these events.

RICHARD J. BROWN

Which was the largest volcanic eruption ever?

Destruction, hence, like creation, is one of Nature's mandates.
MARQUIS DE SADE, 1795

Volcanic eruptions occur on a great range of scales and at widely different frequencies. Small eruptions occur many times a day: Stromboli volcano in Italy erupts a few tens of thousands of cubic metres of magma every 10–20 minutes, providing a fiery night-time spectacle for the hundreds of tourists who tramp up its flanks at dusk. Very large eruptions – *super-eruptions* – occur on timescales of several tens to several hundreds of thousands of years, and have the devastating potential to erupt several thousand cubic kilometres of magma. Fortunately, none of these have occurred in historical times, but this begs the question: how can we understand the nature and effects of such catastrophic super-eruptions if we have never witnessed them?

How can we measure the size of an eruption?
When a volcano erupts explosively, magma (molten rock) is fragmented into ash and pumice (known collectively as tephra) by the rapid exsolution of the gases dissolved within it (p. 72). This tephra is violently ejected at great speed and may form towering, turbulent columns that can reach up to 50 km (31 miles) high in the atmosphere. The tephra is dispersed by winds over huge areas. Gritty abrasive ash falls from these clouds, coating houses, cars and fields and clogging rivers. Some of the tephra fails to make it up into the atmosphere and collapses back to the ground to form extremely hot and fast-moving pyroclastic flows that can travel up to 100 km (62 miles) from the volcano. These flows can deposit ash and pumice layers (ignimbrites) up to several hundred metres thick across the landscape.

In order to understand the size of a particular eruption it is important to be able to correlate the layers of ash erupted by it. All eruptions have a chemical and mineralogical fingerprint that can be analysed and used to correlate layers from the same eruption, and scientists have a range of tools at their disposal for this. Layers of ash can be dated by using the decay of radionuclides in the crystals. At sea, analysis of microfossil populations in intervening mud layers on the ocean bottom can tie down the age of ash layers. All the layers

A small explosive eruption occurs every 10–20 minutes at Stromboli in Italy.

Opposite above *The enormous flooded caldera formed by the Toba eruption of 74,000 years ago measures 100 by 35 km (62 x 22 miles). The whole of Greater London would disappear in such a huge hole.*

Opposite below *Map of the limits of the ash fall from the Toba super-eruption.*

associated with a particular eruption must be mapped; in the case of super-eruptions, these deposits may occur on more than one continent.

Once an understanding of which ash layers belong to which eruption is achieved, it is then possible to try to estimate the total volume of the eruption, although this is fraught with difficulties. Much of the material may have been deposited out to sea, or buried within the collapsed caldera where it is not visible. Loose tephra is very mobile and is readily stripped during heavy rain. Ancient deposits become covered with thick vegetation or even sprawling cities. Some very ancient super-eruptions (around 450 million years old) are only known from extensive thin ash fall layers dispersed far from the source volume.

Volumes of eruptions are usually reported as 'dense rock equivalent' or DRE values, which is the estimated amount of hot magma that was erupted. It is given in this way because the density of the pumice and ignimbrites varies from eruption to eruption. As magma erupts it vesiculates – the gas in the magma comes out of solution (exsolves) and forms bubbles, which can become trapped as cavities in the fragmented magma to form pumice. The vesicularity, and therefore the density and volume, of pumice varies considerably (from 0.5 to 1.5 g/cm^3), as does the density of ignimbrite. In some cases, ignimbrites are deposited so hot and thick that the bubbles in the pumice collapse and flatten (welding), increasing the density of the ignimbrite. The volume of the ash and pumice deposits must be carefully calculated to take account of such variations and then converted back into a DRE value to allow comparison with other eruptions and deposits. Volcanic eruptions are ranked on a logarithmic scale of erupted mass known as the 'volcanic explosivity index' or VEI. The largest VEI 8 events have DRE values of more than 1,000 cubic km (240 cubic miles).

The great Toba super-eruption

The eruption of Toba 74,000 years ago ranks as one of the largest ever on Earth: a VEI 8 event, it lasted for up to two weeks and erupted 2,800 cubic km (672 cubic miles) of magma, creating a huge caldera, 100 x 35 km (62 x 22 miles) in size and generating pyroclastic flows that sped out from the volcano and deposited ignimbrites over an area equivalent to 20,000 sq. km (7,722 sq. miles). Vigorous ash plumes rose from the tops of these pyroclastic flows and dispersed a DRE-equivalent volume of 800 cubic km (192 cubic miles) of ash. This ash formed a layer which blanketed an area of at least 4,000,000 sq. km (1,545,000 sq. miles) – from the South China Sea to the Arabian Sea – and is still 2.5 m (8.2 ft) thick in southern India, 2,500 km (1,553 miles) from the volcano.

The Toba eruption is thought to have had a major impact on global climate, accelerating the

Examples of large explosive eruptions (over 'volcanic explosivity index', or VEI, 5) and their recurrence intervals.

VEI	Minimum volume magma (km^3)	Minimum volume pyroclastic deposits (km^3)	Example of eruption	Recurrence interval
5	0.4	1	Mount St Helens, USA, 1980	10 years
6	4	10	Krakatoa, Indonesia, 1883	50 years
7	40	100	Tambora, Indonesia, 1815	200–1,000 years
8 (low)	450	1,000	Taupo, New Zealand, 26,000 years ago	120,000–200,000 years
8 (moderate)	1,000	2,500	Toba, Sumatra, 74,000 years ago	300,000 years
8 (high)	>3,000	>5,000	Fish Canyon, USA, 27.8 million years ago	less than 1 million years

transition into glacial conditions. Some scientists have argued that the catastrophic climate change associated with the Toba super-eruption caused a marked reduction in the early human population.

Effects of super-eruptions

The number of fatalities from volcanic eruptions has risen through time even though the number of volcanic eruptions has remained more or less constant, the result, no doubt, of the huge increase in the population of the Earth over the last few centuries. More and more people are living within killing distance of volcanoes. In order to understand the potential impact of a future super-eruption, it is instructive to examine the effects of smaller events.

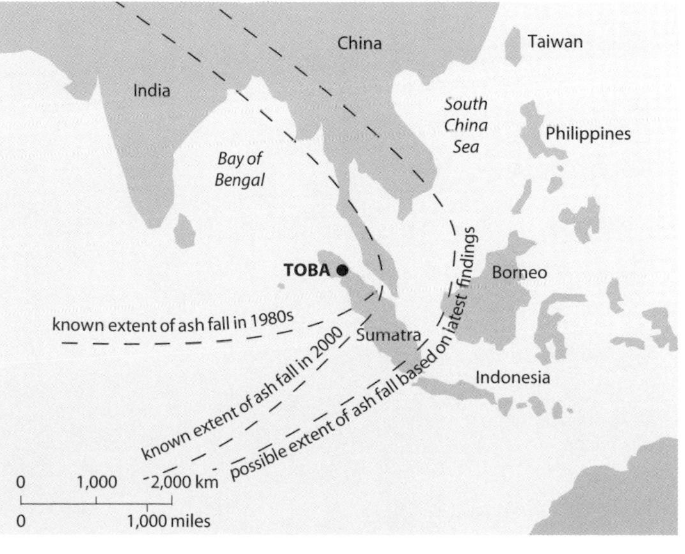

WHAT IS IT LIKE TO BE CAUGHT IN A LARGE ERUPTION?

The 1883 eruption of Krakatoa volcano, Indonesia, spewed out only around 12 cubic km (3 cubic miles) of magma (VEI 6), but it still ranks as one of the largest in modern history. It caused widespread devastation and the death of 36,000 people. This is an account from one lucky survivor: *'Suddenly it became pitch dark … ash [was] being pushed up through the cracks in the floorboards, like a fountain.… I felt a heavy pressure throwing me to the ground … all the air was being sucked away and I could not breathe.… I realized the ash was hot … the hot bite of pumice pricked like needles.… I noticed for the first time that [my] skin was hanging off everywhere, thick and moist from the ash stuck to it … I did not know I had been burnt.'*

A colour lithograph of the eruption of Krakatoa, Indonesia, in 1883.

Consequence	No. of events	No. of people
Death	260	92,000
Injured	133	16,000
Homeless	81	291,000
Evacuated	248	5,282,000
Any incident	491	5,596,000

Approximate number of people affected in different ways by volcanic eruptions in the 20th century (values rounded to the nearest 1,000). Each event may have had more than one consequence.

Volcanic eruptions have a large number of ways of killing people. Those close to the volcano may be wiped out during caldera collapse, for instance during the eruption on Santorini, Greece, in the mid-2nd millennium BC (possibly the source for the legend of Atlantis), or people may be killed by falling rocks (volcanic bombs), crushed by earthquake-damaged buildings or swept away and burnt to death in pyroclastic flows.

At a greater distance from the volcano, people may die as their roofs collapse under the weight of falling ash. Tsunamis (p. 86) can be generated if an eruption occurs near or in a lake or the sea, bringing death many thousands of kilometres from the volcano: most of those killed by the 1883 eruption of Krakatoa were swept away by tsunamis.

For many years after an eruption rain can loosen tephra and create fast-moving lahars (debris-filled floods of ash, boulders and trees) that can inundate huge areas. In 1985, 25,000 people died when an explosive eruption melted snow and ice on Nevado del Ruíz volcano, in Colombia, creating a devastating lahar that wiped out the town of Armero. It is sobering to note that in many modern eruptions, up to two-thirds of the total fatalities occurred more than 1 month after the event.

The climatic effects of a large eruption are perhaps the most worrisome and can persist for up to 10 years. Climate models indicate that loading the atmosphere with large amounts of dust and sulphur gas (volcanic aerosols) may lead to major perturbations in climate chemistry, further degrade the ozone layer and result in significant surface cooling (by about 3–5 °C or 5.4–9 °F) because they reflect solar radiation back into space. In the year following the largest historical volcanic eruption, the eruption of Tambora, Indonesia, in 1815, global temperatures dropped by 1 °C (1.8 °F) and persistent vibrant sunsets were visible over Europe ('The Year Without a Summer'). These were vividly captured in the paintings of the British artist J. M. W. Turner. Persistent climate perturbations may lead to crop failure and mass starvation across large parts of the planet. The modern world is at a greater risk than that faced by our ancestors because of our globalized trading infrastructures and our critical dependence on air travel and satellite communications. The effects of a super-eruption on world financial markets could be severe.

It is difficult to say with any degree of certainty when or where the next super-eruption will occur; we only know that it will. Super-eruption sites are generally situated where tectonic plates collide or where hot magma from deep in the Earth's mantle rises up beneath a continent. Past super-eruptions have occurred in Italy, New Zealand, the Andes, Japan and the US. There is still much left for us to do to understand super-eruptions and their regional and global effects, but mankind must treat this threat seriously, since sooner or later one will occur and we will all feel the impact.

The towering column from the 1991 VEI 6 eruption of Mount Pinatubo in the Philippines.

The mystery of tsunamis

With speeds that can exceed 700 kilometres per hour in the deep ocean, a tsunami wave could easily keep pace with a Boeing 747.
Frank A. González, 1999

Below left *The 2004 tsunami at Koh Pu, Thailand: the waves pile up at the shore because they are slower in shallow water than in deep water.*

Below right *The mechanism of a tsunami-generating earthquake: when an oceanic plate pushes underneath a continental plate, the resulting earthquake causes the entire body of water in the ocean to vibrate.*

The term tsunami consists of two Japanese words meaning 'harbour' and 'wave' – a perfect description of the tsunami phenomenon. Japanese fishermen gave the name to a type of wave which could mysteriously destroy their harbours while they had been out at sea and had noticed nothing unusual.

In deep water, a tsunami disguises itself so well that it is imperceptible, both from aeroplanes and to ships at sea. During a tsunami on Hawaii in April 1946, ships standing off the coast watched tremendous waves breaking onshore, but could not detect any change in sea level at their offshore locations. It is only today, with modern technologies, that we are able to observe tsunamis in the open seas, but even so they are still very small. For instance, the most recent catastrophic tsunami, that of 26 December 2004 in the Indian Ocean, which killed around 220,000 people, reached a height of only about 1 m (3.25 ft) in the open ocean, as measured by satellites. On the west coast of Sumatra it piled up to more than 30 m (98 ft) high in places. It is this sudden occurrence of tsunamis at coastlines, with no apparent warning, that makes them exceptionally dangerous.

What turns a small wave in the open seas into a monster at the shores?

In addition to their relatively small height, another reason why tsunamis are hard to observe in the open ocean is their large wavelength. The distance from one wave peak to the next might be 100 km (62 miles) or more, and so the time between the passage of two wave peaks can be over an hour. Tide gauges, sea-level observation satellites or deep-ocean pressure buoys are needed to detect this. Wind-generated waves, in contrast, have wavelengths of only about 100 m

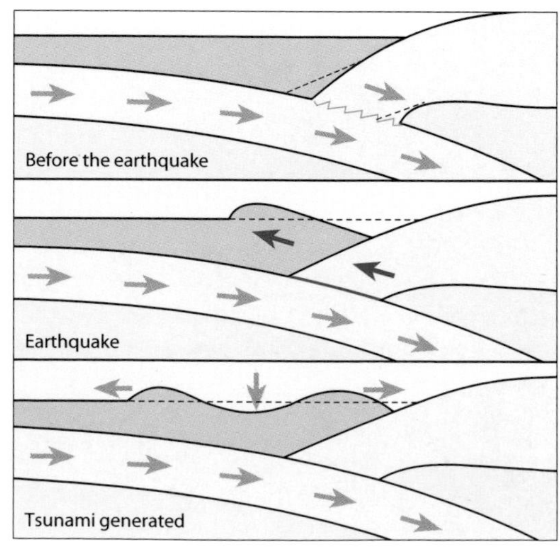

(328 ft) and periods of 10 seconds – these can easily be detected by any observer. Also, wind waves influence only the surface of the ocean – submarines can avoid being shaken by storms by diving just a few tens of metres. Tsunami waves, in contrast, move the entire water body down to the bottom of the ocean. Such a movement is hard to notice on a ship, although it might cause a mysterious sudden blur of the water. So why are the waves so high at the shores and yet so small in the open ocean?

The explanation lies in the relation of the speed of the tsunami wave to the depth of water. In the deep ocean the wave travels at about about 700 km (435 miles) per hour (the speed of a jet liner), but it slows down to just a few tens of kilometres per hour in shallow coastal waters. This leads to an increase of the wave height as the back of the

Too late for escape: fifteen seconds later all these people on Hat Rai Lay Beach, southern Thailand, were caught by the tsunami of 2004; some did not survive, though the photographer did, with his camera.

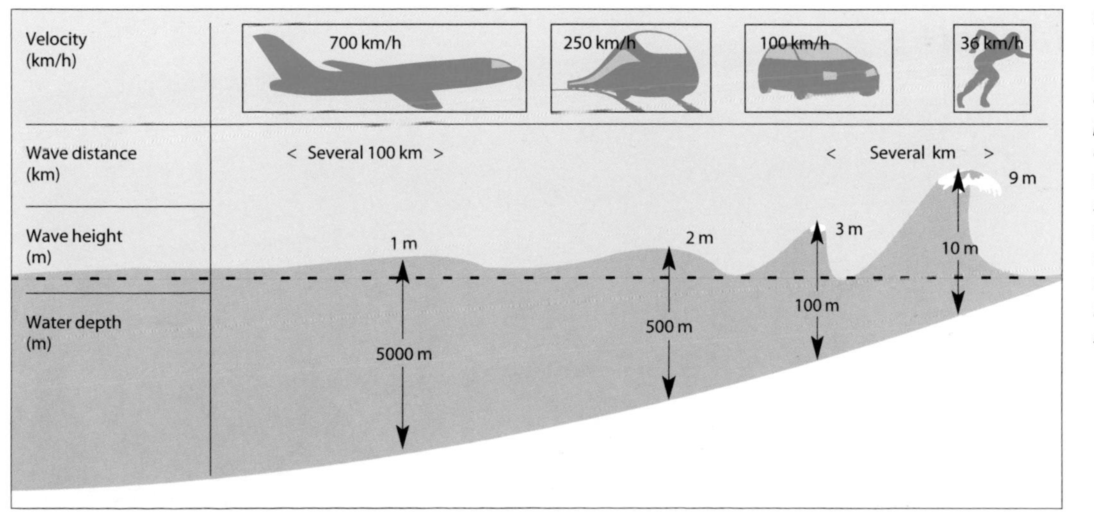

Diagram showing the speed of tsunami waves, and also how they pile up at the shore as the wave, travelling at great speed in deep water, pushes up the front of the wave which is much slower in shallow water.

A wood engraving of 1887, depicting the effects of the 1755 Lisbon earthquake and associated tsunami.

wave, travelling at great speed in the deep water, pushes up the slow front of the wave in the shallow water. As a consequence of the great wavelength, a tsunami is more like a long-lasting flood than the familiar surf ordinarily seen breaking on the beach.

What causes a tsunami?

Tsunamis originate as sudden movements of large masses of water somewhere in the ocean. Such motions could be caused by abrupt vertical displacements of the sea floor as a result of earthquakes, or by large underwater landslides, volcanic eruptions or large meteorites. The vast majority of tsunamis are caused by earthquakes.

Small, local tsunamis are relatively frequent – several occur every year. These are created by earthquakes with magnitudes of around 7 on the Richter scale. The devastating tsunamis are generated by earthquakes with magnitudes of greater than 9. Fortunately, such events are very rare, with only one or two happening per century. This is another reason why tsunamis usually catch people by surprise. Several generations pass between really large, catastrophic tsunamis, and it is human nature to forget unpleasant things.

Before the 2004 disaster, the previous catastrophic tsunami in the Indian Ocean was caused by the eruption of Krakatoa in 1883 (p. 84), which killed more than 36,000 people. Three generations later, this event had been forgotten and people were again completely unprepared. Other examples of large historical tsunamis include those caused by the Lisbon earthquake in 1755 (more than 60,000 casualties) and the Messina earthquake in 1908 (more than 30,000 casualties).

About 7,000 years ago a huge underwater landslide off the coast of Norway (the Storegga slide) caused a 20-m (66-ft) high tsunami in Scotland. And about 65 million years ago a giant meteorite hit the shores of Mexico, causing the extinction of the dinosaurs and many other species on Earth (p. 42). As a result, a super tsunami washed over much of North America.

Is it possible to receive warnings?

Natural events such as earthquakes or tsunamis will always occur and cannot be prevented by humans. Prediction has up to now also been impossible. However, the extent of the disaster inflicted on people can be reduced. Early warning is today considered the most important part of mitigating the catastrophic effects of tsunamis. This approach is based on the fact that information about a tsunami can be propagated much faster than the tsunami wave itself. The potential warning time between the observation of a tsunami near its source and the time when it hits a more distant place varies from minutes to several hours, depending on the distance from the source.

The first tsunami-warning system was installed in Hawaii by the United States after the Alaska tsunami in 1946, which killed several hundred people. A tsunami caused by a very large earthquake in 1960 in Chile killed about 200 people in Japan. Although its travel time from Chile to Japan was 22 hours, no warning was given.

A tsunami-warning system in the Indian Ocean region in 2004 could have saved many thousands of lives – there was a theoretical warning time for Thailand of about one hour, and of about three hours for India. However, the opportunity provided by these warning times was not exploited. Unfortunately, history shows that governments install warning systems only after catastrophes have occurred. Since most tsunamis are caused by earthquakes, present-day warning centres concentrate on near real-time earthquake information, which is obtained from modern global or regional seismic networks.

Sea level observations have up to now been used mainly only for either confirming or downgrading tsunami warnings based on earthquake data. In future, however, direct observations of small sea level changes, probably from space, might prove to be the most reliable technique.

What should we do?

Scientific methods, however, are not the only, or major, approaches available to an early-warning centre. Human and social aspects are just as important. Information about an advancing tsunami must be transmitted quickly to the people who are in danger, and they must have a basic knowledge about tsunamis and what to do. Tsunami-warning centres need at least 15 minutes to issue a reliable first warning. More time will pass before the warning reaches the people on the beach through official channels. This is still too long for local tsunamis. Automatic mobile phone services might be a solution in many cases. The fastest way, however, is simple observation. According to a tsunami expert: 'If you can see the ocean and you feel the earthquake, run like hell away from the water'.

The need for such basic knowledge about tsunamis is illustrated by the example of an English schoolgirl during the Sumatra tsunami. She noticed unusual waves in the sea, remembered what her teacher had told her and alerted others on the beach, saving many lives.

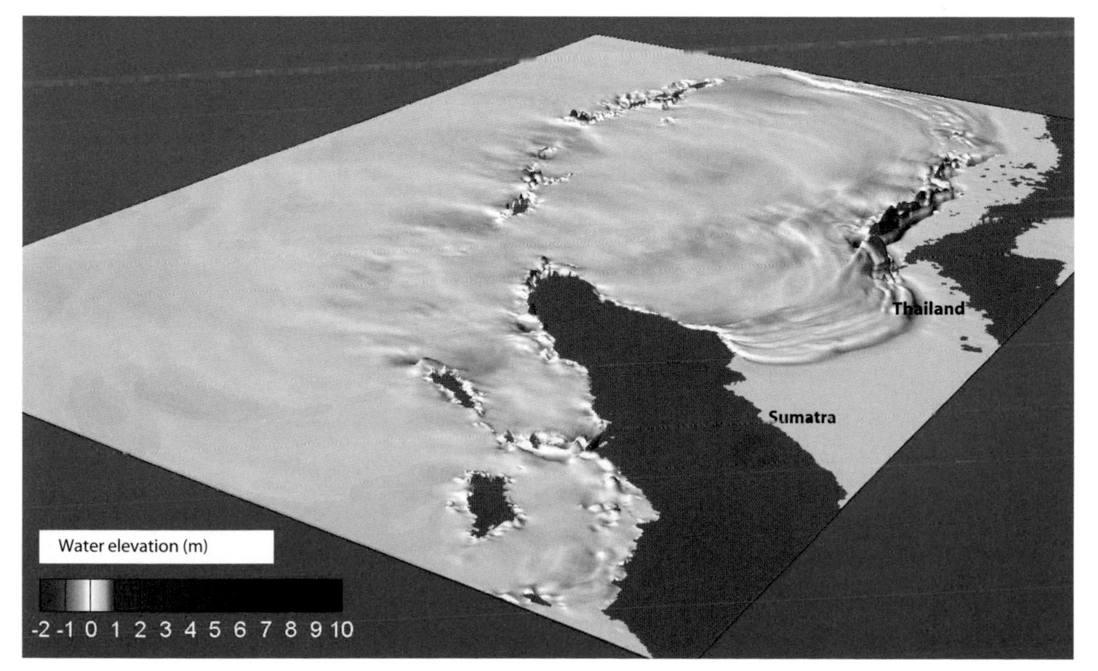

Model of the propagation of the Sumatra tsunami in 2004, 2 hours and 30 minutes after the earthquake occurred. The coast of Thailand had not yet been reached.

PHILIPPE CLAEYS

Asteroid and comet impacts on Earth

... les irréductible gaulois ne craignent qu'une chose que le ciel leur tombe sur la tête ...
... the indomitable Gauls fear only one thing – that the sky will fall on their heads ...
RÉNÉ GOSCINNY, ASTÉRIX LE GAULOIS/ASTERIX THE GAUL, 1961

Collision is a major process in the evolution of planets, as attested by the 1994 crash of comet Shoemaker-Levy 9 on Jupiter. On solid planetary surfaces, these collisions result in topographical features called impact craters. Fresh craters are characterized by an almost perfect circular shape, as visible on the Moon. On Earth, the size of the approximately 170 known impact craters varies between a few tens of metres (Sikhote Alin, in the Russian Far East) to more than 180 km (112 miles; Chicxulub, off the coast of Yucatan, Mexico, p. 42). So far, no crater has been identified on the floor of the oceans.

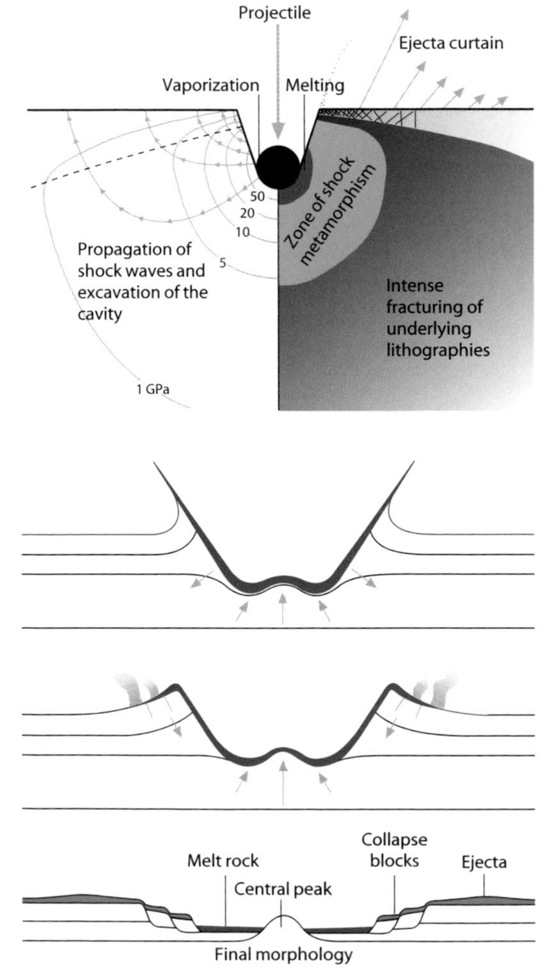

Formation of a complex impact crater: the pressure of the shock waves expanding from the impact point is expressed in gigapascals (1 GPa = 10.000 bars). The excavation phase, when depth reaches to two times the projectile diameter, is followed by the rebound of the compressed underlying rocks and the collapse of the crater margin, leading to the final morphology.

Origin and frequency of impact craters

The formation of a crater by the impact of an object at hypervelocity is in fact a complex process comparable to nuclear explosions. The transformation of the projectile's kinetic energy releases a huge amount of energy ($=\frac{1}{2}mv^2$, where m and v are a projectile's mass and velocity). The impact velocity, determined by celestial mechanics, ranges between 11 and 72 km (7 and 45 miles) per second: 11 km/s is the escape velocity of the Earth–Moon system and a body is simply accelerated to this speed by the planet's gravitational pull. The maximum speed for a comet (72 km/s) corresponds to the escape velocity of the solar system at a distance of 1 astronomical unit added to the Earth's orbital velocity (30 + 42 km/s).

The impact of a 1.5-km (1-mile) object – smaller than most asteroids – releases the same amount of energy as the whole Earth in a year (from heat flow, earthquakes, tectonic activity, volcanism, etc., combined). Asteroid or comet impacts have never been directly observed on Earth, despite possible speculations regarding mythological events. In 1908, an object around 50 m (165 ft) in diameter exploded several kilometres high in the atmosphere above Siberia. This 'Tunguska event' devastated more than 2,000 sq. km (772 sq. miles) of uninhabited forests. Such an explosion over a city would have wiped it off the map.

Morphology and formation of impact craters

On Earth, there is an approximate ratio of 1 to 20 between the size of the projectile and that of the resulting crater. A crater less than 4 km (2.5 miles) in diameter has a simple bowl-shape and a depth equivalent to ⅓ of its diameter. An elevated rim marks the crater edge and its inner flanks contain fractured and collapsed rocks. Above this size, a more complex crater forms: the concavity decreases (1/6 of its diameter) and uplift occurs in the central zone as the compressed rocks bounce back when the pressure is released. Much larger structures are composed of several elevated concentric rings. Multi-ring craters more than 1,000 km (620 miles) in size are known on the Moon, Mercury, Mars and Venus. The largest terrestrial craters – Chicxulub (65 million years old), Sudbury (Canada, *c.* 250 km/155 miles; 1,850 million years old) and Vredefort (South Africa, *c.* 300 km/186

The well-preserved Barringer crater (around 1.1 km/0.7 miles in diameter) in Arizona was formed some 50,000 years ago by the impact of an iron meteorite. The elevated rim is clearly visible.

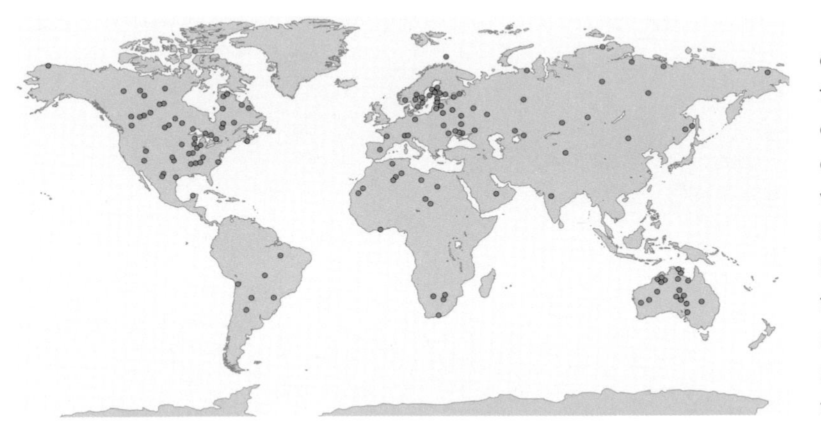

Location of the known terrestrial impact craters. Most are found on old continental shields in the geologically well-studied regions of Europe, Canada, Australia and Russia, while many remain to be discovered in South America, Asia and Africa.

miles, 2,023 million years old) – also all probably belong to this category.

A crater forms by the propagation of intense shock waves at the collision point. These can reach velocities of over 10 km (6.2 miles) per second, and pressures equivalent to several million times atmospheric pressure. Their passage vaporizes, melts, deforms, compresses, fractures and pushes away large volumes of rocks, excavating a cavity. In most cases, the projectile is completely vaporized and the upper layers of the target rock melt. Deeper in the rapidly evolving cavity, the rocks are displaced upwards with sufficient velocities to be ejected from the crater. A cloud of vapour, melt-particles and dust grows above the impact point, rising into the atmosphere like the mushroom cloud of a nuclear explosion.

The South Pole of the Moon showing multiple well-rounded craters; only the few projectiles hitting the planet surface at an angle of less than 10° will produce an elongated crater. Several craters are larger than 1,000 km (620 miles). The lunar stratigraphy is based on the crater record.

Fragments of the target rocks are also expelled on more oblique trajectories as a curtain of ejecta that lands at a distance equivalent to several radii of the crater from the rim. The shock waves quickly lose energy as they spread out. When their velocity decreases to that of sound in rocks (5–8 km or 3–5 miles per second) they are transformed into seismic waves. Gravity and rock mechanics take over at this point, leading to the final crater morphology. A small crater forms in a few seconds, whereas a 100-km (62-mile) structure takes up to 10 minutes. Such a process cannot be reproduced in the laboratory, and so our understanding of it is derived from theoretical calculations, experiments on the behaviour of shock waves in rocks, the study of mini-craters obtained by shooting at a variety of targets and field observations at preserved impact structures.

Identification and age of impact craters

Impact craters are usually recognized by the presence of topographic or geophysical anomalies. Seismic profiles outline the bowl-shape silhouette of the crater, allowing an estimate of its size. Within a crater, the low-density fractured rocks induce a negative anomaly in the local gravity. A crater can also be unambiguously identified by the presence of a meteoritic component and/or shock-metamorphosed minerals. As a high-pressure shock wave passes through minerals, it irreversibly modifies their crystallographic structure, leading to the formation of typical and highly diagnostic microscopic defects. The only other process known to induce shock metamorphism in minerals is a nuclear explosion.

In rare cases, actual fragments of meteorites are preserved close to the crater (Barringer in Arizona). Otherwise, meteoritic contamination is detected by measuring higher than usual concentrations of elements in the platinum group, including iridium. Another approach is to use the isotopic ratios of osmium and chromium, which differ significantly in meteorites and the Earth's crust.

Outside the crater rim, the presence of tektites – impact glass – also represents evidence of an event. Impact glass can be differentiated from volcanic glass by its low water content and chemical

Projectile diameter	Impact frequency (1 event per)	Examples and released energy (equivalent TNT)
< 50 m (< 165 ft)	Detection frequent	Burns in atmosphere
50 m (165 ft)	200 to 400 years	Tunguska, *c.*12 megatons
100 m (328 ft)	1,000 to 5,000 years	Wolf Creek, Pretoria, Barringer, *c.* 15 megatons,
500 m (1,640 ft)	0.1 to 0.5 million years	Zhamanshin, Bosumtwi, Mien, *c.* $11x10^3$ megatons
1 km (0.6 miles)	1 million years	Ries, Rochechouart, *c.* $9x10^4$ megatons
5 km (3 miles)	10 to 50 million years	Popigai, Manicouagan, *c.* $1x10^7$ megatons
10 km (6 miles)	100 to 500 million years	Chicxulub, Sudbury, Vredefort, *c.* $1x10^8$ megatons
> 10 km (> 6 miles)	Probable?	Precambrian ejecta in Australia & Africa?
> 1,000 km (> 620 miles)	At least 1	Origin of the Moon

Estimated frequency of the impact of asteroids or comets on the Earth based on scaling up the occurrence of very small objects (micrometeorites) commonly falling through the atmosphere and crater counts and ages on the Earth and the Moon.

composition, reflecting the melting of the rocks. Tektites are small (about 1 cm/0.4 in), rounded or elongated glass particles ejected at high velocity from the crater and ultimately deposited over a vast geographic area, up to several thousand kilometres from their source, an effect known as a 'strewn field'. Four strewn fields are known from the Cenozoic (or Cainozoic, after 65 million years ago and still ongoing). One, the central European (*c.* 15 million years old), was ejected by the Ries impact in Bavaria all the way to the Czech Republic; the most recent is the Australasian (*c.* 0.7 million years old), extending all over the Indian Ocean – the source crater for this remains unknown.

Most craters are younger than 250 million years. Small structures (less than 5 km/3 miles) erode rapidly and have largely disappeared, except very recent ones. Geological processes quickly erase the traces of ancient impacts, although the large Sudbury and Vredefort craters are Precambrian in age. Ejecta deposits are also difficult to identify in ancient sediments, except at the Cretaceous–Tertiary boundary. However, thick Precambrian ejecta layers occur in South Africa and West Australia, aged between 2.64 and 3.46 billion years old. Some of these layers may have been formed by projectiles larger than 20 km (12 miles) in diameter – twice the size of the one responsible for the dinosaurs' extinction. It is thus plausible to speculate that these very ancient periods of Earth history were marked by a higher occurrence of large extraterrestrial objects on Earth. Indeed, some 4.5 billion years ago, the Moon was formed by a gigantic impact between the young Earth and an object the size of Mars.

Impacts of small and large objects on Earth will continue from the asteroid belt between Mars and Jupiter, as well as from the large number of 'Near Earth Objects' on orbits that can potentially intersect that of the Earth; comets also pose a latent menace. Constant monitoring of all these rogue bodies is an absolute necessity.

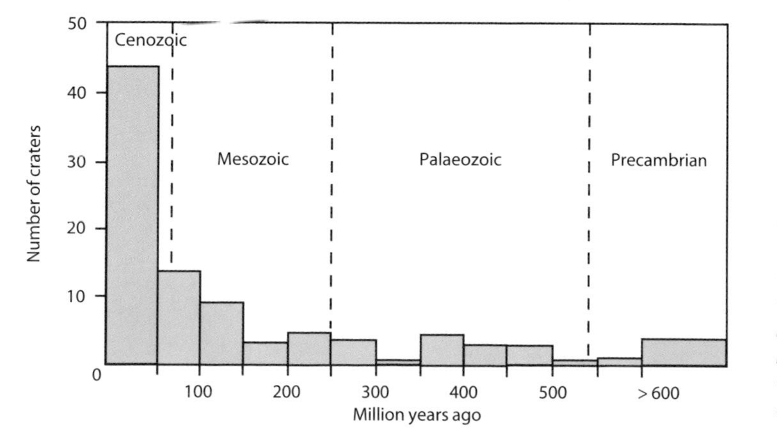

Distribution of known impact craters in the terrestrial geological record. The bias is clearly towards younger structures due to the rapid obliteration of old ones by geological processes such as plate tectonics, erosion and sedimentation.

MARCIO ROCHA MELLO

Where does oil come from?

Oil is the blood and gas is the oxygen of modern civilization!
DAVID CURRY, 2008

Oil was found at the surface in China around 4,000 years ago, and human beings have since discovered many uses for this sticky, black substance. Early civilizations used oil as a medicine, as a weapon of war, for mummification, as cement in the construction of the fabled Gardens of Babylon and the Tower of Babel as well as the Great Wall of China; the Aztec and Inca civilizations used it as a road surface.

Around the 16th and 17th centuries, petroleum started to achieve commercial importance in the pharmaceutical, waterproofing and public lighting industries. It was in the 19th century that the great petroleum revolution took off, when, in 1859, Colonel Edwin L. Drake, in Pennsylvania, USA, drilled the first oil well using a mechanical drilling rig and produced 19 barrels of oil per day.

Edwin L. Drake drilled the world's first commercial oil well in 1859, launching an oil boom in western Pennsylvania. The oil was found at shallow depth (just 21 m/69 ft below the surface), allowing its production using very simple drilling tools.

By this time the industrial revolution was already well under way, and oil became one of the most precious commodities on Earth.

It is extraordinary that, despite such ubiquity and global importance, the origin of oil has ever been mysterious. Early ideas suggested that oil came from non-biological sources. This *abiogenic theory* was dominant through the 19th century, and was strongly supported by the Russian chemist Dmitri Mendeleyev, as well as by 20th-century Russian and Ukrainian scientists, because it explained the presence of petroleum in metamorphic (baked) rocks at depths of up to 7,000 m (23,000 ft). They postulated that this oil originated inorganically from carbon-rich material deep in the Earth's mantle under very high temperature and pressure. The oils then migrated to the surface through faults and fractures.

The great 20th-century Austrian astrophysicist Thomas Gold argued that oil and coal originated from primeval methane trapped at depth when the Earth was formed, or from populations of bacteria that live far below the surface. His influence was so great that he persuaded the Swedish government to invest millions of dollars to drill deep below granite rocks in search of oil. Not much was found, and the abiogenic theory for the origin of oil has been largely rejected: the current view is that only 0.00001% of oil is of abiogenic origin.

In 1933, the German chemist Alfred Treibs manufactured petroleum in the laboratory from animal fat at high temperature and pressure, a practical demonstration of the *biogenic theory* that oil might have come from the remains of living organisms. In the 1950s, advances in

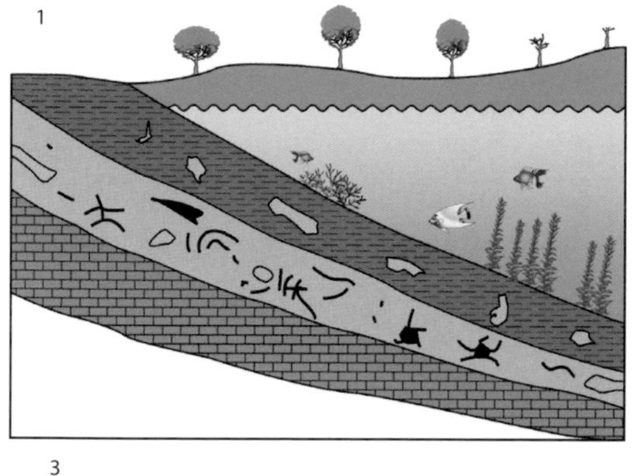

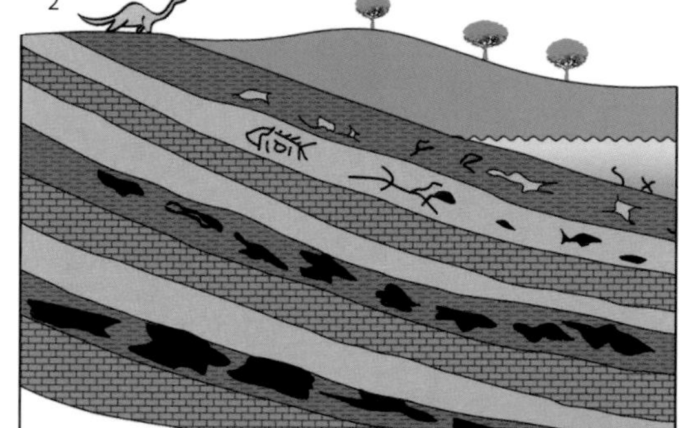

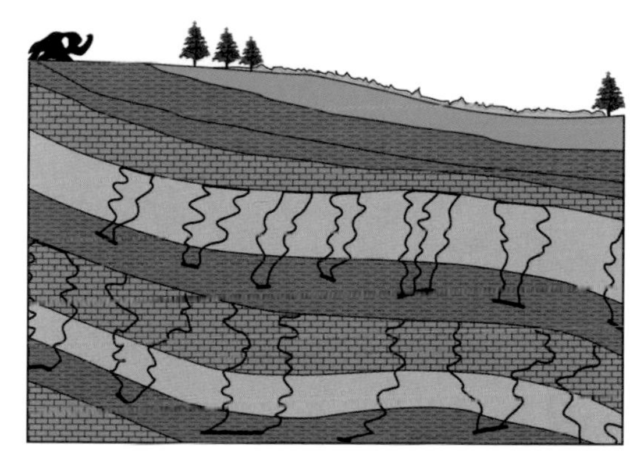

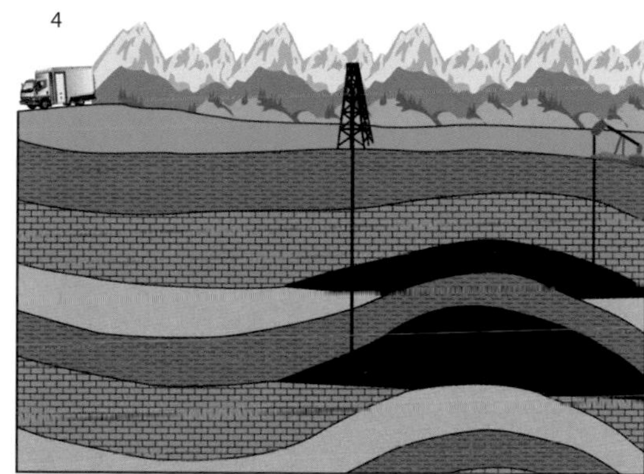

The sequence of oil formation.
1 *Hydrogen-rich remains of algae, higher plants and bacteria settle on the bottom of lakes and seas where anoxic water conditions help to preserve their organic components.*
2 *Through burial by the successive accumulation of overburden, the organic matter is transformed by high temperatures into oil and gas (hydrocarbons).*
3 *Oil and gas are fluids lighter than water and therefore migrate upwards by buoyancy through the sediments, using fractures and faults until they reach an impermeable seal, such as clay.*
4 *Tectonic movements form structures that act as a trap and accumulate the oil and gas in reservoirs such as porous sandstones and carbonates – the targets for oil and gas exploration.*

organic chemistry showed that the components of oil were biological compounds. In the last 20 years, molecular geochemists using new analytical instruments have built a molecular code for petroleum that provides great detail about its formation from the original plant and bacteria precursors.

Organic chemists can sometimes identify the precise biological sources of particular oils. For example, some oils contain the compound Oleanane, which is a marker of angiosperms, the flowering plants. Other organic markers in oil indicate a source from diatoms, single-celled planktonic algae. Remarkable components of some oils are diamondoids, fairly stable carbon compounds that resemble diamonds. Diamonds themselves are of course formed at depth from inorganic sources (see p. 77), but diamondoids in oil carry clear chemical signatures that they formed from lipids (fat-soluble molecules) in the membranes of bacteria.

Petroleum, then, is formed from the remains of algae, higher plants and bacteria that sank through the waters of ancient seas and lakes to the dark, anoxic ('without oxygen') bottom. The lack of oxygen inhibited decay and the original hydrogen and carbon were preserved as the organic remains accumulated. These seabed and lakebed deposits became deeply buried and were heated to temperatures over 100°C (212°F), generating oil and gas. Oil and gas are fluids, lighter than the water present in the pores of the rocks, so they migrate upwards, by buoyancy, until they reach an impermeable layer of rock, normally clay, which forms a seal, and they accumulate in the trap below. These trapped accumulations are the target of petroleum geologists.

Although the question of the origin of oil now seems to be completely resolved, some mysteries remain. The view that oil and gas might exist at great depth has been supported recently by the discovery of very deep gas accumulations around 10,000 to 12,000 m (33,000–40,000 ft) below the Earth's surface. The question now is whether such deep sources of oil and gas will ever account for more than a tiny fraction of global reserves.

Evolution

Evolution is at the heart of modern biology and medicine, and yet it seems to be a hugely mysterious and highly controversial topic. The theory of evolution was famously proposed by Charles Darwin in his *On the Origin of Species*, in 1859. Other biologists before him, notably the Comte de Buffon and Jean-Baptiste Lamarck in 18th-century France, were already convinced that life had not been created in an instant, as set out in the Bible. They used words like 'development' and 'evolution' to refer to the long history of change, and saw a procession of life, from simple to complex. Darwin's insight was that evolution occurred by natural selection.

Natural selection is the principle that only the organisms best adapted to their environment tend to survive and transmit their genetic characteristics to succeeding generations, while those less adapted tend to be eliminated. Darwin's ideas have proved remarkably prescient and durable, even though the scientific basis for genetics was unknown to him, given the technology available in his time.

The evidence for evolution is all around us, and Darwin presented much of it, based on the anatomy of plants and animals, fossils, biogeography and animal behaviour. One problem he wrestled with was the evolution of the eye. Darwin's opponents argued that the eye is so complex and consists of separate, but interacting parts, that it must have been created in one step. His solution, and the explanation today, is that there are many intermediate stages on the way to the complex eye of vertebrates, beginning with simple cells that can detect light and dark.

A model of the structure of the DNA molecule, in front of the output from gene sequencing. The discovery of the structure of DNA in 1953 revolutionized our view of life.

When Darwin studied a particular orchid from Madagscar he predicted that there must be a moth with a proboscis around 30 cm (12 in) long to pollinate it. The hawkmoth Xanthopan morgani *was discovered later.*

The discovery of the structure of DNA in 1953 is celebrated as one of the key scientific advances of the 20th century. But how is the information in our genes translated into the proteins which do all the work in our cells? This is achieved by the genetic code – like the Morse code system for encoding a message – and it is the same in a bacterium as in an elephant, and us.

Darwin also had the remarkable insight that the true pattern of evolution was a branching tree. Beginning with a single species billions of years ago, this great tree split and branched upwards through time. The tree can be reconstructed from various sources of evidence, from external physical characteristics and from genetics: drawing the tree has become a major focus of international research.

But how can one species split into two? Speciation is key to large-scale evolution. New species form in many ways, but most are associated with barriers to interbreeding within a population. So how variable can individuals be and yet belong to the same species? Humans vary remarkably in external appearance, but we are all members of one species, *Homo sapiens,* as shown by our genes. There is still debate about whether evolution happens at the level of individual organisms, or whether genes themselves are the drivers – the 'selfish-gene' viewpoint.

The links between evolution and development are also being intensively studied. Even before Darwin, German scientists remarked how embryos changed from rather simple, apparently similar forms early in development to acquire the particular features of the adult. Many patterns, such as the five fingers of vertebrates, are deeply embedded, if hard to account for. The study of regulatory genes is now leading to fascinating insights into how genes control the patterning of the body as it develops in the egg.

CARL ZIMMER

The evidence for evolution

Nothing in biology makes sense except in the light of evolution.
THEODOSIUS DOBZHANSKY, 1973

Scientists have amassed evidence for evolution in much the same way that they have amassed evidence for other scientific theories, such as the germ theory of disease or the theory of plate tectonics. When scientists talk about a scientific *theory* such as these, they do not use the word theory to mean a hunch or a nice-sounding story. Scientific theories are structures of ideas that explain a wide range of facts. To build a theory, scientists survey the observations that have been made in the past and come up with an explanation that joins those facts together. To investigate a theory, scientists draw hypotheses from it and then test them with experiments and observations. If those tests tend to support the theory, then the theory gains strength.

In the early 19th century, for example, several researchers proposed what came to be known as the germ theory of disease. They argued that many diseases were caused by infections of living organisms. Louis Pasteur, Robert Koch and other scientists then developed hypotheses and devised experiments to test them. Pasteur used sterile flasks of broth to demonstrate that microbes did not emerge spontaneously, instead moving around invisibly. Koch later came up with a series of tests that could link a particular species of microbe with a particular disease.

By the end of the 19th century, most scientists and doctors accepted the germ theory of disease. Yet they still did not know how germs infected their hosts or how they replicated once inside. It would not be until the mid-20th century, for example, that scientists discovered that viruses and bacteria carry genes. And it would not be for some years more that they learned that those genes were made of DNA.

This lack of information was not in itself a reason to abandon the theory. Instead, scientists judged it according to the evidence at hand. The discovery of genes, when it finally did arrive, only deepened our appreciation of the germ theory of disease. Scientists learned, for example, that viruses make us ill by inserting their genetic material into our cells, where they hijack the cellular machinery to make new viruses.

But new research can also cause scientists to expand theories. It turns out, for example, that some infectious diseases are not caused simply by the arrival of a living pathogen. In some cases, a harmless microbe that has lived in our body for years may suddenly turn nasty. If certain strains of *E. coli* move from the intestines to the blood stream, they can cause a fatal case of sepsis. The germ theory of disease must also now take into account the possibility that infectious agents do not have to be alive. Misfolded proteins, known as prions, have been linked in recent years to diseases such as mad cow disease.

A Victorian cartoonist poked fun at Charles Darwin for proposing that humans evolved from an ape-like ancestor.

Darwin's theory of evolution

The theory of evolution has a similar history. Charles Darwin published *On the Origin of Species* in 1859, and in it laid out a theory to explain the diversity and complexity of life. He proposed that over time, life evolved. Existing species gave rise to new ones through a process of descent with modification. While Darwin believed that several mechanisms helped drive evolution, he focused much of his attention on natural selection.

Simply put, in any generation some individuals will have more offspring than others. In some cases, that success is the result of variations in certain traits – thick fur, for example, or large body mass. If these beneficial traits are hereditary, they will become more common from one generation to the next. Over the vast expanses of time since life began on Earth, Darwin proposed, natural selection had produced complex features such as flowers and eyes (see p. 104).

Darwin first came up with the basic outline of his theory in 1837, but he did not immediately publish it. Instead, he spent much of the next 20 years patiently gathering evidence. He observed that many patterns in life are what you'd expect from evolution. He noted, for example, the varied forms that vertebrate limbs can take, from slender bird wings to shovelling mole claws. Despite their differences, they shared many underlying features. The same pattern could even be seen in rudimentary limbs, such as ostrich wings and vestigial leg bones in whales. Darwin argued that this underlying similarity was the result of natural selection acting on one ancestral appendage.

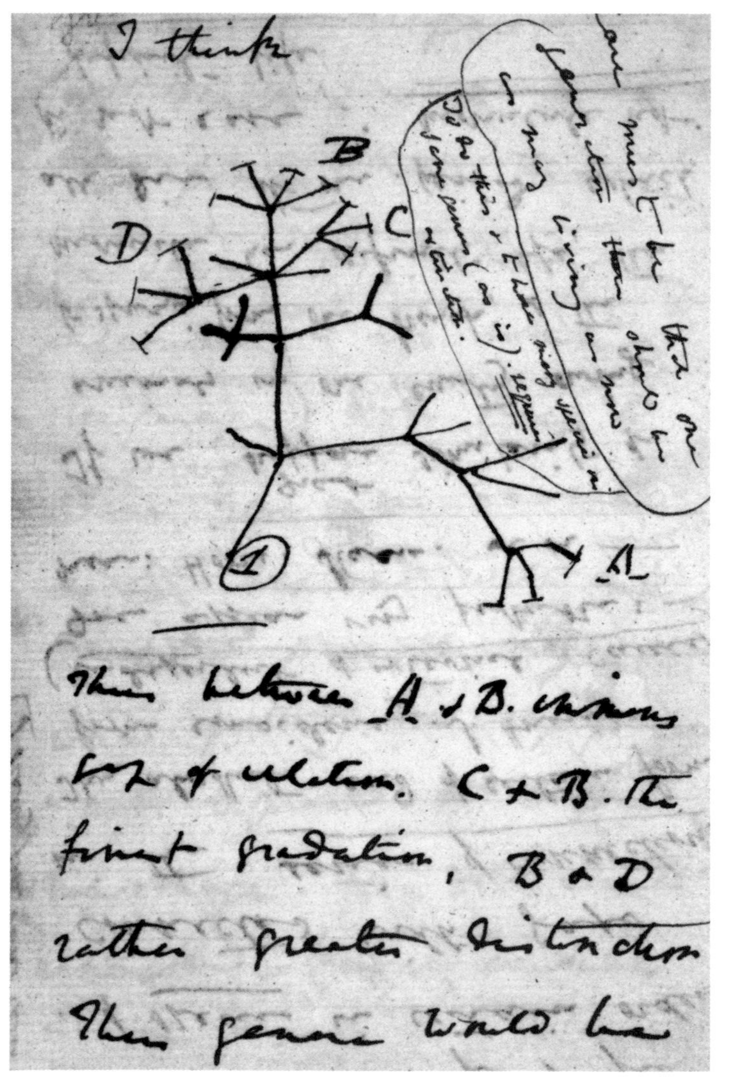

Darwin developed his theory in notebooks. This page, from 1837, shows his first sketch of the tree of life (see p. 118 for the version published in On the Origin of Species*).*

Discoveries after Darwin

As with the germ theory of disease, scientists today can study the theory of evolution in far greater detail than Darwin could. Like Pasteur and Koch, Darwin knew nothing of DNA. But this new research only adds more evidence for evolution. For example, the common ancestry of all living things is inscribed in the genetic code (p. 110). No matter how different living things may look, the dictionaries they use to translate genes into proteins are nearly identical.

Over billions of years, life has branched out into millions of species, and the kinship of closely related species is marked in their DNA. Such closely related species generally have similar genes that encode proteins, but they also share similar segments of DNA that do not encode proteins. Mutations can create disabled versions of genes called pseudogenes. Darwin proposed that humans were most closely related to great apes, and it turns out that chimpanzees and humans share pseudogenes that correspond to working genes in other species.

Geology offers another line of evidence supporting the common ancestry of living things. As new species evolved over millions of years, the surface of the Earth was also changing. Continents drifted and islands rose and subsided. The distribution of species today reflects this history. Many species of flies from the genus *Drosophila* are found on the islands of Hawaii and nowhere else, for example. These islands are arranged in a chain, which gradually formed and disappeared over millions of years. When scientists reconstructed the evolutionary tree of these flies, they discovered that the species belonging to the oldest lineages are found on the oldest islands,

while the youngest islands have the youngest species.

Mechanisms

The theory of evolution explains not just the pattern of evolution, but the mechanisms as well. And scientists can now confirm – on a genetic level – the factors that make natural selection possible. Mutations occur spontaneously, at rates that scientists can quantify. Depending on the circumstances in which an organism lives, a mutation may be harmful, harmless or beneficial. Advantageous versions of genes can spread quickly through a population, followed by further beneficial mutations. It is even possible to study natural selection as it unfolds.

DNA, shown here in its purified form, is the stuff of genes. Scientists can find traces of our evolutionary history inscribed in its sequence.

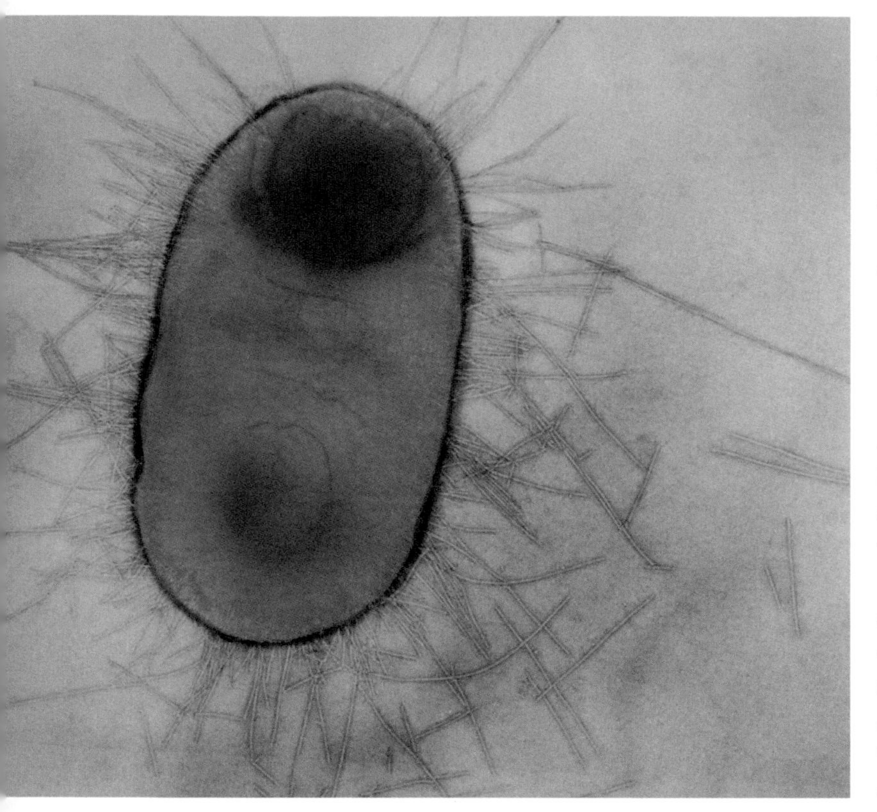

Scientists use fast-reproducing microbes like the gut-dwelling bacteria Escherichia coli *to observe evolution in action.*

Richard Lenski of Michigan State University, for example, has been observing the evolution of the microbe *Escherichia coli*. In the late 1980s he created 12 lines of *E. coli* from a single ancestor and fed them a meagre diet of glucose. Over the course of 40,000 generations, all 12 lines underwent a measurable evolutionary change. In most of the lineages, the microbes got bigger, being on average twice as large as their ancestor. Lenski also froze samples of the bacteria so that he could compare them to later generations. He then found that the bacteria had evolved the ability to reproduce 70% faster than their ancestors. With advances in DNA sequencing, scientists can scan the entire genome of an ancestral microbe and compare it to those of its descendants to pinpoint exactly which mutations caused these adaptive changes.

In the 20th century scientists learned how to analyse the spread of mutations with mathematical precision. This approach to evolution, known as population genetics, has allowed scientists to distinguish between different kinds of change. Under some conditions, a mutation can manage to become common from generation to generation thanks to nothing more than chance. Scientists have developed ways to distinguish this so-called neutral evolution from changes brought about by natural selection. They can even estimate the strength of natural selection on different genes – even on parts of genes. By comparing human DNA to chimpanzee DNA, scientists have identified several thousand genes that have undergone significant natural selection over the past few million years in our ancestors. These adaptive changes have altered genes that are involved in fighting diseases, in producing sex cells, in building the brain, and in many other functions.

The evidence for evolution also comes from studies on the origins of complex features. Wings and arms and other kinds of vertebrate limbs may show signs of a common ancestry, as noted above, but where did that ancestral limb come from? Studies on fossils and DNA agree that the closest living relatives of land vertebrates are a group of fish called lobefins. Their living representatives are the coelacanth and the lungfish. But 370 million years ago, as shown by fossils, there were many species of lobefins swimming around in coastal wetlands. The evolutionary tree of fish and land vertebrates reveals how fins were gradually reworked into legs with toes (p. 131). Studies have shown that these fish had even evolved proper digits long before they came on land.

Scientists do not know the precise series of mutations that made this transformation possible, but many important clues come from studies on how limbs and fins develop on embryos. The same genes govern the development of all of them, and it appears that simple changes – when and where these genes become active – brought about the shift (p. 127).

There are still many questions to be answered about life's evolution, as indeed there are about plate tectonics, infectious diseases and other important theories in science. But all these theories have stood the test of time, because they are such powerful tools for understanding the natural world.

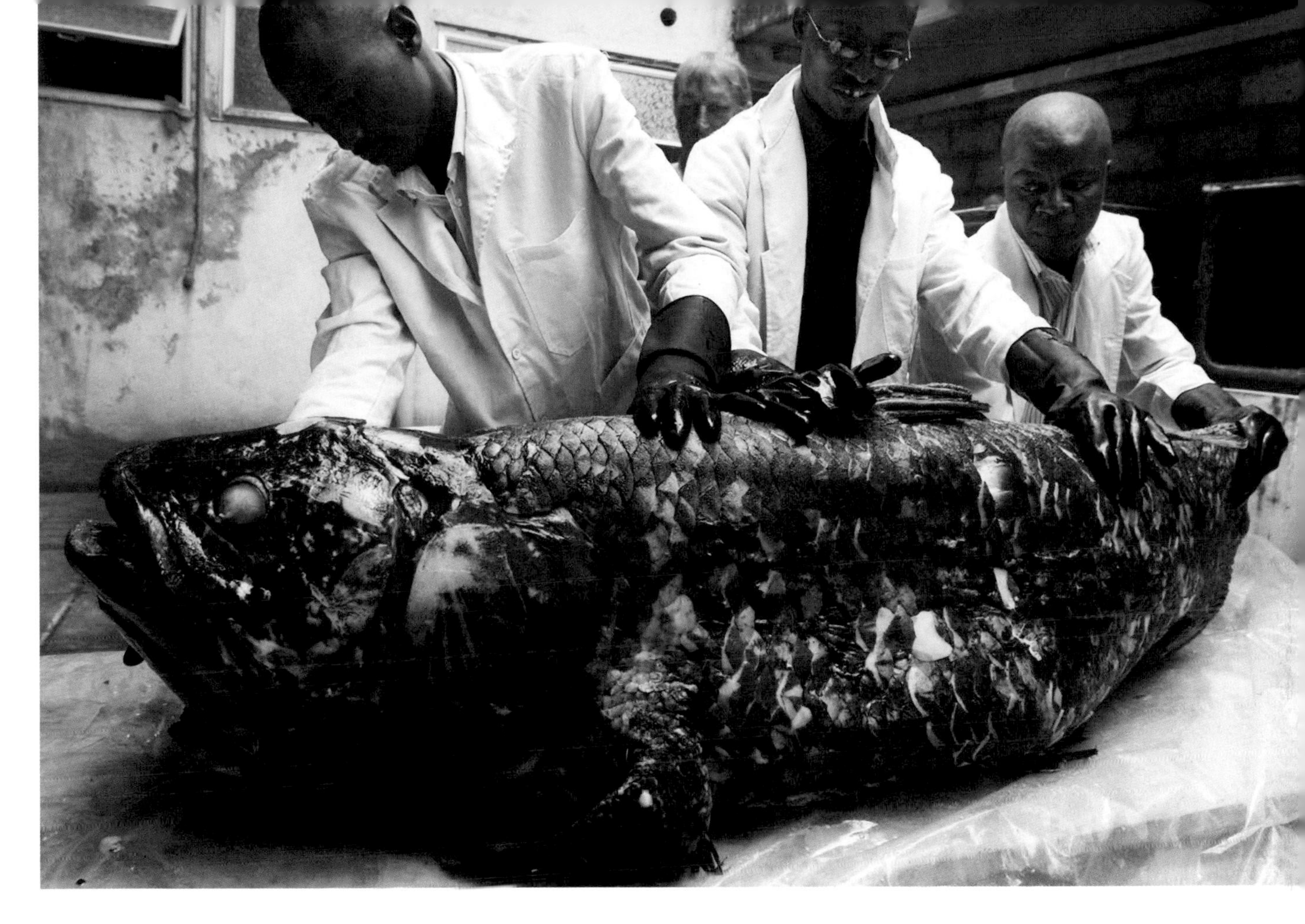

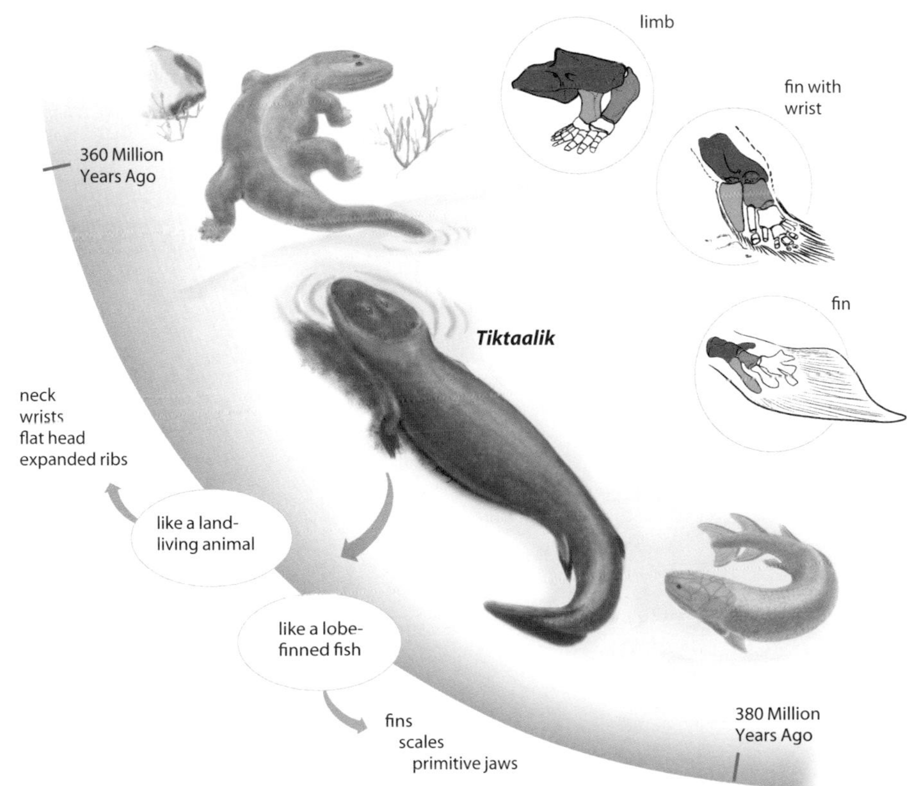

Above *The coelacanth, a fish that lives in the Indian Ocean, is among the closest aquatic relatives of land vertebrates like ourselves – studies of DNA and fossils confirm its close kinship.*

Left *The 370-million-year-old fossil of* Tiktaalik roseae*, discovered in 2006, offers clues to how our ancestors came on land.*

ANDREW PARKER

How did the eye evolve?

ORGANS OF EXTREME PERFECTION AND COMPLICATION:
To suppose that the eye ... could have been formed by natural selection seems,
I freely confess, absurd in the highest possible degree.
CHARLES DARWIN, 1859

Slotting a lens into the body of a digital SLR camera, complete with its light-sensing array, and connecting the camera to a computer for image processing, is a good model of the workings of the human visual system. The human eye and associated visual cortex in the brain mirror the camera in its lens, retina and processing unit. And that's the crux of Darwin's dilemma: the visual system comes in three, unrelated parts. To arrive at such an eye, all three parts must evolve independently and simultaneously. That, at first, does appear absurd.

There are two reasons, however, why this is not illogical. The first involves the way the eye evolved, the second concerns Darwin's thoughts on 'perfection'. Here we look at both.

Eyes detect the peak in the sun's electromagnetic radiation – light – and, because of constraints set by the physics of light, they come in limited, convergent forms. They can be classified as either chamber (or simple) or compound types. Chamber eyes, like ours, have a single entrance for light; compound eyes have multiple entrances. But both produce an image on a retina, as opposed to simple light sensors (that cannot be classified as eyes). In fact the transition between simple light sensors and the first eye during life's history holds hidden significance for eye evolution.

The compound eyes of flies: each facet, armed with its own lens and retinal cells, collects light from a unique sector of the environment. The images formed by individual facets are assembled in the brain like pieces of a jigsaw puzzle.

From light sensors to lenses

The argument that eyes cannot in theory have evolved because their intermediate stages could not function does not hold up. A computer model has shown how a patch of light-sensitive skin could gradually transform into a chamber eye within 2,000 sequential changes, or just 364,000 generations (as a pessimistic estimate). Certainly, there have been enough individuals within a species and geological time to accommodate this. Under continuous selection for improved light perception, the organ first alters to determine between sunlight on or sunlight off, then to sense the direction of sunlight, and finally to form an increasingly well-resolved image. The important point is that each stage is an improvement in light perception or vision over its predecessor.

Even the gradual introduction of a lens does not affect the developmental relationship between structure modification and efficiency of light perception, in this case the resolved number of pixels in the eye. It is only in the very final phase of focal-length tuning that the spatial resolution (or number of effective pixels formed) shoots to extremely high values with minor changes in the eye's structure. But although the evolution of a lens would at some point have provided a 'leap' in the information supplied by the eye, it does not necessarily mean that all information gathered in this way gets used. For instance, human vision under-samples the retinal image over most of the visual field. Maybe the evolution of the central nervous system's capacity for processing visual information lagged behind the available information for a long evolutionary time.

Recent research has centred on photoreceptor evolution based on advances in molecular biology. It was found that the gene encoding opsin – the protein that responds to light in the retina – as well as some developmental genes for light sensors date back beyond the first eye, and were recruited repeatedly during the evolution of different eyes, possibly between 40 and 65 times. Because the same development toolkit was used each time, and is anyway found within all animals, it is no problem to explain why the eye evolved so many times. We can demonstrate the action of these developmental genes in the model organism *Drosophila* (a fruitfly) by turning them on in the wrong places within an embryo, resulting in eyes sprouting on legs and wings in the adult fly (see also p. 127).

The evolution of lenses may owe something to physical phenomena such as molecular self-assembly or crystallization. If an enzyme, such as a liver enzyme, is overproduced within a cell it will crystallize; our own lenses, and even the lenses within some bioluminescent organs (to focus the light generated into a beam), are formed this way. The phenomenon of crystallization enhances the effects of genetic mutation in the evolutionary process. Current research aims to demonstrate this using the living cells that make the unusual calcitic lenses in brittlestars (related to starfish).

Rather than being difficult to explain, the evolution of the eye is now so well understood that it is used as a model for how common genetic tools build complex organs in general.

Perfection and imperfection

That the three elements of the visual system came together perfectly may be difficult to explain, but an imperfect system is less so. As in the under-utilization of information available mentioned above, there are signs of imperfections in all eyes. That the human retina is inside-out is another example – light must pass through nine layers of connecting nerve cells and other tissue to reach the receptors at the deepest level.

Imperfections can also include failure to deliver on visual requirements. Such shortcomings can be revealed within the arms race that exists between the eye of a predator and the colour of its prey. We see cuttlefish resting on the sea bottom, but their predators cannot. Predators, in this case, are dolphins with only 'black and white' vision, rendering the cuttlefish perfectly camouflaged to them. Small primates that eat seemingly camouflaged butterflies are oblivious

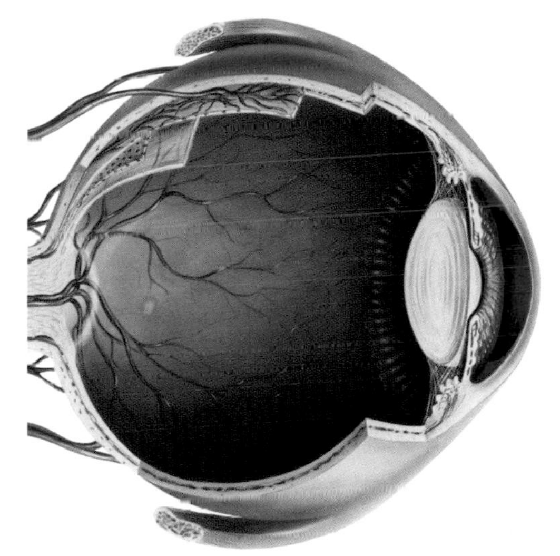

Left *In a chamber eye, for instance that of humans, birds and squid, light is focused by the lens at the front, forming an image on the retina (the focal plane). The information collected passes to the visual cortex in the brain, where it is processed.*

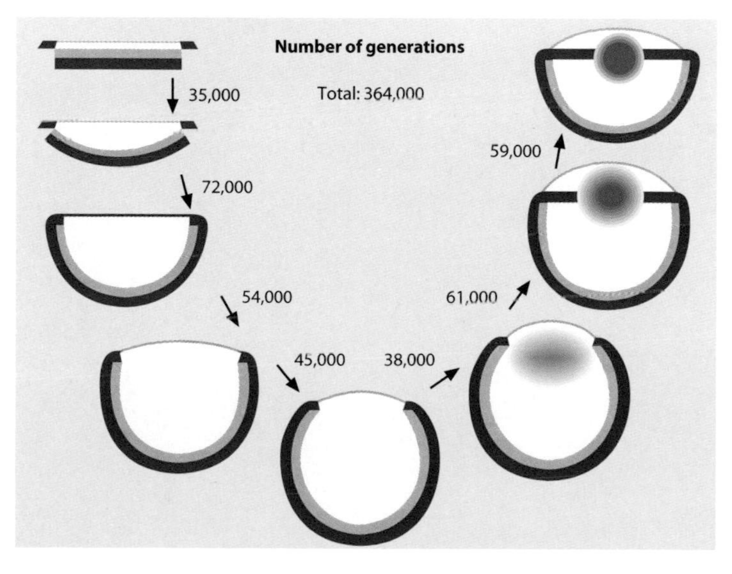

Below *Computer model of the evolution of a chamber eye by Dan-Eric Nilsson and Susan Pelger. A patch of light sensitive cells, with an outer (transparent) protective layer and a background, dark, absorbing layer, evolves under a continuous selection for improved light perception.*

A Cambropallas trilobite fossil from Morocco, around 540 million years old, with eyes protruding from the head shield (the left eye is casting a shadow) – this was one of the first animals on Earth with eyes.

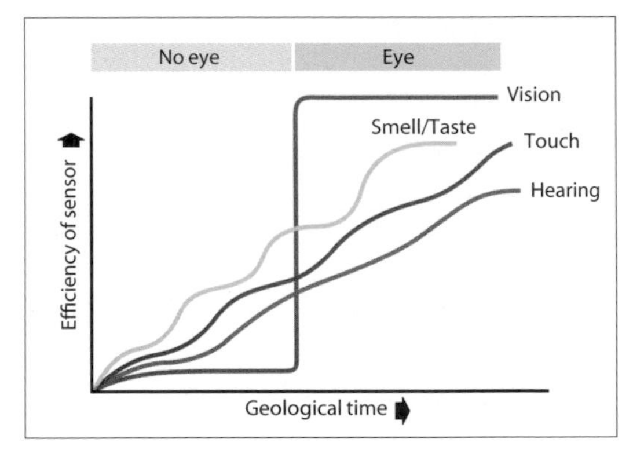

A chart of the quality of light perception (culminating in vision) over geological time, compared with the evolution of other senses. Life on Earth can be divided into two periods – pre-visual and visual.

to their conspicuous patterns – the patterns are in ultraviolet, which only the butterflies can see. The point is that the words 'extreme perfection' in Darwin's phrase are inappropriate. He would have been less troubled by the eye if he had realized so.

The very first eye

The question 'when did the first eye evolve?' is just as significant and consequential as *how* it evolved. The moment the first eye evolved was the point in time that vision was introduced on Earth. And vision today is the most universally powerful sense, in terms of its impact on animal interactions and behaviour.

For the first few hundred million years of multi-celled life on Earth, animals were blind, yet the selection pressures for vision were always immense. What probably held back visual evolution were the high energy costs to build and maintain an eye. That first eye evolved in a trilobite (a predator), around 542 million years ago. Prior to that animals possessed only simple light sensors. This is where the difference between an eye and a light sensor really matters. Light sensors provide information on night and day, or up and down in the water column. Eyes, on the other hand, allow animals to interact with each other using light as a stimulus. With the evolution of the eye, the size, shape, colour and behaviour of animals were revealed for the first time.

The introduction of vision must have been the launch of the most powerful weapon on Earth. And we know that it evolved relatively quickly in evolutionary time. The evolutionary countermeasures that were sparked off within the other, non-sighted, soft-bodied animals of the time, must have been significant. It is interesting that the Cambrian explosion (p. 30), when animals simultaneously, and independently, evolved their hard parts, began at this time. The 'Light Switch Theory' holds that the eye was the cause. Whatever the case, the introduction of vision must have triggered evolutionary chaos to some extent.

Since that first eye, vision has remained on Earth. Although only six of the approximately 37 animal phyla possess eyes, over 95% of all species belong to these. Vision has been a powerful weapon and a successful innovation.

KEVIN PADIAN

Why do so many people not accept evolution?

The purpose of the defense will be to set before you all available facts and information from every branch of science to aid you in forming an opinion of what evolution is ...
Dudley Field Malone, Scopes trial, 1925

In 1925, John Scopes was fined $100 for teaching evolution in Dayton, Tennessee. Scopes lost his case because he deliberately broke an existing law in Tennessee that in effect banned the teaching of evolution. Since then, every major court decision has supported the teaching of good science and eliminated religion and pseudoscience from American public classrooms. Yet polls consistently show that many or most Americans are ignorant of, or opposed to, the theory of evolution. Why?

Fundamentalist opposition

Christian fundamentalism witnessed a resurgence in America a century ago. Its proponents take literally the words of the Bible, as interpreted in contemporary English-language translations. Corollaries of this fundamentalist view include an abhorrence of homosexuality, abortion, the equality of women and the theory of evolution, among other things. The logic behind such opinions drawn from the Bible is often unclear on theological grounds and even self-contradictory, and the positions have been rejected by mainstream Christian denominations. However, the promise of salvation through faith alone has been overwhelmingly enticing for a vast number of Americans in the past three decades – and sufficient, it seems, to abandon rationality in the judgment of scientific concepts.

Christian fundamentalism, with its opposition to evolution, is considerably stronger in America than in other countries, but such opposition is also periodically felt in western Europe, Australia and parts of Asia. Fundamentalist Muslims likewise shun evolution, although more liberal Muslims do not; the pseudonymous theologian Harun Yahya of Turkey publishes prolifically against evolution, and this is one reason why Turkey is at the bottom of a list of 34 developed countries in the acceptance of evolution among its populace (the United States is second to last).

For Christian fundamentalists, the concept of evolution must be wrong because God created the Universe and everything in it in six days – not six metaphorical days, but six 24-hour days. Although the two accounts of that creation in Genesis differ in several sequential respects, neither agrees with the order proposed by the rational enquiry of scientific disciplines.

Anti-evolution books for sale in Dayton, Tennessee, during the trial of John Scopes for teaching the theory of evolution in a school in 1925.

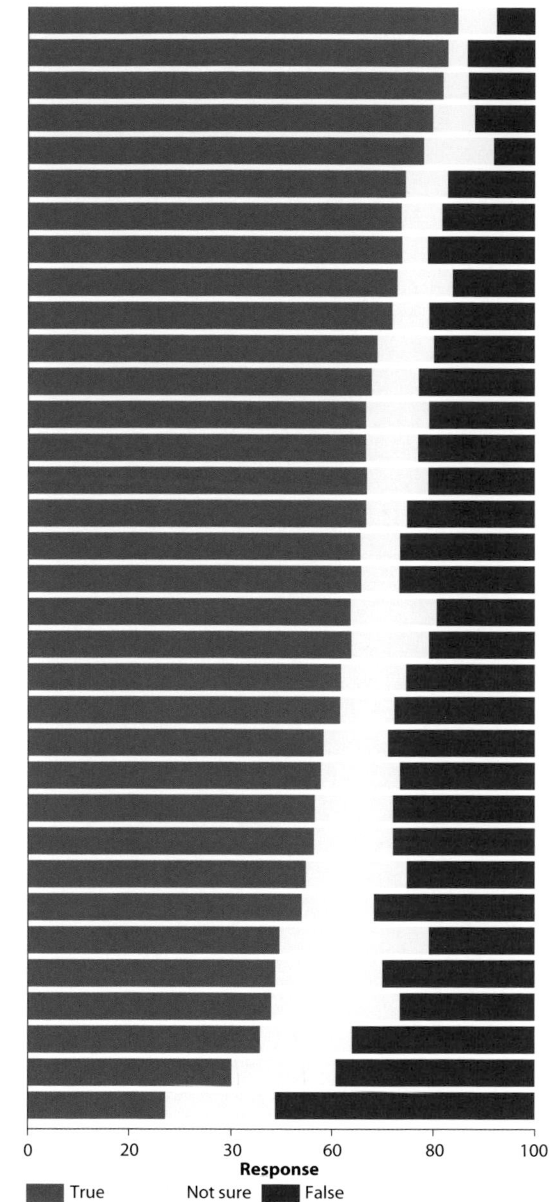

In an article in Science *in 2006 Jon Miller, Eugenie Scott and Shinji Okamoto analyzed the results of surveys from 34 countries on the public acceptance of evolution: only Turkish adults were less likely to accept it than American adults. (The figures after each country are the numbers of adults surveyed.)*

Opposition to evolution originally took the form of a theologically based rationale. In the 1960s, however, a school of polemics developed that attempted to refute the theory of evolution on 'scientific' grounds, such as the notion that it violated the Second Law of Thermodynamics. This approach, known as 'creation science', held that the stories in the Bible were literally true and could be supported by the proper application of 'good' science. Its proponents, however, had virtually no training in or understanding of the scientific concepts they were discussing, even though some had degrees in a field of science, technology or engineering. Because they published no peer-reviewed scientific papers, the scientific community did not take them seriously. But they were prolific evangelists, and very successful among the fundamentalist faithful, who had a generally poor understanding of science.

If people are ignorant or mistrustful of an idea, they are not likely to support it. So it is that the teaching of evolution declined after the Scopes trial. In one textbook of the time, the frontispiece of Darwin was replaced in its new edition by a drawing of the digestive system.

The new Creationists

Proponents of so-called 'intelligent design' (ID), as it has been redefined in the 1990s, are the intellectual and cultural heirs of the scientific creationists of previous decades, although they distance themselves as much as possible from their forebears. ID proponents appear to accept most scientific understanding, but are coy on topics such as the age of the Earth and even the possibility that one species could arise from another. They believe that natural processes operate most of the time, but that occasionally a change is so complex that natural processes cannot account for it. In this case, the intervention of a 'Designer' must be inferred as a causal agent.

Although for many ID appears at first glance to be a reasonable compromise between strong Christian convictions and necessarily non-religious scientific inferences, there are problems. Its notion of an interventionist 'Designer' is at odds with Enlightenment theology, and raises the conundrum of theodicy: if the Designer can intervene, why does He not do so more often, to relieve pain and suffering? If He cannot, then His omnipotence is called into question; if He can, then the power of prayer is doubted.

Proponents of ID have not published a single article in a peer-reviewed journal that demonstrates anything empirical about ID. For these and other reasons, a federal judge ruled in 2005 that intelligent design was not accepted as science by the scientific community, was clearly a religious notion and had no place in science classrooms. The Discovery Institute in Seattle, Washington, is the principal and virtually only centre of ID thought and public relations efforts; since the 2005 court ruling they have sniped at the judge's decision but have been unable to come up with any evidence to support their legal, 'scientific' or educational positions.

Microevolution to macroevolution

Biologists define microevolution as the evolutionary processes that take place in populations of organisms, often leading to adaptations to changing environmental conditions and also to the formation of a new species, or speciation (p. 123). Macroevolution comprises the patterns and processes that occur among groups of unrelated organisms (i.e., not members of the same species) as they evolve over millions of years. Even in Darwin's day, the progression of faunas through the fossil record was well understood, although the mechanism of the formation of new species was not; nor was the amount of time represented by the layers of rocks in the geological column. In the century and a half since, these questions have been largely settled, but the public is still mostly ignorant of the progress that has been made.

The creationist campaign of misinformation about evolution can claim credit for a good deal of the confusion. But it is also true that scientists, educators, textbook authors and government officials have failed to bring the understanding of evolution – notably how major changes occur in evolution – to the public consciousness. We have nearly all the major details, for example, about how vertebrates came on to land (pp. 103, 132), how birds evolved from dinosaurs (p. 35), how the mammalian ear and jaw complex evolved (p. 46) and how humans evolved from other apes (p. 50). Yet these discoveries, and the scientific methods behind them, are largely obscure to the public, especially in America.

A further difficulty is the inability of most people to comprehend the vast amounts of time that scientific evidence provides for the age of life, the Earth and the universe. These discoveries proceed from accepted scientific methods and observations, and are independent of all ideas about organic evolution. For most people, however, a history of life, including humans, measured in the thousands of years is far easier to comprehend. Nevertheless, the natural processes of evolution have been working on a timescale of much greater magnitude, a piece of information which would have delighted Darwin, because he suspected it. Merely because it is difficult to conceive of something does not mean that there is no evidence of its probability.

The six days of creation, as recorded in Genesis, here illustrated in the 12th-century French Bible of St Sulpicius of Bourges.

Disentangling the genetic code

What is true for E. coli *is true for the elephant.*
JACQUES MONOD, 1954

The genetic code provides the link between the genes in our genome and the proteins that are responsible for essentially all of the thousands of chemical reactions in our cells. There are two remarkable facts about the way we, and indeed all organisms, are structured on a biochemical level. First, our genomes are exceedingly stable, which allows DNA (deoxyribonucleic acid – the carrier of genetic information) to be amplified from extinct species tens of thousands of years old. Second, the chemistry of life proceeds extremely rapidly, because specific kinds of proteins called enzymes are able to catalyze reactions. The fastest proteins are limited only by the rate at which they encounter molecules of their substrates, and can accelerate reactions by a factor of 10^{18} – for comparison, 10^{18} microseconds is about 32,000 years. It is the genetic code that allows information about protein sequences to be transmitted stably from generation to generation, while at the same time providing for variations that allow natural selection to occur.

Below left *Pencil sketch of the DNA double helix by Francis Crick, 1953.*

Below right *James Watson (left) and Francis Crick with their model of part of a DNA molecule in 1953.*

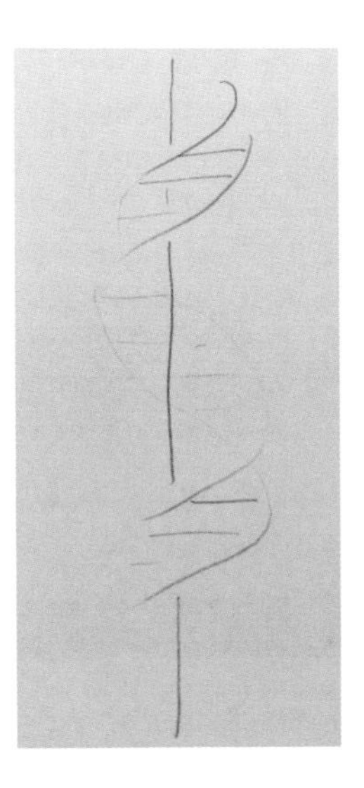

The genetic code

The term 'genetic code' is often incorrectly used to refer to the sequences of DNA that make up our genome, but in reality the genetic code is an encoding scheme for information, like the Morse code encoding of English letters as patterns of dots and dashes, rather than a message transmitted in that code.

The genetic code translates information from the DNA into proteins by establishing a correspondence between various sequences of three consecutive bases (the nucleotides, which, in pairs, make up the 'rungs' of the double helix) in the DNA and specific amino acids in the protein. For example, in all known organisms, the sequence of bases ATG – adenine, thymine, guanine – encodes the amino acid M, standing for methionine. Because there are four kinds of DNA bases (the other is cytosine), there are $4^3 = 64$ different three-base sequences, or codons. As there are only 20 kinds of amino acid typically used in proteins, the genetic code is said to be degenerate. Here 'degenerate' is used in the physicist's sense of having several different states that lead to the same outcome, rather than being somehow degraded from a more precise form.

The dawn of genetic code research came in 1954, the year after James Watson and Francis Crick's discovery of the structure of DNA. George Gamow, a physicist at George Washington University, recognized that the sequence of bases in the

DNA must code for proteins through some translation table, and he suggested that this was achieved by direct templating. In other words, the amino acids would line up against the diamond-shaped holes bounded by four nucleotides: two successive nucleotides on one strand and two on the other. The constraints of base pairing, combined with the geometry of the situation, indicated that there would be precisely 20 different kinds of holes – one for each of the amino acids typically used in proteins. However, this beautiful hypothesis was undermined by the fact that RNA (ribonucleic acid) acts as an intermediary between the DNA in the genome and the proteins that build the cell, and so any explanation based on direct templating between DNA and proteins could not be correct.

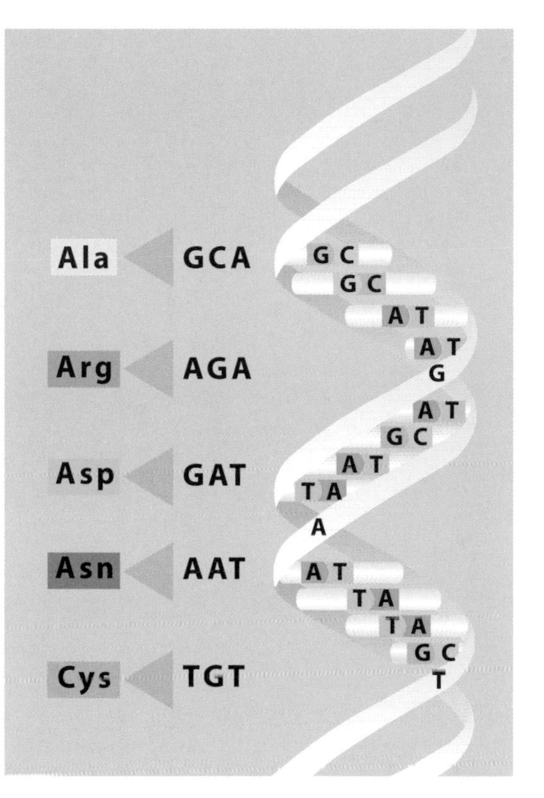

Diagram of DNA and codons, each consisting of a sequence of three bases, and the corresponding amino acids that they relate to.

Reading the genetic code

Shortly after this, several key pieces of research led to our modern understanding of the genetic code. Sydney Brenner and colleagues at the Cambridge Molecular Biology Laboratory were the first to establish that the genetic code reads bases in a set of three at a time, and, significantly, that the deletion or insertion of a single nucleotide would disrupt the entire protein. Second, Noboru Sueoka at Princeton showed that there were strong correlations between the amount of each of the four bases and certain amino acids, suggesting that these correlations could be used to uncover the relationships between the codons and amino acids. Third, Marshall Nirenberg and Heinreich Matthaei at the National Institutes of Health demonstrated that they could achieve translation without the presence of a cell, using purified components of the proteins and RNAs that make up the translation machinery. This allowed the relationships between many codons and amino acids to be determined directly, by synthesizing artificial templates that could act like messenger RNAs (mRNAs).

The full genetic code table was worked out in 1965, just four years after the first codon was matched up to the first amino acid experimentally. Since then, many scientists have been drawn to the puzzle of what caused the present arrangement within the genetic code. One of its striking features is its near-universality: the bacterium *E. coli* uses the exact same relationships between codons and amino acids as we do, at least in the nuclei of our cells. The fact that the same genetic code operates in very different groups of organisms suggested that there might be universal principles that govern which codons are assigned to which amino acids.

How did the genetic code evolve?

Shortly after the code was determined in 1965, Carl Woese at the University of Illinois at Urbana-Champaign observed that similar amino acids tend to be assigned to adjacent codons. He then proposed that the genetic code was organized by direct interactions between codons and amino acids. In 1969, by comparing the actual genetic code to a sample of random genetic codes, Cynthia Alff-Steinberger of the University of Geneva found that it was better than almost any of the random ones at placing related amino acids together. Then, in 1975, J. Tze-Fei Wong at the University of Toronto showed that amino acids

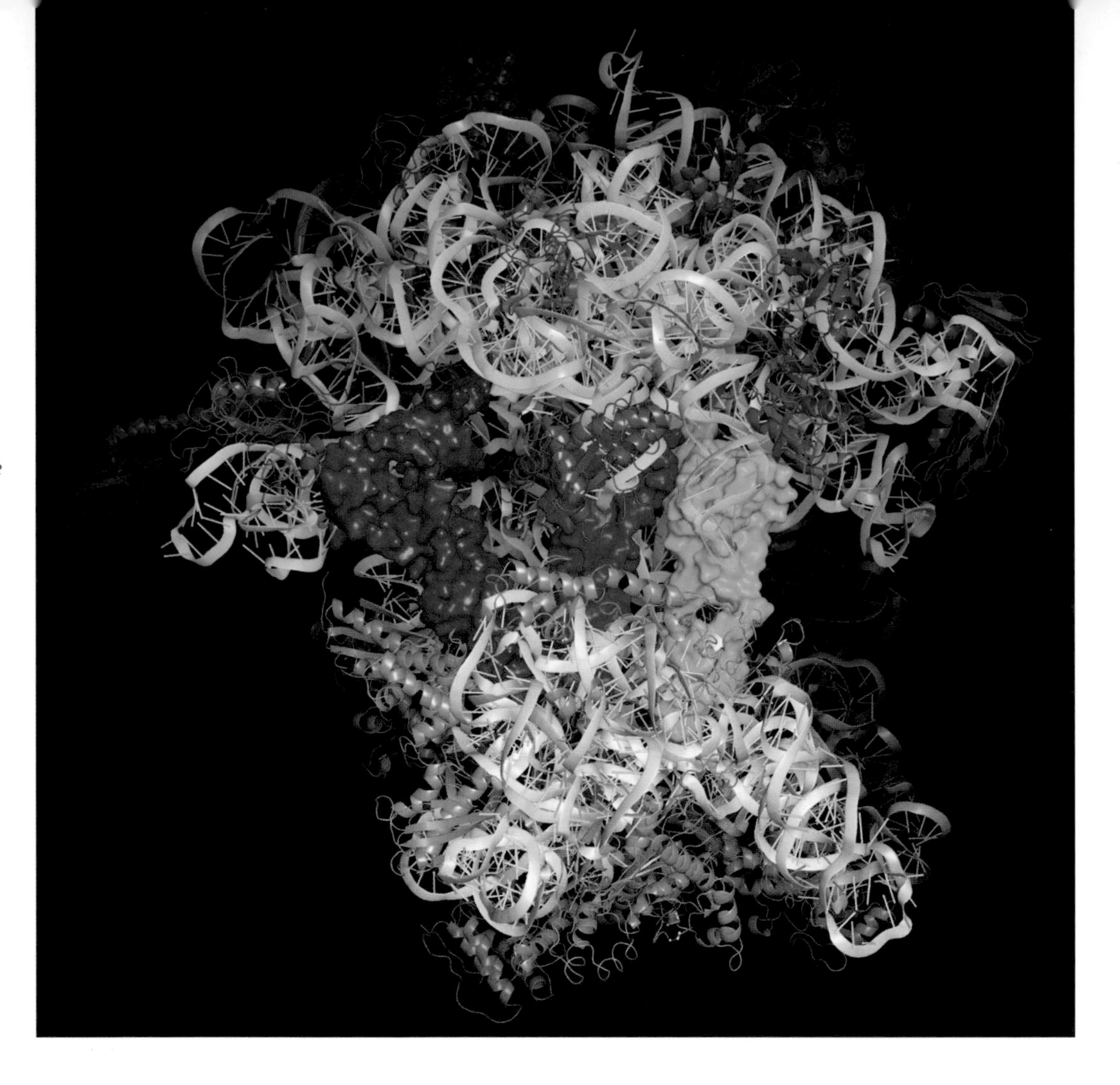

Molecular model of a bacterial ribosome – the place in a cell where groups of codons are translated into amino acids – showing the RNA and protein components in the form of ribbon models.

that were synthesized by similar pathways were also often close together. These three ideas, that the genetic code might stem from direct associations between codons and amino acids, that there was a selection process to minimize errors, or that new elaborations of amino acids were added to the code in a stepwise process, provided three competing models of genetic code evolution.

However, Francis Crick's pronouncement in 1968 that the code was a 'frozen accident' – in that changing the code would disrupt every protein, so that, once formed, primitive organisms would be unable to change it – was also influential. If his theory were correct, no explanation for the genetic code table would be necessary.

More recently, evidence for all three theories has come to light, and the emerging consensus is that several processes probably played important roles in the evolution of the genetic code. Michael Yarus at the University of Colorado, Boulder, and colleagues have shown that, out of vast random pools of RNA sequences those that best bind specific amino acids are the ones that have the codon (anticodon on RNA) sequence for that amino acid at the binding site, supporting the physical interaction theory. Steve Freeland, now at the University of Maryland, and colleagues have shown that the genetic code is highly optimized relative to other possible genetic codes, and, intriguingly, that it may promote the evolution of functional protein sequences faster than other genetic codes.

Finally, a large body of research by numerous investigators has shown how components of the translation apparatus can change to modify codon assignments or add new amino acids: indeed, some organisms, including us, incorporate 'non-standard' amino acids such as selenocysteine into the genetic code. Thus, our ultimate explanation of the genetic code's evolution must explain not just one genetic code, but a range of variations on a common theme.

OLIVER CURRY

Selfish-gene theory

Natural selection is the process whereby replicators out-propagate each other. They do this by exerting phenotypic effects on the world, and it is often convenient to see those phenotypic effects as grouped together in discrete 'vehicles' such as individual organisms.

RICHARD DAWKINS, 1982

The 'selfish gene' is a phrase coined by Richard Dawkins to convey the basic logic of Charles Darwin's theory of evolution by natural selection. Darwin's theory states that if you have a population of entities that vary in their ability to replicate themselves, then, inevitably, the composition of the population will change over time. Entities with features that make them better able to replicate under prevailing conditions will become more common, whereas those less able will become less common.

Given a constant supply of novel variation, this process of differential replication can continue indefinitely, with surviving members of the population accumulating ever more replication-promoting features. Populations of replicators living in different environments will accumulate different features, eventually giving rise to new forms of life (p. 123).

With this theory, Darwin was able to explain how, from simple beginnings, living things had evolved, diversified and come to possess 'innumerable adaptations and contrivances, which have justly excited the highest admiration of every observer'.

Enter the gene

Evolution is often viewed in terms of individuals struggling to survive and reproduce. However, the entities that replicate are in fact 'genes': strings of digital information encoded in DNA. It is genes, not individuals, that undergo random variation (mutation), are copied and become more or less common down the generations. Thus, evolution consists of competition among genes – each acting as if it were trying 'selfishly' to replicate itself by means of the effects that it has on the world. This 'selfish-gene' or 'gene's-eye' view of evolution sheds new light on the design of living things, and alerts us to previously hidden examples of natural selection's handiwork.

Jacket of the first edition of The Selfish Gene *by Richard Dawkins (1976), with a painting by Desmond Morris.*

The peacock's tail – used to attract peahens – is one of the 'innumerable adaptations and contrivances, which have justly excited the highest admiration of every observer' as explained by Charles Darwin in his theory of evolution by natural selection.

Take individuals: it is common to think of individuals as using genes to make more individuals; but from the gene's-eye view of evolution it's the other way around. Genes use individuals to make more genes – the chicken is the egg's way of making more eggs. From this perspective, instead of taking individuals for granted, we have to ask why genes adopted them as their principal means of transport down the generations. The answer is that by working together genes can build bigger and better adaptations – the peacock's tail, the elephant's trunk, the orchid's flower – than they can by going it alone.

It is precisely because genes have a common interest in creating individual bodies that we intuitively see individuals as the principal actors in evolution, and their adaptations appear to promote individual survival and reproduction. But selfish-gene theory reminds us that this need not have been – and indeed is not always – the case: genes have strategies other than the formation of individuals for ensuring that they are passed on into future generations.

Other ways to make a living

It is now recognized that, if they can get away with it, genes will replicate themselves at the expense of other genes in the same body. They litter the genome with superfluous duplicates of themselves (junk DNA), or they barge their way into more than their fair share of eggs and sperm (segregation distorters). Such genes can spread through the population, even though they do so to the detriment of the other genes and the individual that houses them.

Genes can also replicate by helping identical copies of themselves that reside in other individuals. Such individuals are most likely to be genetic relatives. This explains why parents are altruistic to their offspring, and why social insects – such as ants, bees and termites – forego individual reproduction, and even sacrifice their lives, in order to help raise colonies of their siblings (see p. 221).

What is more, genes can replicate themselves by exerting effects beyond the individual in which they are housed – that is, they create 'extended phenotypes', such as animal artifacts. For example, a gene might build a bird that builds a better kind of nest, a beaver that builds a better kind of dam, or a spider that builds a better kind of web, and thereby propagate more genes for nest-, dam- and web-building.

Myths and misunderstandings

Although selfish-gene theory is now mainstream in biology, it has been subject to a number of persistent misconceptions – especially when applied to human behaviour.

First, selfish-gene theory does not imply that individuals want to spread their genes. Evolution is a theory of design, not of motivation. It explains why knees and leaves and fins and livers are designed the way they are, and consequently why they work the way they do. It does not suggest that knees and leaves and fins and livers are motivated by an unconscious desire to spread genes. The same is true of brains. Evolutionary theory explains why brains are wired up in the way that they are; and, equipped with this wiring diagram, it can explain behaviour. At no point do such explanations invoke ulterior genetic motives.

Second, selfish-gene theory does not imply that genes are the sole determinants of behaviour, or that genetic effects are fixed and immune to environmental influence. On the contrary, genetic instructions are usually 'if/then' rules that depend on the environment at every stage. Alligator embryos develop according to the rule 'If the egg is cool, then become male; if warm, then female'. Bones operate according to the rule 'If under stress, then lay down more calcium; if under-used, then reabsorb the calcium'. Sunflowers follow the rule 'If the sun is to the right, then bend right'. And so on. Brains consist of hundreds of thousands of such decision-rules, organized into sophisticated software. Hence, when explaining behaviour, the question is not 'Genes or environment?'. Rather, it is 'Given the evolved design of the organism, which aspects of its environment are important, and why?'.

Finally, selfish-gene theory does not imply that individuals will be selfish. The 'selfishness' of genes is a technical term that means 'self-promoting' or 'self-replicating'. Genes must be selfish in this sense, for the simple reason that 'selfless' genes would remove themselves from the gene pool. But evolutionary theory has identified a number of ways in which selfish genes can replicate themselves by building social, cooperative and altruistic individuals. We have already noted that genes can spread by helping replicas in other individuals – that is, in other family members. Genes can also spread if they dispose organisms to team up with others, to do one another favours, and to settle disputes over resources without coming to blows. Evolutionary biologists have found numerous examples of these cooperative traits in nature, including a range of social instincts in animals (p. 225). Now they are beginning to find equivalent instincts in humans – such as sympathy, trust and generosity – which we call 'moral sentiments'. And all are the result of genes 'selfishly' pursuing their own replication.

So, just as Darwin's initial statement of the theory of evolution solved some of the great mysteries of life – the origin of species, the descent of man, the expression of emotion – modern selfish-gene theory is uncovering new mysteries – from conflict within the genome to altruism between individuals – and contributing to their solution. And, as Darwin envisaged, his theory is laying the foundation for investigating one of the few remaining mysteries in science: human nature.

Organisms are hard-wired to be flexible. This alligator embryo develops according to the rule 'If cool, then become male; if warm, then female.'

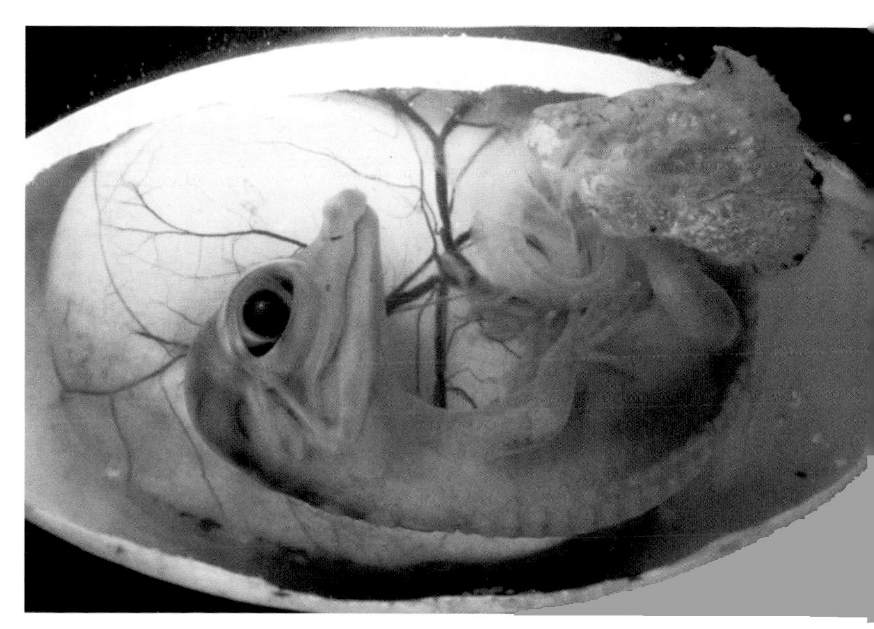

Drawing the tree of life

The affinities of all the beings of the same class have sometimes been represented by a great tree. I believe this simile largely speaks the truth.
CHARLES DARWIN, 1859

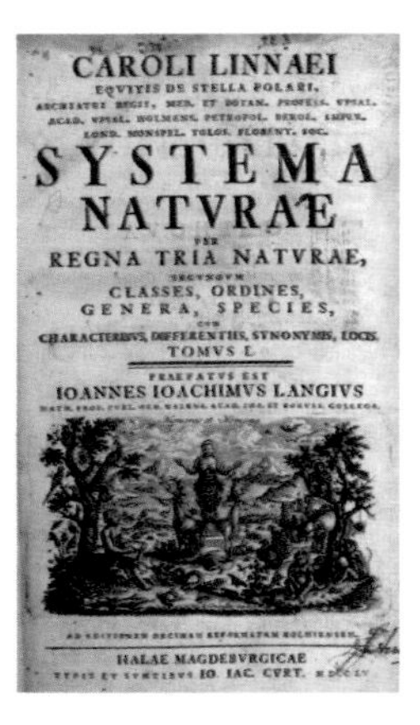
CAROLI LINNAEI
SYSTEMA
NATVRÆ
REGNA TRIA NATVRAE,
CLASSES, ORDINES,
GENERA, SPECIES,
TOMVS I.
IOANNES IOACHIMVS LANGIVS
HALAE MAGDEBVRGICAE

Title page of the classic tenth edition of Carl Linnaeus' Systema Naturae, *published in 1758. This edition is regarded as the starting point for the nomenclature of animals.*

The tree of life is a classic icon of evolution, and the only illustration in Darwin's *On the Origin of Species* was an evolutionary tree. His view, highly original at the time, was that all living species are the top twigs of a great tree, and that it should be possible to track back in time down the tree to see how the species branches progressively join together right back to the root – the single species from many billions of years ago that was the progenitor of all others.

Although this concept may seem obvious now, it has proved hard to establish the true shape of the tree, or even segments of it. Only in the past 20 years have serious strides been made in drawing large sections of the tree to wide satisfaction. This difficulty in establishing the very backbone of evolution is a mystery – why has it taken so long, and why is it so difficult?

The shape of evolution

Darwin's tree is a hierarchical diagram: if each small twig represents a species, the next hierarchical level down is a larger grouping, called a genus (plural, genera); further down are families, then orders, then classes. Long before Darwin, naturalists realized that life was not arranged randomly – there is a pattern. We can all tell a dog from a cat, and the great classifier of life, Carl Linnaeus, called them *Canis* and *Felis*. Linnaeus established the system of Latin binomial (two-name) terms for all living species, thus the cat is *Felis catus* and the dog *Canis familiaris*. Without being an evolutionist in any sense, Linnaeus could see that the wildcat, the lion, the puma and the leopard were all cats, so he put them in Family Felidae, and the wolf, fox, jackal and Cape hunting dog were all dogs, so he put them in the Family Canidae. These families are members of Order Carnivora, distinct from monkeys, apes and humans, Order Primates, or rats, mice and beavers, Order Rodentia.

Each year more species are discovered and named (p. 177), and the complexity of the tree of life becomes clearer. When Darwin looked at the classification he saw evolution. Why should lions, cheetahs, leopards and domestic cats all share a common design if they were not related? He also visualized the branching pattern of the classification as a real family tree. The Genus *Felis* may have arisen a few million years ago, while the Family Felidae must have arisen earlier, the Order Carnivora earlier still.

Darwin also saw that not all species had survived to the present – many had become extinct. As he put it, many twigs of the great tree of life died long ago. The broad outlines of the fossil record were known in Darwin's day, but discoveries since then have fleshed out the timescale of the tree, and shown us that he was right: the order of branching of the tree pretty much matches the order of occurrence of fossils in the rocks.

In *On the Origin of Species* Darwin presented two key ideas: that the process of evolution was natural selection, and that all living things are related to each other through common descent, meaning that any pair of species shares a *common ancestor* if you delve deep enough down the tree.

Drawing the tree

At one level, drawing the tree is easy. We can link the various species of either cats or dogs as close

MAMMALS, MORPHOLOGY AND MOLECULES

We know that a whale and a bat and a primate are all mammals, but how do they relate to each other? Perhaps surprisingly, the classification of living mammals has long been mysterious. The picture became clearer in 1997, when Mark Springer and colleagues discovered the Afrotheria, a clade consisting of African animals, which linked the elephants (Proboscidea), hyraxes (Hyracoidea) and sirenians (Sirenia) with aardvarks (Tubulidentata), tenrecs and golden moles. The last three had all been assigned various positions in the classification of mammals, but their genes show they shared a common ancestor with the elephant-hyrax-sirenian group. After 1997, everything began to fall into place.

It was discovered that the South American placental mammals form a second, separate, major group, the Xenarthra. And the remaining mammalian orders form a third major clade, the Boreoeutheria ('northern mammals'), in turn split into Laurasiatheria (including insectivores, bats, whales, carnivores and ungulates) and Euarchontoglires (primates, rodents, rabbits).

So, in the course of two or three years, several independent teams of molecular biologists solved one of the outstanding puzzles in the tree of life. The key mystery remaining is this: if the Afrotheria is really a clade, then there must be some obscure anatomical feature shared among them all! The hunt goes on.

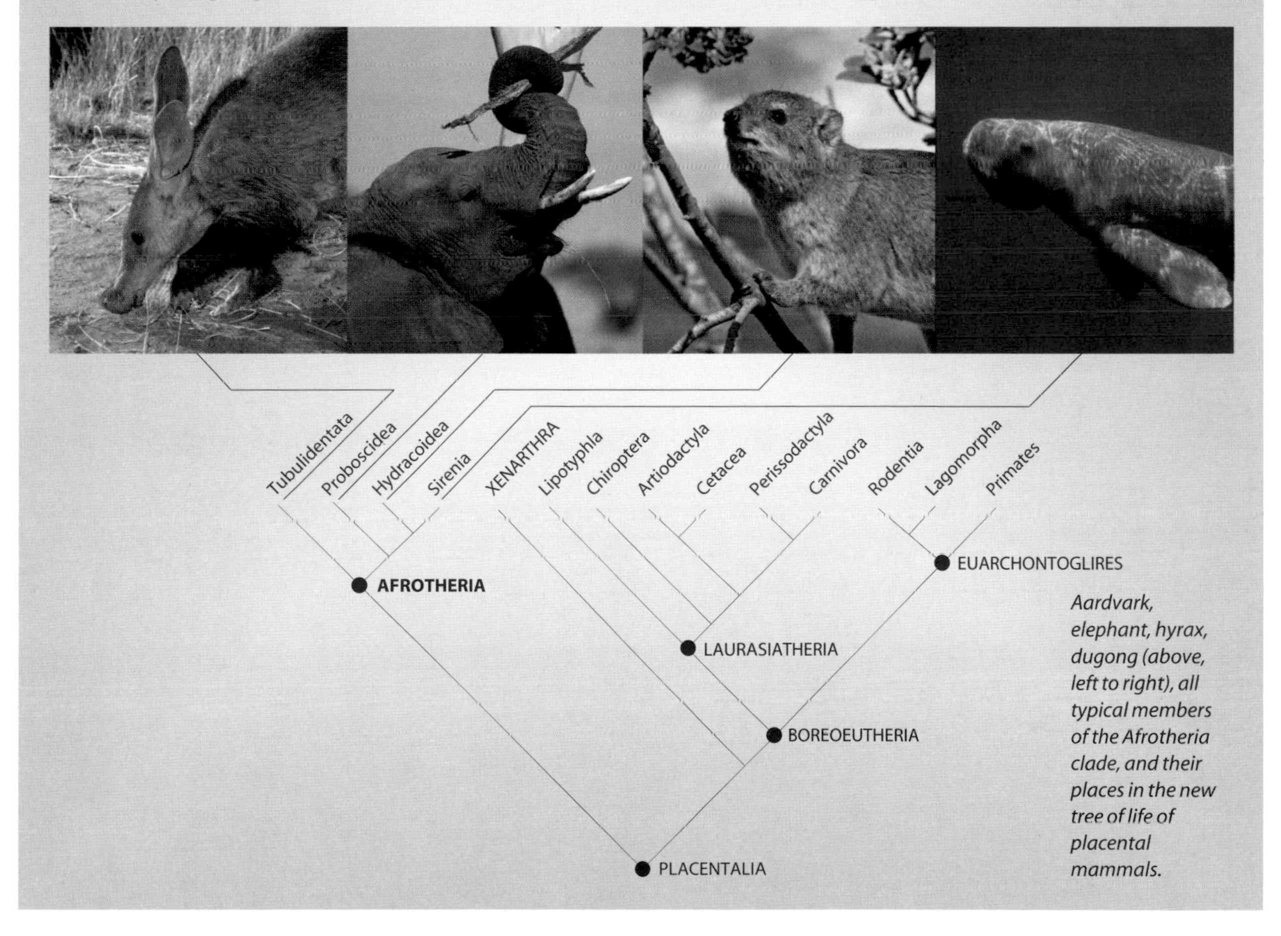

Aardvark, elephant, hyrax, dugong (above, left to right), all typical members of the Afrotheria clade, and their places in the new tree of life of placental mammals.

relatives based on their *morphology* – their external appearance and anatomy. Detailed study of their bones and muscles shows both the shared structures and also which species are closest. For decades, *systematists*, the biologists who study the evolution and classification of species, did this work based on their observations of general physical resemblances.

The method was improved in the 1960s and 1970s with the rapid introduction of *cladistics*, a toolbox of observational and analytical methods. Cladists catalogue critical characters that match

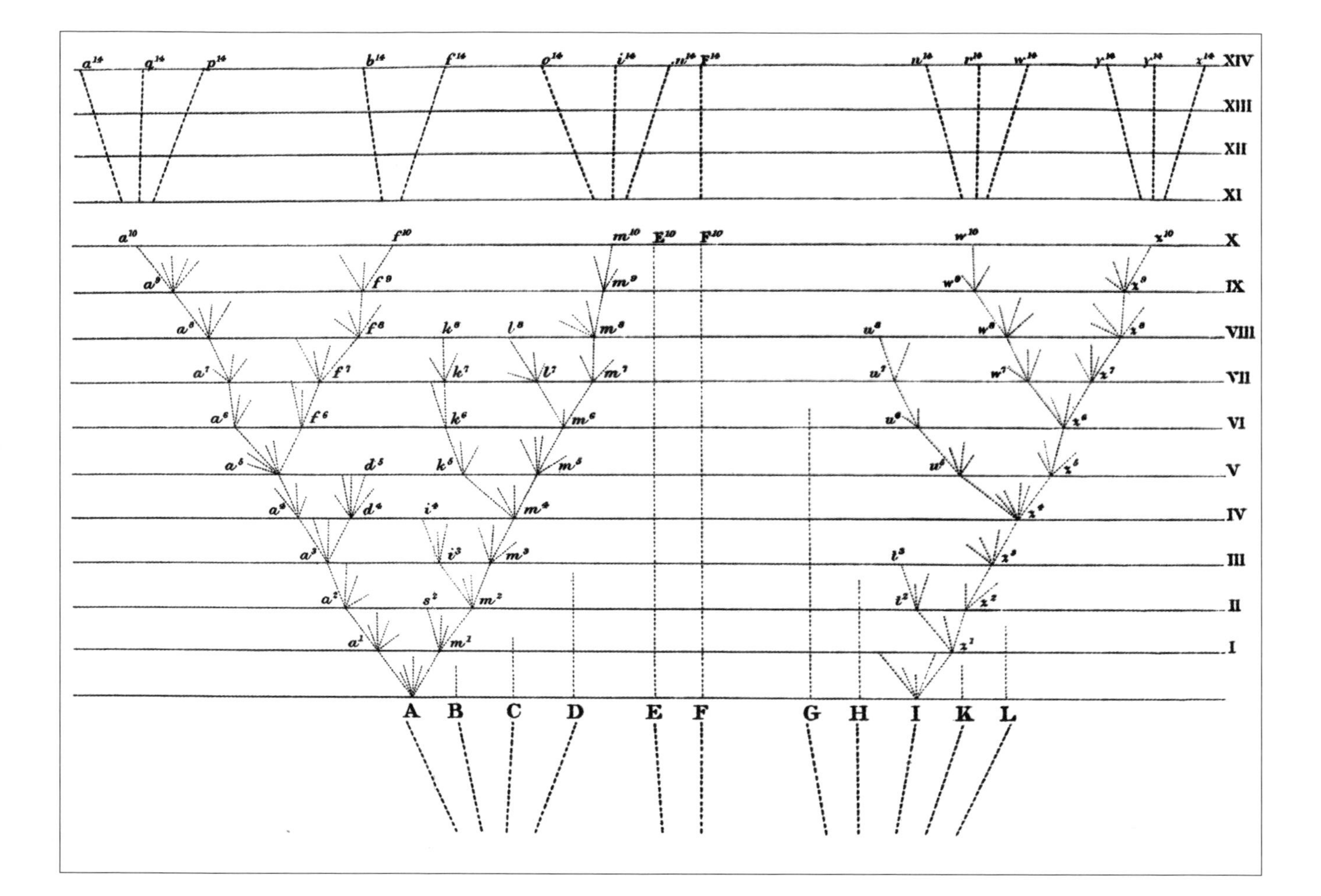

The famous (sole) illustration from Darwin's On the Origin of Species *(1859), one of the first ever representations of the tree of life, showing 11 initial species (A–L), two of which (A and I) are founders of groups that branch into many species, and the others of which die without issue.*

particular branching points in the tree – many of them exceedingly obscure anatomical points – for dozens of species, and run vast multivariate statistical analyses to draw the tree that best fits the data. Cladistics provides clearcut solutions to sectors of the tree of life. But how do we know whether the results are correct or not?

The real revolution has come with molecular sequencing. The DNA in every cell of every organism provides the genetic sequence. The sequence of bases in the DNA is unique to each species, and indeed comparisons of sectors of DNA – genes – between different organisms show that the degree of difference is proportional to the date of their last common ancestor.

The molecular clock hypothesis is based on the premise that the DNA mutates at a steady rate. It is therefore possible to make comparisons between any pairs of species to allow the sequence and rough timing since their last common ancestor to be determined.

The importance of the development of cladistics and molecular sequencing is that systematists can now investigate the same sector of the tree using two independent methods. If they find the same answer, then that is probably correct. The answer to the question why the tree of life has not been completely worked out may now be clearer – the modern era of tree exploration began only in the 1970s and 1980s, and there are many millions of species to study. Each analysis may take many months or years of study, and the computations are huge, but work so far is expanding the possibilities. In 2007, the first complete tree of all 4,550 species of mammals was published. We can look forward to such analytical solutions to ever-greater sectors of Darwin's tree of life, but the work will not be finished in our lifetimes.

MICHAEL J. BENTON

Human genetic variation

Natural selection is a mechanism for generating an exceedingly high degree of improbability.
RONALD A. FISHER, REPORTED BY JULIAN HUXLEY, 1936

People have always wondered why every human being seems different, and yet at the same time so similar. Genetic inheritance is obvious in the way that children look like their parents, inheriting perhaps their hair colour or nose shape from one parent, and their height or propensity to baldness from the other. There are more than 6 billion people on the Earth today, and countless more have existed in the past: can all those people really have had a completely different genetic composition, or might each of us find someone identical to us somewhere, or at some time in the past? And how does genetic make-up vary, if at all, between the different hierarchical levels, from families, through nations and ethnic groups, to the major continent-wide races?

These questions have been major discussion points for centuries, and answers to some are at last being provided by modern genetics. But research exercises such as the human genome project are often so huge and so technically demanding that a consensus view is achieved only slowly, and can then sometimes change dramatically. One key question that was once thought settled has produced surprises recently, and that is the degree to which humans differ genetically from our nearest cousin, the chimpanzee.

The myth of the 1% difference

Since 1975, geneticists have suggested that humans and chimps differ by perhaps 1% of their genetic make-up. This rather low figure was in line with the genetic differences between many other pairs of closely related genera. After all, genetic and fossil evidence suggested that humans, *Homo sapiens*, and chimpanzees, *Pan troglodytes*, diverged perhaps 5 million years ago. Even so, a 1% genetic difference seemed rather modest in the light of the obvious physical and mental differences between humans and chimps – but we had to try to accept the revelation that we were more closely related than we might like to think.

Every human being is different in appearance and different genetically, unless they are identical twins.

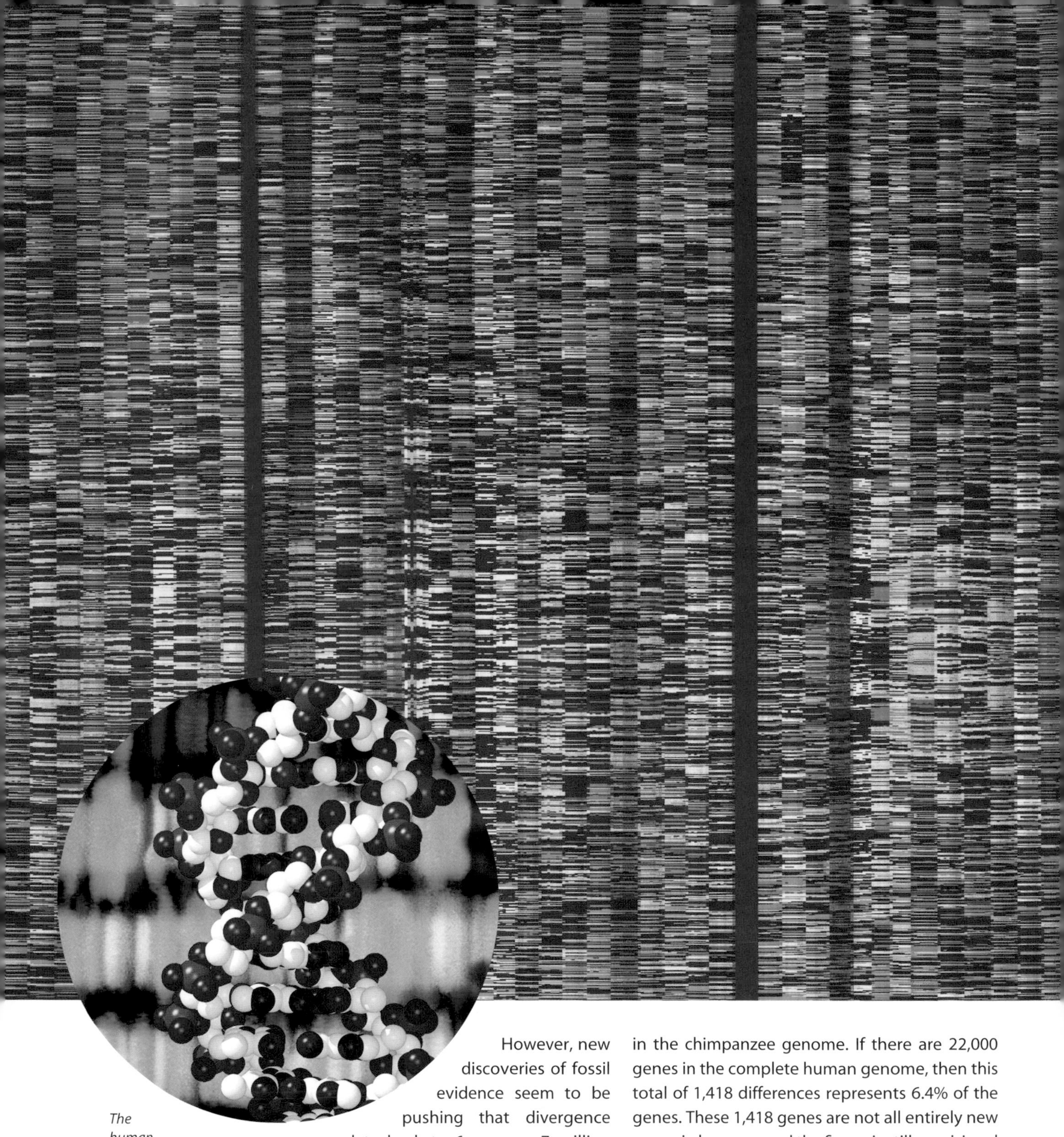

The human genome, seen as colour bands for each gene in an array, behind a model of the famous double helix structure of DNA.

However, new discoveries of fossil evidence seem to be pushing that divergence date back to 6 or even 7 million years ago (see p 50), and recent DNA studies have shown that the genetic difference between humans and chimps is actually more than 1%. In a study published in 2006, Jeffery P. Demuth of Indiana University and colleagues showed that we have 1,418 genes that do not have equivalents in the chimpanzee genome. If there are 22,000 genes in the complete human genome, then this total of 1,418 differences represents 6.4% of the genes. These 1,418 genes are not all entirely new genes in humans, and the figure is still provisional because the chimp genome is not yet as well documented as the human genome.

Human diversity

The complete human genome was sequenced in 2003, by teams in 18 countries. The project began in 1990, and was expected to last 15 years, but was in fact completed in 13, partly as a result of unanticipated advances in technology that speeded up the sequencing, and partly because of competitive pressures between government teams and commercial firms. Three billion DNA bases were sequenced, and it was discovered that humans have 22,000 genes – a figure that had been uncertain previously, and fewer than many had predicted.

It has long been known that genetic variation explains some of the more obvious differences among people, such as eye and hair colour, and blood group. Genetic variation also plays a role in whether a person has a higher or lower susceptibility to certain diseases such as cystic fibrosis and sickle cell disease. There is also a major genetic component – although it is rather more complex and may result from the actions of several genes in combination and involve environmental factors – in many conditions such as diabetes, cancer, stroke, Alzheimer's disease, Parkinson's disease, depression, alcoholism, heart disease, arthritis and asthma.

With all these genetic variables, and so many physical traits, what is the likelihood that you could meet someone else who shares precisely your genetic make-up? Of course, identical twins do just that: they are 100% identical to each other genetically because they arose from the same fertilized egg in the womb that split into two individuals at an early stage. But the chance of meeting another human being who is identical to you, but is not an identical twin, is actually zero.

Any random pair of humans are likely to differ in at least 6 million of the 3 billion DNA base pairs that we all possess, which is only 0.2% of the genome. Much of the genome is 'fixed' in the sense that endless variations are not possible because the genes have a specific job to do, and they cannot mutate endlessly. Equally, there are vast numbers of repeated genes within the genome that can undergo mutation without their performance being affected.

This means that the scope for genetic differences between humans lies somewhere between 6 million and 3 billion DNA base pairs. The exact figure is not known, but is vastly huge and is more than the number of atoms in the Universe. And this is the minimum estimate, so the likelihood you will meet a genetically identical human is effectively zero.

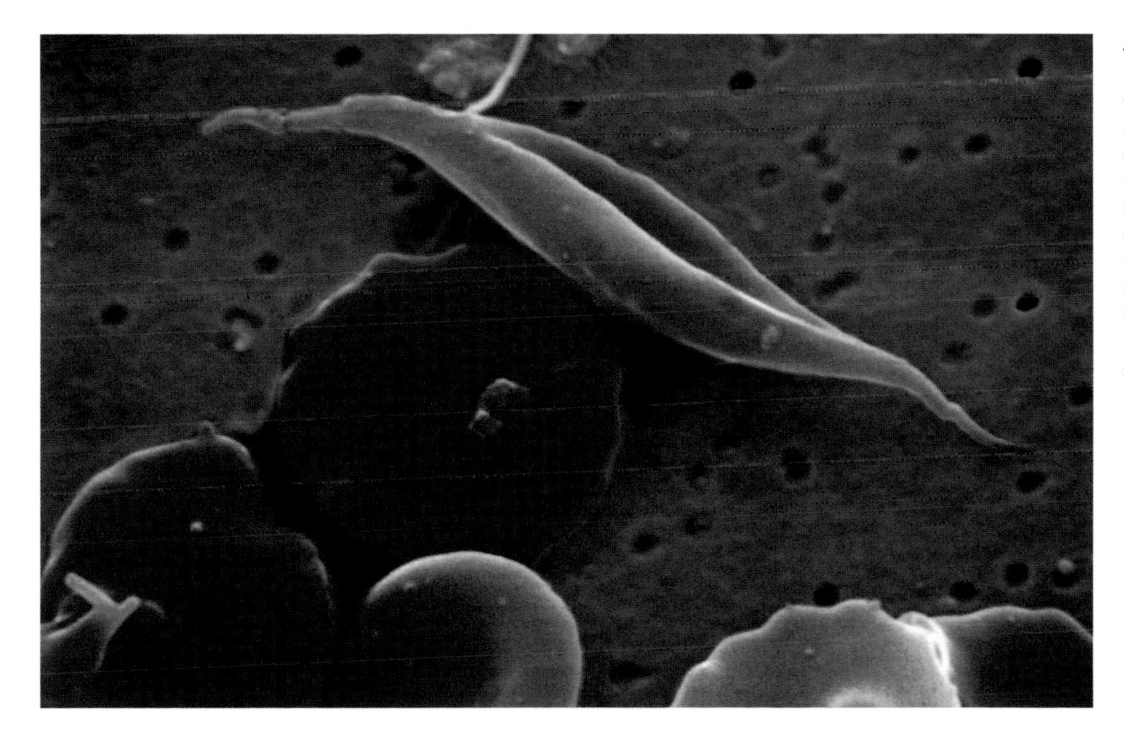

A microscopic image of red blood cells, showing the normal appearance (below), and a collapsed 'sickle cell' (top) that cannot carry oxygen efficiently. Susceptibility to this condition is genetically linked.

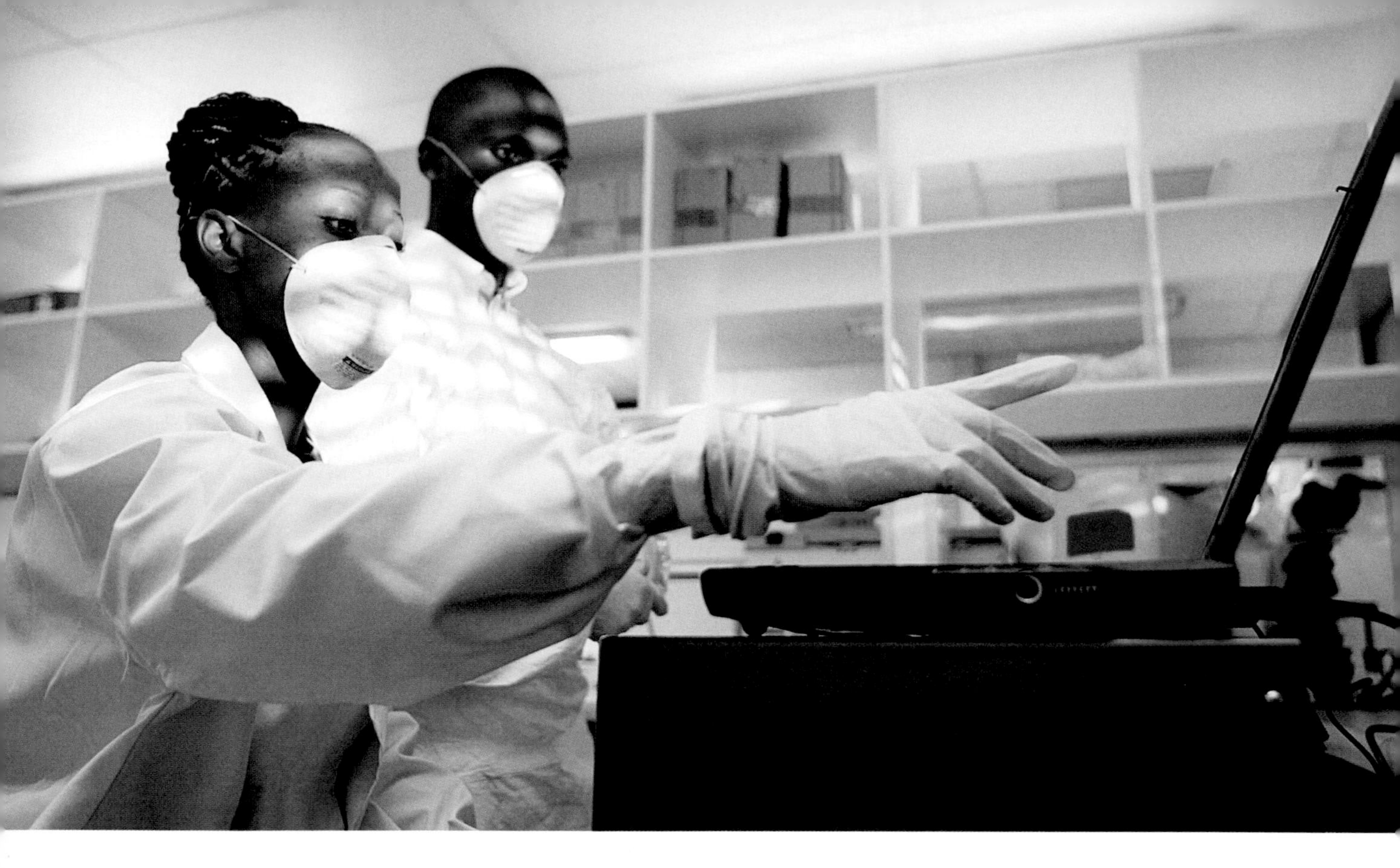

An HIV laboratory in Botswana where blood samples are screened for antibodies to the human immunodeficiency virus; this dreadful disease kills millions, but some people have mutated genes that give them some immunity to HIV infection.

Human genetic variation and group identity

According to current estimates, more than 90% of observed genetic differences in humans occur within populations, and less than 10% between populations. This 10% of genetic variation between populations may be divided into two parts: about 9% comprises the differences between local, regional, and national groupings, and only 1% appears to characterize the major continental groupings, namely races.

Two important consequences arise from this, relating to attitudes to race and also to medical developments. For centuries, humans have noted differences in skin colour and facial characteristics and used them to define the human races. Perhaps the fact that no two anthropologists could agree on just what were the races – were there four, five, six or even more? – indicates that these subdivisions are not as clearcut or as real as once thought. Now we know that major ethnic groups differ in only 1% of their genomes, it becomes clear that the historical attempt to define archetypal human races is futile – they do not represent fundamental and major divisions within the species *Homo sapiens* and have more to do with our preconceptions than with any reality.

Intra-group genetic variation is also important for medicine. If one ethnic or regional group is more or less susceptible to a particular disease, then this should provide a key in investigative studies to determine appropriate responses. For example, the *CCR5* gene is a receptor used by the human immunodeficiency virus (HIV) to enter cells. There is a variant of this gene, *CCR5-delta32*, that offers some protection against HIV infection and progression, and yet its occurrence differs substantially among ethnic groups: about 25% of Americans of European origin have this version of the gene, while it is virtually absent in other ethnic groups. This might then explain some differences in susceptibility to HIV infection between ethnic groups, and will affect future recommendations for medical care and drug development.

New work by Henry Harpending of the University of Utah, and colleagues, published in 2007, shows that humans have evolved faster in the past 5,000 years than before, and that variation between geographic races might be increasing. They point to the overall much larger population size today, and the likelihood that new mutations might therefore arise, and that environments and diets are changing fast, so inducing more change.

How do new species form? 28

... speciation is not a single phenomenon.
MENNO SCHILTHUIZEN, 2001

Members of the same species reproduce with each other, giving them a shared genetic and evolutionary history that is distinct from other species. Because every species today shares the same basic genetic code, they must all have evolved from a single common ancestor in the past. Some process must therefore have led one species to become millions: this is the process of speciation.

Two rather distinct types of events increase the number of species on the evolutionary tree of life: hybrid speciation and lineage splitting. For both types, the challenge is to understand three things: what stops different groups of individuals reproducing with each other; how groups of individuals diverge genetically; and how they diverge ecologically. For a new species to retain its distinctiveness, its members must stop reproducing with other species, which may itself result from genetic divergence. Ecological divergence can aid both, and is necessary to prevent a new species becoming extinct through competition with its parent or sister species.

Hybrid speciation

It has been estimated that as many as 11% of plant species, but probably fewer animal species, have evolved as a result of the natural hybridization of two existing species. For hybrid speciation to be successful, in addition to the three standard challenges listed above, another problem presents itself: when two species hybridize, their offspring are often infertile, because the genetic differences between parents make it difficult for hybrid offspring to form sperm and eggs properly.

One way of overcoming this problem is chromosome doubling in the hybrid, which occurs naturally at a low rate but can be encouraged by chemical treatments. This gives each chromosome a complementary one to enable sperm and eggs to form properly. Chromosome doubling also creates genetic divergence, and makes successful reproduction with other groups impossible. The cord grass *Spartina anglica* originated in this way in southern England in the 19th century as a hybrid of *S. maritima* and *S. alternifolia*, and many other examples are also known.

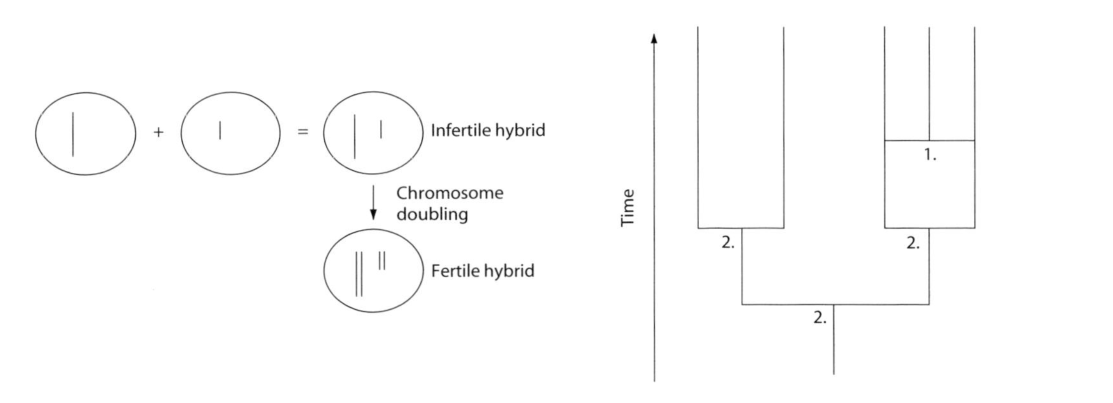

Far left *Hybrid speciation through chromosome doubling.*

Left *An evolutionary tree showing two types of processes giving rise to new lineages: 1 hybrid speciation; 2 lineage splitting.*

Top and above
The hybrid sunflower Helianthus anomalus *and one of its parent species.*

However, chromosome doubling is not the only way in which hybrid speciation can occur. In the United States, three hybrid sunflower species, *Helianthus paradoxus*, *H. anomalus* and *H. deserticola*, have formed by different means from two parent species, *H. annuus* and *H. petiolaris*. Here, the hybrid species have distinctive chromosome arrangements which suggest that chromosomes have been broken down and then re-assembled. Such rearrangements have a similar effect to chromosome doubling, in that they isolate the hybrids reproductively from their parents. Genetic differences have also led to the hybrids occupying different ecological niches from their parents: *H. paradoxus* prefers saline habitats while *H. anomalus* and *H. deserticola* live in arid environments. These ecological differences further reduce the chances of back-crossing with parents and also prevent the extinction of each species through competition with their parent species.

Because of the rarity of chromosome doubling in animals, it used to be thought that hybrid speciation occurred almost exclusively in plants, but recently some convincing hybrid animal species

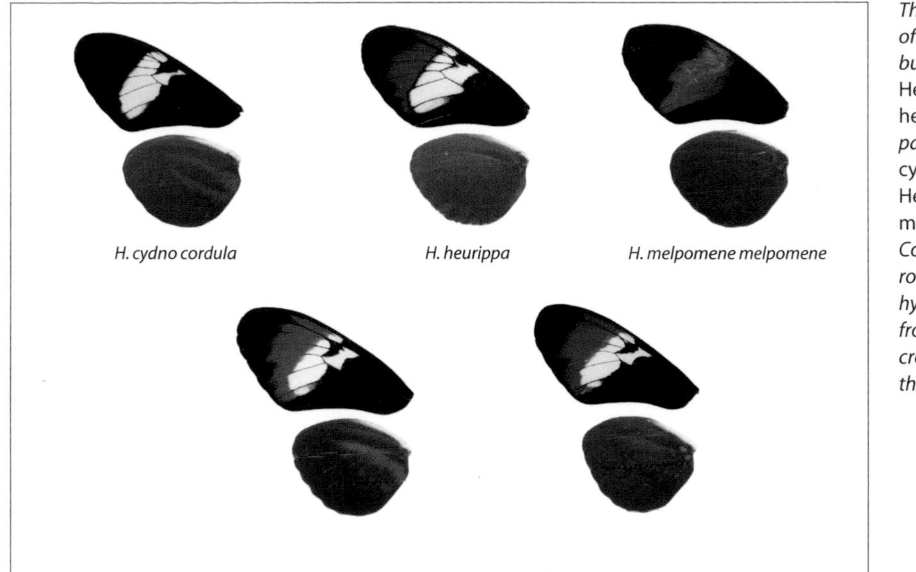

The wing patterns of the hybrid butterfly species Heliconius heurippa *and its parents* Heliconius cydno *and* Heliconius melpomene, *from Colombia (top row). Bottom row: hybrids formed from laboratory crosses between the parent species.*

have been identified. In Colombia, the butterflies *Heliconius cydno* and *H. melpomene* have given rise to a hybrid species, *H. heurippa*, identified by its mixed genetic make-up, and also because similar hybrids can be created in the laboratory. The hybrid has a wing pattern intermediate between its parents, and since the butterflies choose mates on the basis of wing pattern, this deters mating between the hybrids and their parent species.

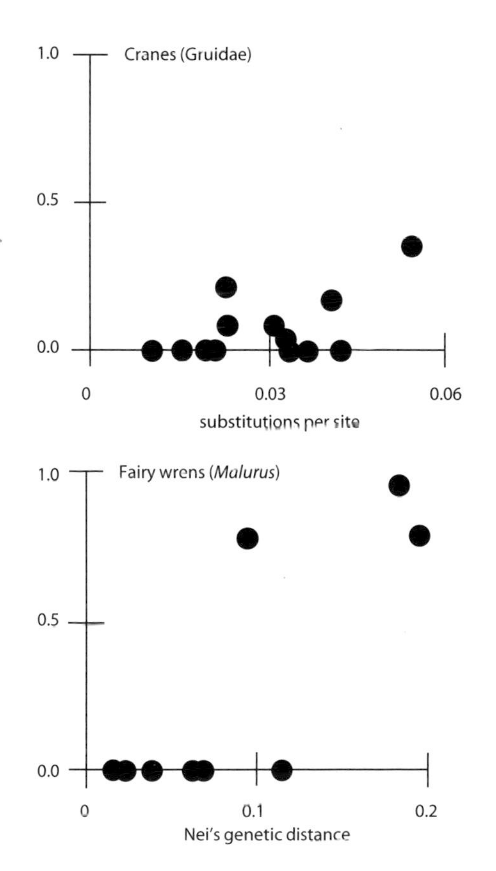

The degree of geographic range overlap (vertical axis) between pairs of species that originate at the same time, plotted against measures of time since splitting (horizontal axis). The older the date of splitting (the further to the right), the higher the degree of geographic overlap. This positive relationship between time and overlap indicates the action of allopatric speciation.

Lineage splitting

If two species can give rise to a third by hybridization, how does a single species split into two? The same three challenges apply: what stops reproduction with other groups, and how do genetic and ecological divergences occur? Unlike in hybrid speciation, where all three processes occur relatively quickly, during lineage splitting one or more of these processes may be delayed. Lineage splitting generally relies on two processes to different extents: geographic isolation between populations of the original species, and natural selection on ecology or the mating system.

Allopatric speciation occurs when the level of geographic isolation is high, and this is thought to be the commonest form of lineage splitting. Evidence comes from comparing the overlap in the geographic ranges of closely related species. This is very low in young species, as would be expected in allopatric speciation, but decreases as species age. When the level of geographic isolation is high, populations are prevented from interbreeding. Ecological and genetic differences between the populations can then simply accumulate

Two examples of the reef fish Halichoeres bivittatus*: individuals of this fish inhabiting the same type of environments are always closely related despite wide geographic separation, whereas individuals that inhabit spatially close but different environments are distantly related. This suggests that speciation has been parapatric.*

gradually, and strong selection pressure may not be needed for this.

When new species form at the edges of the continuous geographical range of a single population, this is known as *parapatric speciation*. Because some mating occurs between individuals from different locations, the strength of selection required to create new species is higher than in allopatric speciation. Groups living in different locations adapt to their particular environments, and although interbreeding between them may occur, the offspring from such crosses are not as well adapted to either environment and leave few progeny, effectively creating a reproductive barrier between populations. Recently, evidence from evolutionary trees and geographic ranges has suggested that much speciation in continuous and species-rich habitats, such as the Amazon rainforest and coral reefs, may have occurred by this process. Thus, parapatric speciation may be reasonably common.

Speciation in the total absence of geographic isolation, known as *sympatric speciation*, is much more controversial. Strong selection against intermediate forms is required for new species to be created in the absence of geographic barriers. Although the evidence does not yet allow us to say that sympatric speciation is common, several likely cases have been identified in recent years. Many of these involve species inhabiting small discrete habitats such as crater lakes (fish) or oceanic islands (palms). Others involve plant-feeding insects, with species shifting the host plant they feed on, thus creating reproductive and ecological isolation simultaneously. Perhaps the most intriguing examples are found in the cichlid fish of the East African Great Lakes, one of the greatest vertebrate radiations with some 1,200 species. In these lakes, closely related species exist which have different colours but which have the same ecology and a high degree of range overlap. Mate preferences for extreme colours are strong, and this appears to have created a reproductive barrier between species, largely in the absence of genetic or ecological differentiation. Later, the ecologies of the different species diverge.

Where next?

Because it holds the key to understanding how biodiversity is generated, speciation science is one of the most active areas of evolutionary biology. Theoreticians scrutinize conditions that favour speciation and devise ways of identifying different mechanisms. Field biologists map geographic ranges and ask how species and individuals are related. Laboratory scientists test the degree of interbreeding and genetic divergence between populations and species, and observe what conditions promote them: all, like Darwin before them, are attempting to understand the great mystery that is the 'Origin of Species'.

PHILIP DONOGHUE

Exploring the links between evolution and development

29

In order to achieve a modification in adult form, evolution must modify the embryological processes responsible for that form. Therefore an understanding of evolution requires an understanding of development.

T. J. Horder, 1989

Animal life is extraordinarily diverse. Roughly 35 major groups ('phyla') of animals can be distinguished, mostly on the basis of the profound differences between them. How did such diversity emerge? One intriguing fact is that while adults can be very different from one phylum to another, this is less true of their earlier developmental stages – the egg, embryo and juvenile. At first, all eggs and embryos are very similar, and the different organisms only achieve their distinctiveness as development proceeds.

The relationship between development and evolution has a long history of study. Darwin used embryology as one of the five principal pillars of evidence on which he based his argument that evolution had occurred and also that all organisms were related by genealogical descent.

In the later 19th century, the German biologist Ernst Haeckel interpreted this phenomenon of diverging embryologies as organisms recapitulating their individual evolutionary histories during their embryology, with each stage of development representing a sequential step in evolution. However, this theory has long since been discarded, primarily because of evidence that any step in an organism's embryology can evolve, not just the final one. And rather than recapitulating evolution, variations in embryology are viewed as the raw material, not the record, of evolution.

Regulatory genes

Given the importance of embryology in evolutionary change it is perhaps surprising that it does not feature in standard syntheses of evolutionary theory. It was not until the discovery of specific genes that regulate development, as well as technological advances that facilitated their study, that embryology was integrated into mainstream evolutionary biology. This was heralded by the analysis in fruitflies (*Drosophila*) of the effects of mutations in developmental genes, specifically of members of the *Hox* gene cluster. Mutations of these genes lead to dramatic changes in the specification of parts of the fly's anatomy, producing legs in place of antennae, and extra pairs of wings.

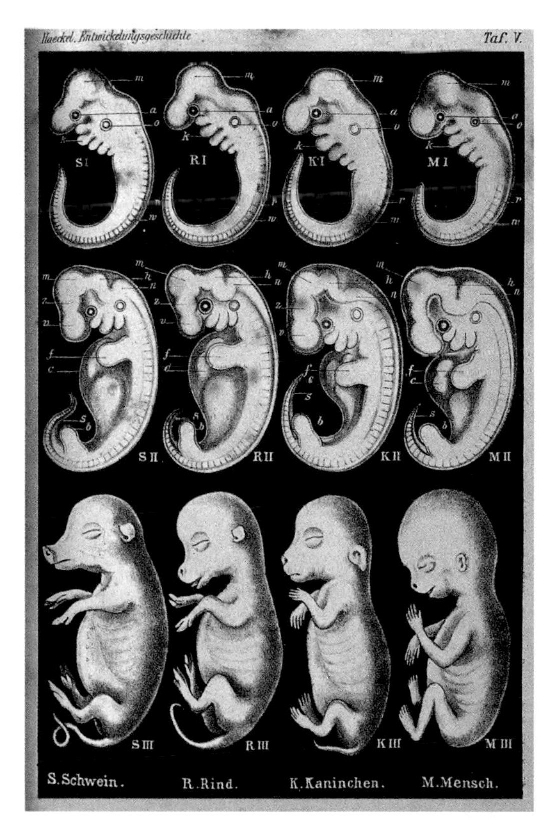

Plate from Ernst Haeckel's Anthropogenie *(first edition 1874) comparing different embryos – pig, cow, rabbit and human – demonstrating the principle that animals become distinct as embryology proceeds.*

Right *A normal fruitfly,* Drosophila melanogaster *(left), with two wings, and a bithorax mutant (right), with four, the result of failure of expression of the* Ultrabithorax *transcription factor.*

Opposite above *Development of the limbs in a mouse embryo showing the gradual emergence of the digits from a rudimentary limb.*

The discovery of regulatory genes brought a new perspective to developmental biology, most especially in understanding the mechanics of embryology. It also revealed that not only are these genes shared by almost the full range of animal diversity, but that they are so structurally and functionally conservative that the same gene from a mouse could be used to rescue induced genetic deficiencies in as distant an evolutionary relative as a fruitfly.

Developmental regulatory genes perform multiple roles, and for the vast majority of them these roles are still being worked out. By far the most completely understood are the *Hox* genes, which most especially perform the role of establishing the main head–tail axis within the developing embryo, as well as positions along it. The areas of expression (where the gene is being transcribed) of individual *Hox* genes can be much greater than their areas of influence. This is because more posterior *Hox* genes suppress the influence of their more anterior counterparts. Where they do have influence, they perform the role of a transcription factor, encoding a protein that binds to other genes, regulating their activity, either positively or negatively. Effectively, they act as traffic wardens within developing embryos, promoting the activity of genes in one cell but not another.

If *Hox* genes can confer the identity of specific body parts, it is easy to see how mutations in these genes can lead to the transformation of one body part into another (homeosis): if a posterior gene is not expressed, then its role is overtaken by its more anterior counterpart. Mutant fruitflies with two pairs of wings, rather than one, have long been known to occur naturally. This happens when the *Hox* gene in the segment posterior to the standard wings, which would normally suppress wing development, is not expressed. Thus, the second segment also produces wings. Perturbation (disturbance) of other *Hox* genes can lead to changing the identity of limbs, such as legs in place of antennae. Similar phenomena are encountered in distantly related animals. For instance, in vertebrates, variations can occur in the relative numbers of different classes of vertebrae, such as lumbar and cervical. However, while these are clear cut examples of the relationship between genes and morphology, it should be

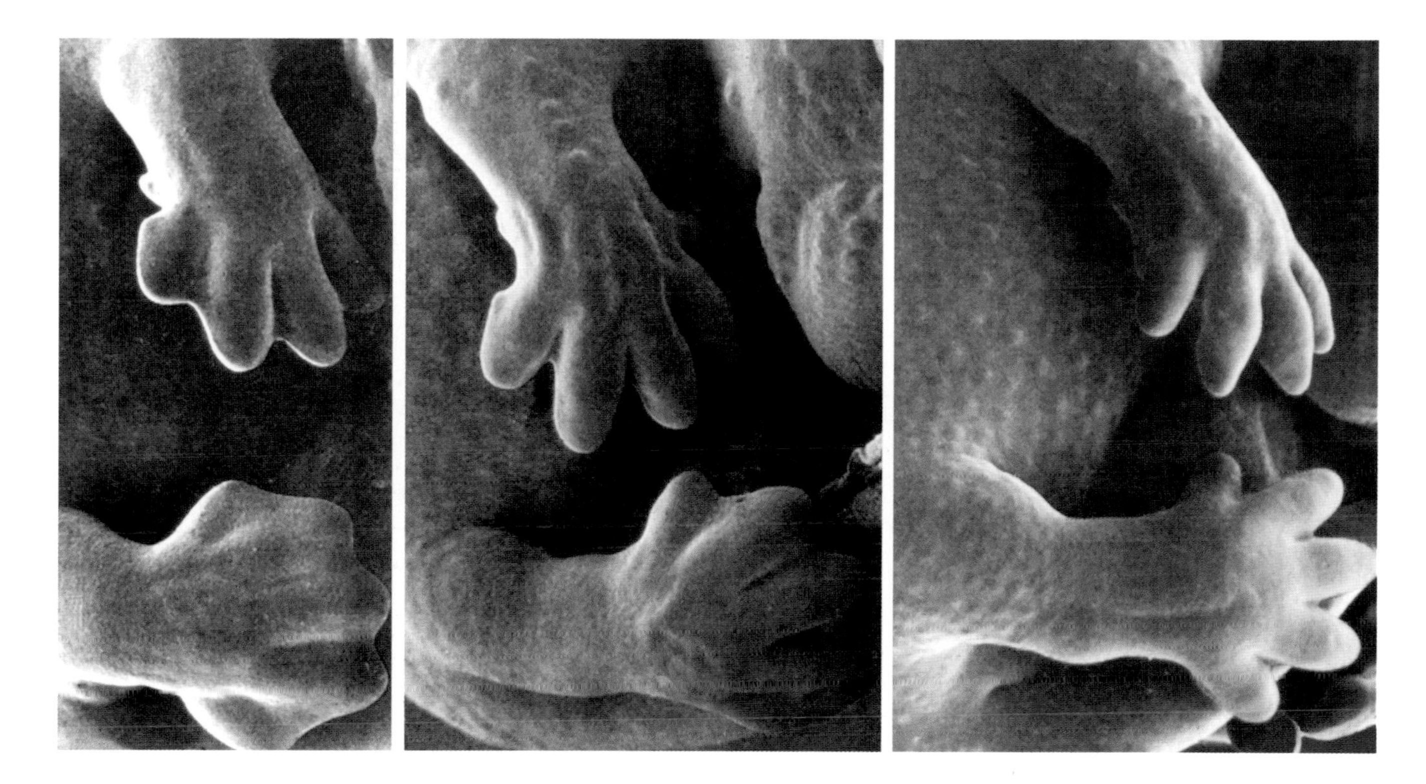

remembered that the role of these genes is indirect and there are many intermediate steps. Thus, the effects of changing *Hox* gene expression is often less obvious.

Few other classes of regulatory genes are as well characterized in terms of function and evolution as the *Hox* gene. What we do know, however, is that there is a toolkit of regulatory genes common to all animals, from sponges to worms, humans and fruitflies, and that these genes are more often than not deployed in the same way in disparate organisms. For example, the gene *Pax6* is so tightly linked to the specification of eye development that in experiments it is possible to initiate eye development in areas of the embryo where eyes should not normally occur by introducing the *Pax6* protein. Other regulatory genes that have been directly connected with anatomical development include *Tinman* and heart development, *Caudal* and gut development, *Distalless* and limb development.

Regulatory genes and evolution

The connection between regulatory genes and the specification of anatomical features led to the idea that the diversity of complex, bilaterally symmetrical animals (bilaterians) may be explained by the evolutionary origin of this genetic toolkit. However, analysis of simpler animals such as sponges and jellyfish has revealed that they, too, possess the genetic toolkit. Indeed, they share with humans a number of genes that fruitflies do not, indicating, somewhat ironically, that there has been significant gene loss among some

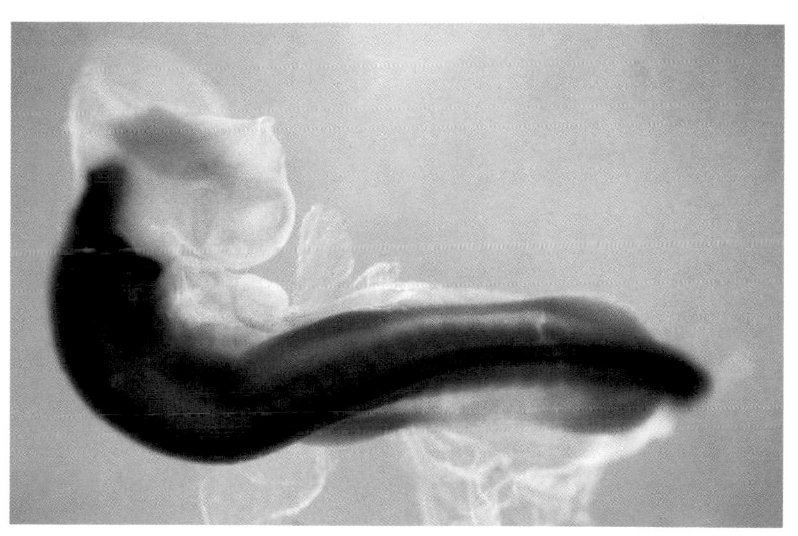

Above *A stage 16 chick embryo, removed form its yolk sack, with a molecular label to highlight where the* Hoxa-2 *gene is being transcribed into messenger RNA.*

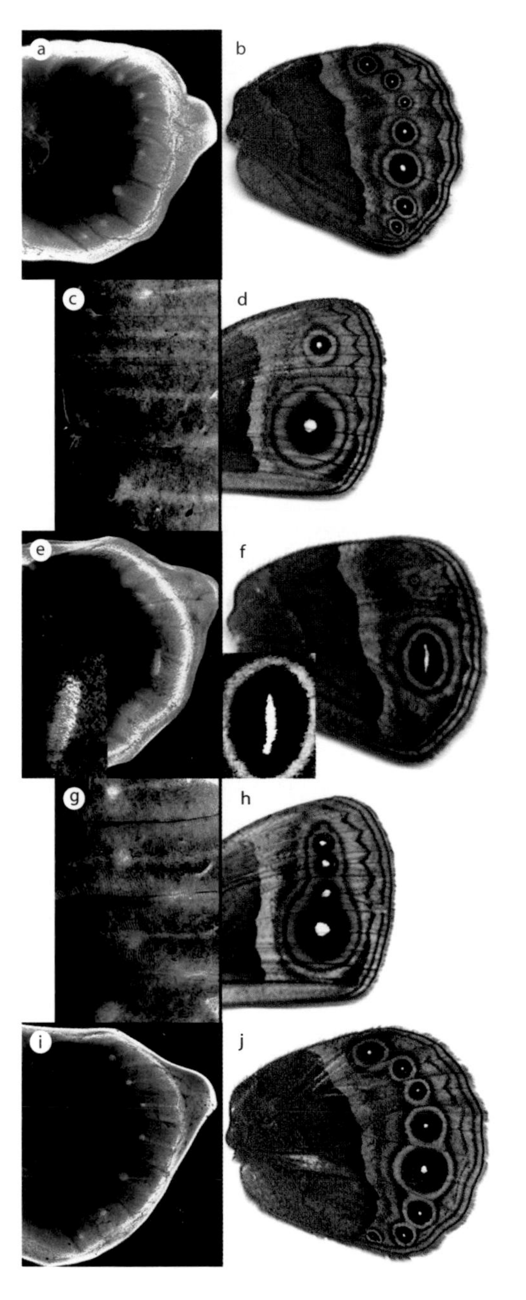

Right *Expression of the transcription factor* Distalless *(Dll) in the wing of the butterfly* Bicylcus anyana. *The first pair (a–b) show the natural pattern of* Dll *expression and eyespot phenotype; the other pairs are a variety of mutants*

Far right *Two forms of the stickleback fish with (top) and without (bottom) pelvic fins and a fin girdle, resulting from different alleles in the transcription factor* Pitx1.

bilaterians, most especially along the lineage leading to fruitflies. It also indicates that the genetic toolkit cannot explain the complexity of animal diversity.

So it appears that animals are more than the sum of their genetic parts – but not too much more. Much of the complexity of organisms is effected within the embryo through the interaction of cascades of regulatory genes in complex genetic circuits. Discovering the architecture of these circuits is extremely labour intensive – the function of every component interaction (not merely every gene) has to be validated experimentally. Nonetheless, from what little is already known it is clear that there is a significant amount of evolutionary conservation of the circuit architecture, even among the most distantly related complex animals.

It is tempting to think of these regulatory gene circuits as key innovations that facilitated new thresholds of complexity in animal evolution, but our understanding of their own evolution is extremely scant and certainly insufficient to contemplate stages in their evolution. It is far more likely that complexity in organisms evolved gradually through selection acting on developmental variation. So far, however, the role of developmental variation in models of selection has not been widely considered, although there have been pioneering studies on specific traits like eye spot patterns on butterfly wings, pigment patterns in fruitflies, and fin skeletons in stickleback fishes. These have revealed that developmental variation is not limitless – the same patterns are converged upon, but this is because certain alleles (one of a pair of alternative genes that occur at a specific locations on a chromosome) within highly conserved genes act as hot spots for the recurrent evolution of those features.

As we are learning more about the genes that direct development, it becomes increasingly clear that genetic conservation is the rule, not the exception. Indeed, the surprising fact is that so much organismal diversity could have emerged in the face of so much genetic conservatism.

MICHAEL COATES

Are five fingers essential?

What can be more curious than that the hand of a man, formed for grasping, that of a mole for digging, the leg of a horse, the paddle of a porpoise, and the wing of a bat, should all be constructed on the same pattern, and should include the same bones, in the same relative positions?

CHARLES DARWIN, 1859

The presence of five fingers and toes is a classic feature of vertebrate anatomy, and the persistence of this pattern, deep within embryonic development, has long been a challenge to comparative anatomists and evolutionary biologists. There is no easily identified adaptive and/or functional explanation as to why it should be so. Instead, it appears to be a textbook instance of evolutionary contingency: once the pentadactyl (five-digit) pattern was established in an ancestral form, natural selection tinkered with the constituent parts, stretching and shrinking bones and cartilages, but left the underlying network of connections, the groundplan, unaltered. This begs questions about what conditions preceded five digits, and what might be responsible for the long-term stability of this arrangement.

Pentadactyl limbs: the evolutionary context

It is now reasonably clear that pentadactyl limbs evolved before the divergence of the ancestral lineages leading to modern amphibians (frogs/toads, salamanders and the limbless caecilians) and amniotes (birds, mammals and reptiles), a parting of evolutionary ways that occurred at least 340 million years ago, in the Lower Carboniferous Period. However, having five or fewer fingers and toes is no longer assumed to be the most primitive condition for vertebrate limbs.

Earlier and undoubtedly more primitive examples have been discovered with sets of six, seven and even eight digits. Evidence is found in exquisitely preserved fossil tetrapods (four-limbed vertebrates – the amphibians, reptiles, birds and mammals) from the end of the Devonian Period, some 360 million years ago. Furthermore, the skeletal remains of these Devonian tetrapods reveal a whole series of associated fish-like characteristics, including well-developed tail fins and substantial gill supports, showing that limbs with digits evolved first in animals that were primitively aquatic, and that the evolutionary transition from fish to tetrapod, and fin to limb, occurred before the vertebrate colonization of dry land.

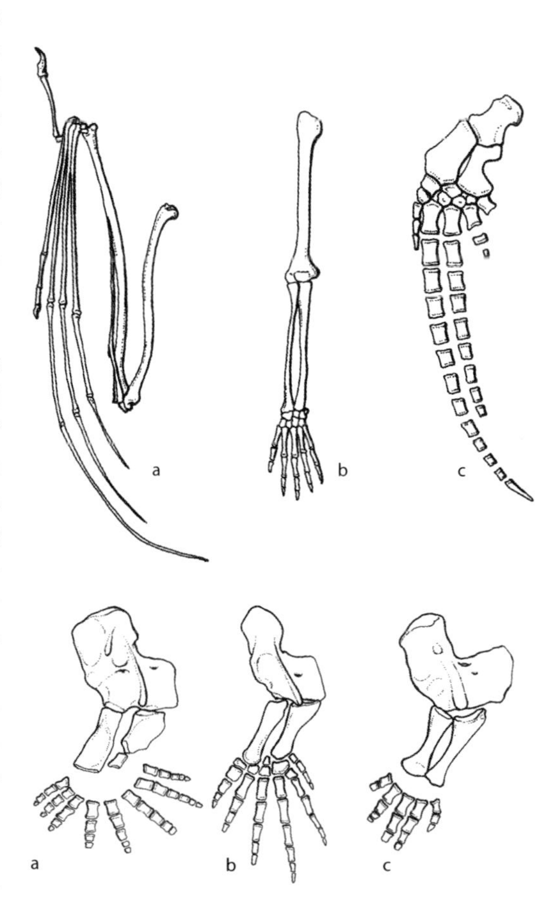

Left *Limb skeletons of modern tetrapods with the same underlying pentadactyl (five-digit) pattern. The forelimbs of a bat (a), a human (b), and a dolphin (c) all have the same basic suite of bones, even though they perform different functions.*

Below left *Forelimbs of primitive tetrapods showing patterns of eight, six and five digits: (a)* Acanthostega, *from the Upper Devonian of East Greenland; (b)* Tulerpeton, *from the Upper Devonian of Russia; (c)* Greererpeton, *from the Carboniferous of the eastern USA.*

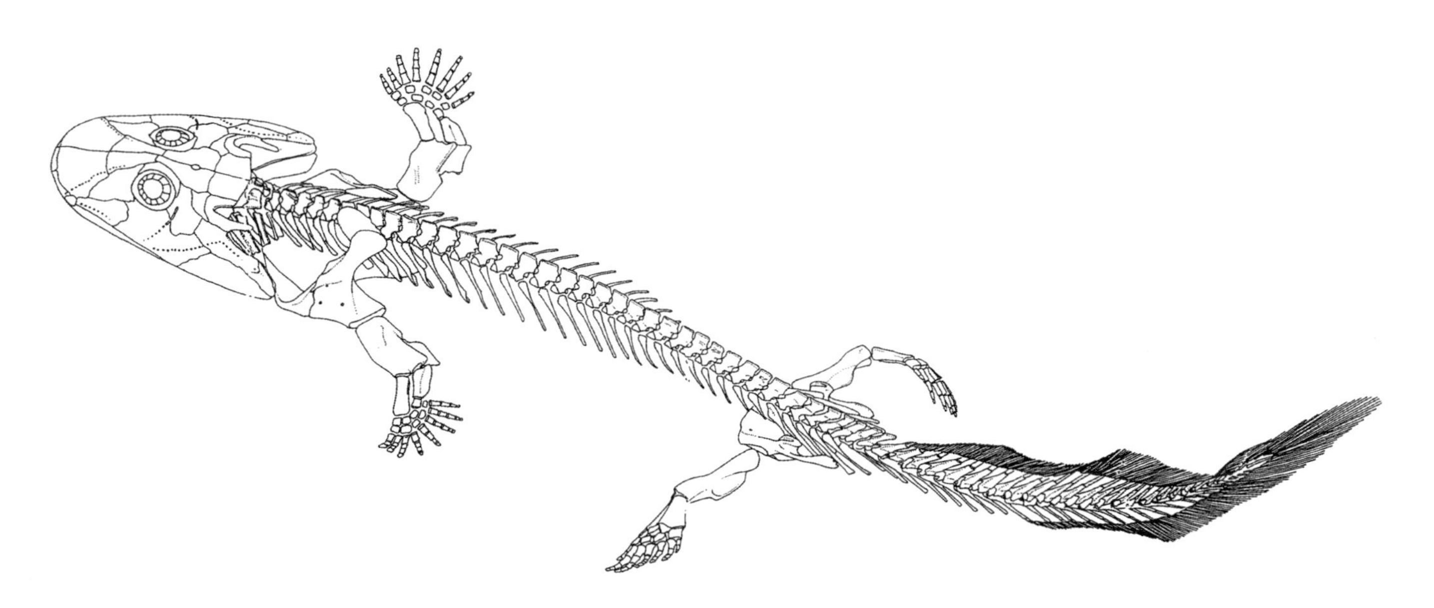

The complete skeletal anatomy of Acanthostega, *one of the most primitive tetrapods. It has broad, paddle-shaped limbs with digits, associated with an otherwise fish-like skeleton, including a wide tail fin and well-developed gill arches.*

This raises questions about the adaptive advantage of limbs rather than fins in an aquatic environment. Clues to this puzzle are provided by the extraordinarily limb-like fins of certain fishes alive today, such as the batfish, frogfish and aptly named handfish. Limb-like fins can be used for stealthy advances on prey, to negotiate restricted spaces, or simply to maintain position in the face of strong tidal or river currents. However, none of these examples shows the pentadactyl ground-plan of tetrapod limbs. Natural selection has had to work from a different set of initial conditions, and while superficial resemblances are striking, underlying structures are different.

It is quite possible that five-digit tetrapod limbs evolved with the shift to life on dry land. The reduction to five digits was accompanied by the origin and increasing complexity of wrist and ankle joints, and these complex articulations might have evolved in response to the new functional demands placed on weight-bearing limbs. Terrestrial limbs have to gain purchase on a wide variety of surfaces, provide the platform for an efficient push-off and allow for rotation between the hand or foot and the lower and upper limb bones. Intriguingly in this respect, in the very few instances found in the fossil record of secondarily evolved limbs with more than five digits, the phenomenon is associated with aquatic habits. The most obvious examples of this are in the paddles of ichthyosaurs, extinct fish-like marine reptiles that lived more than 65 million years ago.

Pentadactyl limbs: the developmental context

In fact, five fingers might not be essential at all, in the sense of being absolutely necessary, but instead represent a condition that once arrived at has persisted within embryonic development. So, if five fingers are no better or worse than four or six, could the precise number relate to some aspect of embryonic development?

Given the immensity of evolutionary time and the extraordinary variety of vertebrate body forms, the rarity of six-digit limbs among living vertebrates points to some sort of developmental constraint. Moles' paws and pandas' thumbs are classic instances in which strangely re-modelled wrist bones serve as sixth digits and represent baroque solutions to the apparently straightforward task of growing an extra finger. Exceptions exist, such as Hemingway's cats (sometimes called 'mitten cats' because of the unusual breadth of their six-fingered paws), and historical figures such as the occasional Philistine giant and Anne Boleyn. Nevertheless, these examples serve mainly to emphasize the general rule of the five-digit pattern.

Limbs with six or more digits can be generated in the laboratory by manipulating gene expression in mouse embryos, and it appears that some of these changes in gene activity might reflect the gradual recruitment of gene functions (rather than the evolution of new genes) that occurred during the developmental evolution of limbs from fins. When the activity of genes responsible for patterning embryonic hands and feet is reduced (down regulated), one result is an increase in digit number, resembling conditions known in the earliest tetrapod limbs.

New research hints at a strange association of gene patterning. During embryonic development, copies of the same genes are used repeatedly for establishing different features of the anatomical pattern, and this may have multiple effects on more than one physical characteristic, a phenomenon called pleiotropy. Pleiotropic effects might well underpin the prevalence of pentadactyl limbs.

Hand-Foot-Genital syndrome is a rare condition in which, as the name implies, the genitourinary tract and the limbs are malformed. This appears to be a pleiotropic effect, where the genes responsible are within the set involved in digit patterning. There is no reason for this linkage at gene level, but it suggests why the pentadactyl limb has been so stable through evolution. If mechanisms involved in regulating the development of fingers and toes include those also involved in our reproductive success, then the side effects of changes to these genes' activity could be evolutionarily devastating.

The spotted handfish, Brachionichthys*: this endangered relative of the angler fishes is just one example of several living species with extraordinarily limb-like fins used for underwater walking and station-holding.*

Biogeography & Environments

The Earth and the life on it are inextricably intertwined. Naturalists have long noticed that plants and animals vary greatly on different continents, and that there are marked contrasts between the equator and the poles. For one thing, life is much more diverse in the tropics than in icy polar regions. The explanation for this may seem obvious, but in fact is not as straightforward as we might think.

It was Alfred Russel Wallace, co-discoverer of natural selection with Charles Darwin, who first put biogeography on a scientific basis. He organized the observations he made in South America and Southeast Asia into a global framework. Famously, while working in Malaysia, Wallace had crossed from Borneo to Sulawesi (Celebes) and back again, and noticed a profound difference in the plants and animals on either side, which convinced him of the distinct divisions of modern floras and faunas.

Additional layers were soon added to the basic mapping of the Earth. First, the acceptance of continental drift during the 20th century provided a rich vein of historical evidence for modern plant and animal distributions. At last it was clear why marsupials are found primarily in Australia and South America – there was once a land connection across Antarctica. Australia is the biggest island of all and has been remote for 50 million years or more. Its past and present unique fauna of emus, kangaroos and wombats is still the source of remarkable insights into geographic change and evolution.

Biologists are fascinated by deserts and polar ice caps because they represent the extremes. It

Biologists still debate why diversity is greatest in tropical regions, such as in this rainforest, where food supplies are not always richest.

A desert landscape, with little apparent life, but even here some plants and animals can survive. In an environment without water, desert plants and animals use a variety of different strategies to survive.

might seem impossible that life could survive in the soaring temperatures and severe drought found in deserts, and yet certain plants and animals have developed a variety of strategies and ingenious adaptations to cope successfully with such conditions. Deserts are not fixed, and they have moved and changed through time. Studying their past fluctuations may help evaluate the impact of climate change in the future.

Similarly, our image of the poles is one of icy wastes, populated by few animals, yet they too have not always been as we see them now. Fossil evidence shows that forests once flourished here and marine animals swam in ice-free seas. And even today, when biologists drill down through the Antarctic ice they find microscopic organisms in the form of bacteria and fungal spores at great depths and extremely low temperatures.

Equally amazing are the discoveries of life on black smokers – vent chimneys along the mid-ocean ridges which pour out black sulphurous fluids. The teeming communities of strange creatures there have radically altered our view of life on Earth, which was once all thought to depend ultimately on the energy derived from the Sun. It was discovered that the microorganisms metabolize sulphur in a process called chemosynthesis, and the worms, clams and snails feed on the microbes. Other so-called extremophiles have been found living in acid conditions, in salt solutions, and in temperatures above the boiling point of water and way below freezing point. Such extraordinary microbes can perhaps give us clues about life on other planets.

Islands are a laboratory of evolution. Remote islands harbour unique plants and animals, many of them now under threat of extinction. Studies of island biology have revealed key evidence about the potential of evolution: Darwin realized this when he studied the different species of finches on the Galápagos islands, with their unique range of beak shapes and feeding habits. A classic case of adaptive radiation, Darwin's finches continue to illuminate how species multiply today.

Numbers of species in the tropics and at the poles

31

The equatorial regions are then, as regards their past and present life history, ... a more ancient world than that represented by the temperate zones, a world in which the laws which have governed the progressive development of life have operated with comparatively little check for countless ages.
ALFRED RUSSEL WALLACE, 1878

Since the earliest days of natural history exploration it has been clear that tropical and low-latitude regions contain far more species of plants and animals than any other part of the Earth's surface. The contrast between, say, a scene in the Amazonian rainforest, with lush vegetation and teeming with abundant insects, birds, and animals, and one at the tip of the Antarctic peninsula, where, on a rocky and sparsely vegetated shore, there might be a few penguins, could not be more extreme.

If you take an area of tropical rainforest and compare it with a similar-sized tract of temperate forest you will invariably find that it contains far more species of plants, birds and insects; very similar high–low latitude contrasts can also be demonstrated in the sea. But why should this be? From the human perspective of being a warm-blooded mammal it might seem obvious that life proliferates where it is hot, but mammals currently comprise less than 0.4% of all known animal species and it is clear that many organisms have adapted quite successfully to what we would regard as extreme environments (p. 171).

Patterns of tropical diversity

It has been estimated that tropical rainforests, which comprise only 6% of the Earth's land surface, could contain more than half of all known organisms. Some 250,000 species of vascular plants (flowering plants, ferns, mosses, etc.) have been described globally, and of these 170,000

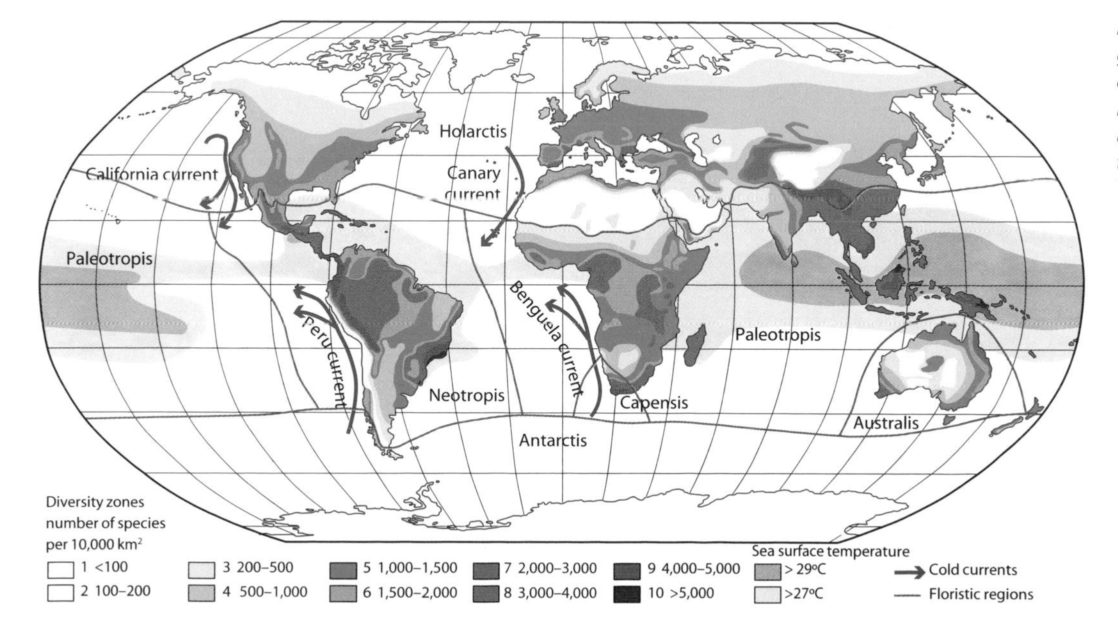

Map showing the global biodiversity of vascular plants. Note the high but uneven spread of tropical values.

Lowland rainforest of the Danum Valley on Borneo – one of the most diverse habitat types in the world.

(68%) occur in the tropics and subtropics. Of the estimated 1.7–1.8 million species on Earth (p. 177), 76% are animals, and 56% of these are terrestrial insects. It has been calculated that 1-ha (2.5-acre) plots of Panamanian rainforest could contain as many as 18,000 species of beetles – this figure can be compared with an estimated total of 24,000 species for the whole of the USA and Canada.

Very similar taxonomic diversity contrasts can be demonstrated in the marine realm too. The tropical Indonesian region contains at least 3,000 species of shallow-water fish, compared to well under 300 species in both polar regions; comparable figures for gastropods (marine snails and sea slugs) are 7,500+ species for Indonesia, 388 for the Arctic and 450 for the Antarctic. There are well over 1,000 species of decapod crustaceans (crabs and lobsters) in the Caribbean, but only 37 in the Arctic and 24 in the Antarctic. Many tropical species are small, many are rare, and many have yet to be formally described.

Steep gradients of decreasing taxonomic diversity from the tropics to the poles are one of the most striking features of life on Earth. However, they do not occur in all groups (certain freshwater invertebrates, seaweeds and planktonic organisms are notable exceptions). There is also evidence to suggest that the differences can be asymmetric between the northern and southern hemispheres. It is also important to recognize that diversity is not uniformly high across the entire tropics – there are distinct peaks and troughs longitudinally, with notable diversity gradients between.

The role of energy

Primary productivity – the energy of the Sun captured by plants and converted into carbon

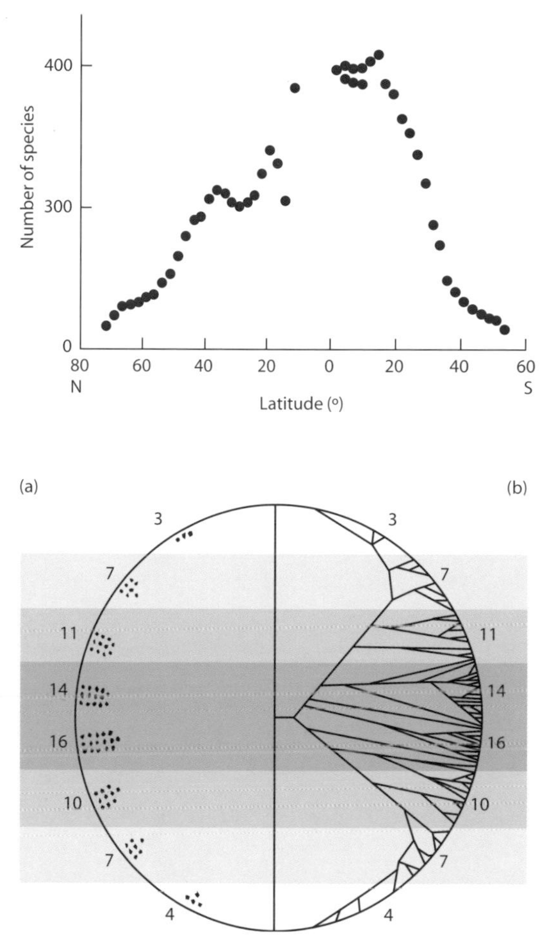

Left *Plot of the number of mammal species at different latitudes. Note the peak in the tropics, although the pattern is not symmetrical between the two hemispheres.*

Below left *Two approaches to explaining species richness. A standard approach (on the left) is to look for correlations between the numbers of species of a group at a particular location and variables in the environment (temperature here). The right side shows how many common taxa may have evolved in tropical latitudes but then spread only slowly through time towards the poles. In the past the tropics were more widespread in distribution and the polar regions more restricted.*

compounds – is the basic resource that fuels life on Earth. Therefore, as long as available water and minerals for plants, and food and water for animals, are not limited (as in a desert), on simple theoretical grounds we would expect there to be a much greater proliferation of life in the tropics. There is a problem with this apparently logical line of reasoning, however, as it might only lead to an increase in biomass; something else must be involved to convert a large number of individuals into a large number of species.

One concept that has been discussed intensely is that of evolutionary speed. If a general acceleration of physiological processes with increased temperature is added to the observation of faster genetic mutation rates and shorter generation times (i.e. the time taken to produce offspring) in the tropics, then it is possible to postulate a faster net rate of natural selection – and thus evolution. In other words, there is a more rapid rate of species production in the tropics than at the poles. Some important evidence that just such a process is in operation has recently come from a comprehensive study of both living and fossil planktonic marine organisms. Absolute rates of both DNA evolution and individual speciation vary exponentially with temperature, thus giving strong support to the concept of the tropics as a persistent 'species factory', or evolutionary cradle.

Whether such a process also operates in higher organisms, such as birds or mammals, has yet to be demonstrated. The relationship between taxonomic diversity and energy is complex and may be fundamentally different between plants and animals: whereas the former are dependent upon radiative energy (photons in sunlight), the latter rely on energy from food released during

Above *A coral reef and fusiliers at the Wakatobi Dive Resort, Sulawesi, in the Indian Ocean. Coral reefs are by far the most diverse habitat type in the marine realm, and may contain as many as 1 million species.*

metabolism. The precise influence environmental temperature might play in either is unclear.

The role of area and time

The tropics span a huge area of the Earth's surface, from 23.5°N to 23.5°S. Within this belt, mean annual temperatures show very little variation with latitude. They also comprise the largest area of available habitats of any biological region (or biome), and as such have the greatest potential for biological diversification. The reasoning here is simple: speciation is promoted through the greater likelihood of species being geographically isolated in such a vast region, and extinction is retarded as there are 'more places to hide' in times of stress. Although a positive species–area relationship has been demonstrated on various scales and in many different regions, some caution has to be exercised in its application. Take coral reefs, for example. They comprise by far the richest habitat in the marine realm and yet, in total, cover less than 0.1% of the Earth's surface.

Both Darwin and Wallace appreciated that the great antiquity of the tropics was a likely further reason for their unusually high diversity. And indeed we can now demonstrate an almost continuous tropical fossil record that stretches back into at least the Late Palaeozoic era (i.e., 300 million years ago), and this can be compared with a geological record of modern polar climates that is no more than 30–40 million years. We also know that the modern latitudinal temperature gradient is no more than 15 million years old.

Tropical habitats also covered a much more extensive area in the geological past and many modern groups of plants and animals can be traced back to an origin within the Late Cretaceous–Early Cenozoic greenhouse world. One idea – the 'tropical niche conservatism' hypothesis – is based on the concept that such groups have simply had insufficient evolutionary time to radiate fully into temperate and polar regions.

There is then no simple answer to the question of why there are more species in the tropics than at the poles. Factors such as energy, area and time are undoubtedly important, but there could be others too. The sheer habitat complexity of reefs and rainforests may promote intense resource partitioning, and it may be that competitive interactions have led to a form of tropical evolutionary arms race: first predators predominate, and then their prey. Glacial climate cycles may also have promoted intense speciation through successive expansions and contractions of range. Speculation is intense, for this remains the central problem in contemporary biodiversity studies.

Right *A fossil fish,* Aipichthys velifer, *from the Upper Cretaceous. The fossil record of many key tropical taxa is still poorly known, but the great antiquity of the tropics is probably one reason for the diversity found there.*

MICHÈLE L. CLARKE

The evolution of deserts

Without water, a desert is nothing but a grave. First I saw a lovely lake with trees standing on its furthest bank in mid-Gobi, but the caravan leader only smiled and spoke indulgently as one might speak to an ignorant child: 'That's not water', he said, 'that's glitter sand'.

MILDRED CABLE WITH FRANCESCA FRENCH, 1942

The warm deserts of the world are characterized by a lack of water – they are climatically controlled environments where limited rainfall and seasonally or permanently high evaporation rates cause a negative water balance. Aridity is responsible for the familiar image of a bare, rocky environment populated by sparsely distributed, spiny, drought-tolerant vegetation and a largely nocturnal animal life (p. 144).

Travellers and early scientists who traversed individual deserts described them as hostile, unchanging, barren and 'sterile' environments. However, within many of the driest deserts there is significant evidence for the impact of environmental change during the recent geological past. Remnants of ancient lakes, with relict shorelines, lake beds and freshwater fossils, can be found in a number of locations, including the hyper-arid desert environments in Namibia, the Sahara and Death Valley. Such evidence for much wetter past conditions within lands that are today arid raises the question as to whether many of the characteristic desert landscapes may, in fact, have evolved under a very different climatic regime. Given

Remnants of a lake which once sustained trees, at Dead Vlei, in the hyper-arid Namib Sand Sea.

concerns about future impacts of climate change, current research seeks to help us better understand the role of climate in desert evolution.

What is a desert?

Although there is no agreed scientific definition for the term 'desert', all deserts have climate regimes where loss of water by evapotranspiration from rocks, soils and vegetation exceeds that supplied by rainfall. While it is relatively straightforward to measure rainfall, quantifying water loss from desert surfaces, plants and trees is difficult. Measurements are complicated and depend on a range of factors such as shade, aspect, colour and albedo (reflectiveness), as well as how much water is available to evaporate. Thus, rather than directly measure water losses, estimates of potential loss are often calculated from other more easily quantifiable parameters such as solar radiation, wind speed and humidity. The ratio of water supply to estimated water loss can then be used to determine the extent of aridity at a given location within a desert, with classifications into hyper-arid, arid, semi-arid or dry sub-humid. The word desert is often, although not exclusively, associated with hyper-arid and arid drylands.

Because of the harsh environments of deserts, coupled with the scarcity of human settlement, meteorological conditions are often interpolated across vast areas of rugged terrain, thus oversimplifying the complexity of desert conditions. Furthermore, different international agencies (for instance UNEP and UNESCO) use different measures of aridity, and so identifications will depend on which definition is used. Identifying and accurately mapping the margins of deserts is therefore largely arbitrary. This has important implications for evaluating the accuracy of concerns about the spreading of desert margins.

Causes of aridity

Interactions between oceanic or atmospheric circulation patterns and global land masses cause areas of aridity and create the current deserts of the world. Cold ocean currents caused by the upwelling of deep water on the western seaboard of large landmasses such as southern Africa and the Americas mean that sea surface temperatures are low and little evaporation occurs. This limits moisture availability to the coastal zone and can cause hyper-arid conditions to persist inland – hence the Namib and Atacama deserts.

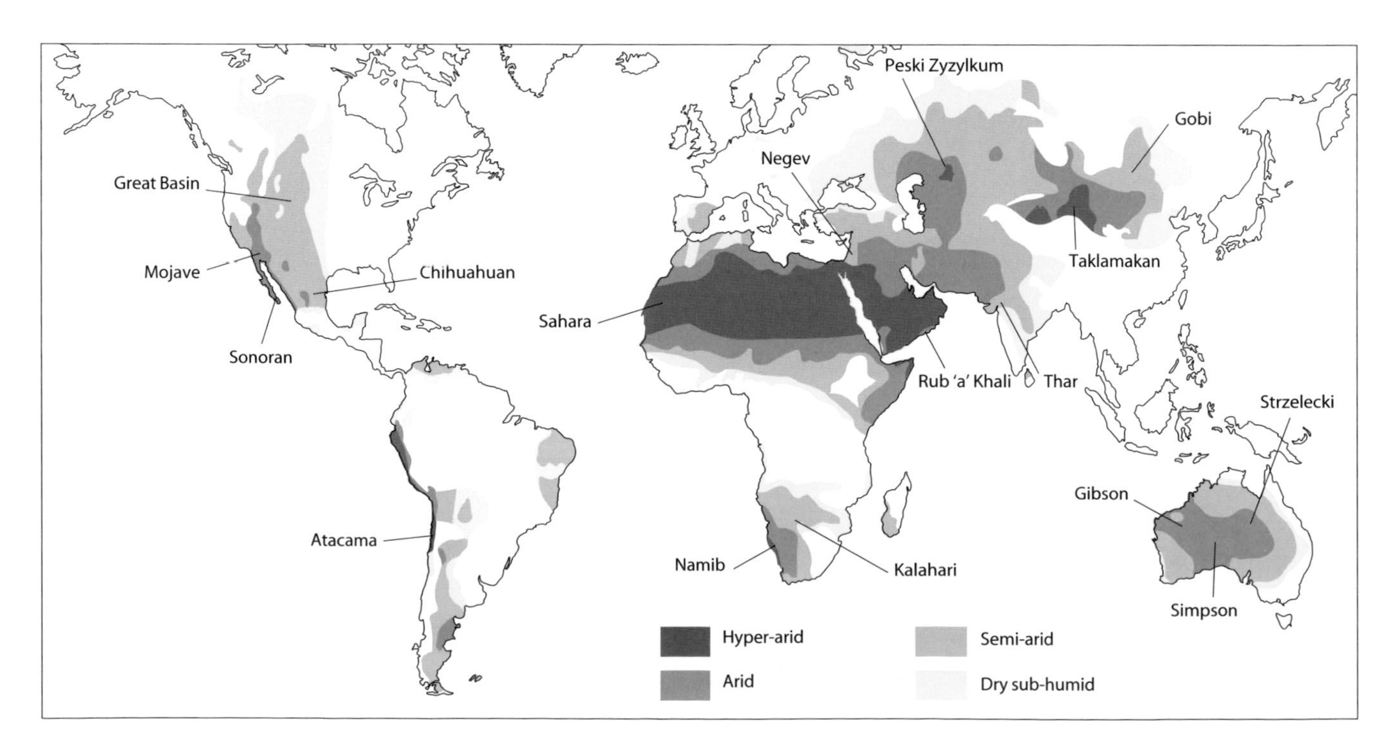

The global distribution of deserts and arid lands, based on the United Nations Environment Programme's aridity classification.

An alluvial fan in Death Valley, California, currently a hyper-arid desert. These fans develop from massive erosion and outpouring of sediment during occasional rainfall in arid desert regions where there are insufficient plants to bind the soil.

Many deserts, for instance the Sahara, Arabian, Thar, Kalahari and Australian, lie at latitudes between 10 and 30–40° north and south of the equator, where atmospheric circulation is dominated by sub-tropical high pressure and subsiding, relatively cool, air. The general stability of this area of high pressure means that rain-bearing cyclonic disturbances are usually deflected elsewhere or have only seasonal impact.

Continentality is another important factor, especially for the deserts of central and east Asia, as distance from the nearest ocean will determine the availability, or lack, of atmospheric moisture. Topography plays a role too: the deserts of Asia and North and South America are in the rain-shadow of high mountains – moisture falls as rain on windward slopes causing a deficit downwind.

Climate change

The effects of climate change have the potential to shift the patterns of oceanic and atmospheric circulation, bringing rain to desert landscapes and aridity elsewhere. As our understanding of inter-annual variability in ocean-atmospheric circulation gets better, our appreciation of the role of natural phenomena, such as El Niño (p. 266), in driving the evolution of desert features, also improves. El Niño is geographically limited and does not affect all global deserts, but it does serve as a useful analogy for the vulnerability of deserts to short-term changes in oceanic and atmospheric patterns.

Desert evolution: an arid myth?

Persistent rains have considerable impact on deserts. In addition to flooding and the formation of lakes, the normally stable surface environment changes as mobilized flood sediments are blown by wind action, creating sand dunes and dust storms. Increased moisture also stimulates plant growth (if briefly) and promotes rock weathering. The dormant desert comes to life and for several months after the rains, desert features evolve rapidly – dunes, gullies and alluvial fans develop new features and new channels, rocks rapidly break down, cacti flourish. In fact, it is likely that under normal conditions, deserts are dormant landscapes, waiting for the next rains to develop their natural features. The implication is that climate change is responsible for what we mistakenly interpret as arid features. Mildred Cable was right: deserts are a grave and need water to evolve.

BRAD C. FIERO

33 How do plants and animals live in the desert?

...life nowhere appears so brave, so bright, so full of oracle and miracle as in the desert.
EDWARD ABBEY, 1968

Deserts, with their stark beauty and extreme conditions, have moulded a fascinating flora and fauna, armed with myriad adaptations to make desert living not only possible, but highly successful. These adaptations include anatomical, physiological and behavioural means of expiration, evasion and endurance (the three E's of desert existence).

For some plants and animals, ironically, desert survival requires timely death (Expirers). Seeds and juveniles develop into adults when conditions are relatively mild. As conditions become more severe, adults die, but only after producing progeny in a form that can endure the harsh conditions (e.g. tough seeds or eggs). Annual wildflowers and ephemeral insects have mastered this approach to long-term survival.

Others (Evaders) avoid the harshest conditions by changing locations, body temperature (going dormant) or activity time. Some animals enter deserts only for a brief period when conditions are favourable for them and then migrate to escape the harshest extremes. Some stay year round but enter burrows or cavities that offer refuge from

Being nocturnal and migratory helps the lesser long-nosed bat thrive in the desert when its food supply (columnar cacti and agave nectar) is plentiful.

heat, cold and death by desiccation. They can do this for hours, days or even months, often going torpid to conserve energy and water. Many desert species become nocturnal during the summer to avoid the daytime's water-hungry inferno. Lesser long-nosed bats in the Sonoran Desert do all three.

Finally, there are the species that are out there, experiencing it all the time (Endurers). These plants and animals must endure both heat and water stress.

Adaptations to heat stress

The dry desert air rapidly gains and loses heat, requiring plants and animals to endure temperature extremes. Desert plants (e.g. saguaro cacti) protect themselves from freezing temperatures by using nurse plants (taller plants) that provide insulation against cold, clear nights. Desert animals avoid freezing temperatures by seeking shelter in burrows and cavities, and by insulation provided by fur, feathers and scales. Both plants and animals defend themselves from high temperatures by both reducing the input of heat into and dissipating heat out of their bodies, and by tolerating higher than normal body temperatures.

Animals can reduce heat input by selecting cooler microclimates (shade or burrows) or cooler activity times (nights or cool seasons). Insulation, shading (by spines/hairs, etc) and reflective surfaces (using waxes or hairs) also reduce heat input, as do changes in posture – the sidewinder rattlesnake lifts most of its body off the hot sand –

The annual wildflowers of the Sonoran Desert flourish when conditions are mild and then die, leaving their tough seeds behind. The saguaro cactus, with its widely spread, shallow roots is able to capture even tiny amounts of moisture, and its body can swell to store vast quantities of water.

Right *Although the fennec fox is the smallest in the dog family, this nocturnal burrowing African fox has the longest ears relative to body size, allowing it to dissipate heat better and hear its prey in the dry, desert air.*

Opposite *The sidewinder rattlesnake has a unique method of raising all but two points of its body off the ground while moving in order to minimize contact with the very hot sand.*

and orientation – vertical leaves minimize the surface area exposed to the midday sun.

If reducing heat input is not enough, then heat must be dissipated by evaporation (at the cost of water loss) or by having long appendages (such as the long, vascularized ears of the jackrabbit), or small bodies or body parts (e.g. leaflets) to radiate heat. Animals may pant or urinate on themselves (vultures for instance) for evaporative cooling.

If these prove insufficient, then there is no choice but to tolerate higher body temperatures. Some plants and animals can survive body temperatures that would kill humans – antelope squirrels up to 43°C (109°F), some cacti to 60°C (140°F) and some grasses to an incredible 70°C (158°F). This is accomplished through the use of specialized enzymes and heat-shock proteins.

Adaptations to water stress

Deserts are defined by the lack and unpredictability of water (p. 141). This dearth of water selects for plants and animals that capture, store and conserve water well, or that can tolerate dehydration. Desert plants capture water through their roots, which may reach many metres down to tap a deep water table or extend less than a metre in depth but many metres wide to soak up sporadic rainfall before it evaporates. Leaves and shoots may condense and funnel water vapour to roots. Desert animals know where to find water or acquire it from their food, and can even capture water vapour using special anatomy that condenses and funnels water to their mouths, as found for instance in some lizards and beetles.

Once obtained, animals may store water in fatty deposits in their tails and other tissues, like the gila monster lizard. Water can be stored in the roots, stems and leaves of plants (plants that do this are called succulents). Water storage defends against desiccation, but then itself requires defence against thieves (by spines and poisons), toppling over (ribs) and problems caused by expansion-contraction (pleats). A saguaro cactus may reach over 12 m (40 ft) tall, its vertical ribs allowing it to expand until up to 90% of its weight is water.

Water must then be used conservatively. This can be achieved by minimizing the loss of water through the outer surface (cutaneous water loss), through urine and faeces (excretory water loss) and through gas exchange (respiratory water loss). Plants lose most of their water through their stomata (pores) during gas exchange, but have a variety of means to minimize this loss.

Other water conservation methods include waxes on the outer surfaces of some plants and arthropods that stop water from leaking out. Animals concentrate their urine (e.g. kangaroo rat) or convert urea to uric acid, which uses 10 times less water, and can also aestivate (go dormant) during hot/dry spells to conserve water.

As with heat, if water storage and conservation methods are insufficient, then plants and animals must tolerate dehydration. Humans can tolerate water losses of 12% of body mass, but many desert plants and animals can tolerate water losses of between 30% and 50% or more. Some organisms (for instance brine shrimp and nematodes) survive dry periods almost completely dehydrated and lifeless, and then spring back to life when water becomes available again.

Considering these and other adaptations, deserts provide a beautiful laboratory in which to witness and study life's inventiveness in conservation and endurance.

JANE FRANCIS

Has there always been ice at the poles?

We must always remember with gratitude and admiration the first sailors who steered their vessels through storms and mists, and increased our knowledge of the lands of ice in the South.
ROALD AMUNDSEN, 1912

Opposite above *Antarctica today – home to penguins and spectacular icebergs.*

Opposite below *Axel Heiberg Island, in the Canadian Arctic, is treeless tundra now, but the black lines are the remains of layers of fossil forests.*

Below *The fossil leaf of a flowering plant that grew in Antarctica about 80 million years ago.*

Our current vision of the polar regions is of pristine ice and snow, populated by animals such as penguins and polar bears that are adapted to low temperature extremes. Today, the large landmass of Antarctica sits over the South Pole, with a giant ice cap up to 4 km (2.5 miles) thick. The northern polar region is covered by the Arctic Ocean, which has frozen to form the Arctic ice cap. On a geological timescale, however, it is only relatively recently, from about 40 million years ago, that the high latitudes became glacial; for at least 200 million years before this the polar regions were warm and ice-free, fuelled by high levels of atmospheric carbon dioxide.

Fossil evidence

Evidence for past polar warmth is found in rocks and fossils, of both marine animals such as ammonites and plesiosaurs that lived in polar seas, and dinosaurs and plants that lived on the land. Paradoxically, among the most common fossils found in these now snow-covered regions are leaves, tree trunks and pollen, the remains of vegetation that once lived at high latitudes. These fossils hold clues about the nature of the climate at the time, and about the type of vegetation that was able to survive so close to the poles.

Some of the most spectacular remains are of fossil forests, with many tree stumps still in their original positions of growth. In the Transantarctic Mountains in Antarctica large petrified tree stumps of Permian age (around 251 million years old) are preserved in coal seams that were once large peat swamps. At that time Antarctica was part of the large southern landmass of Gondwana, and geological evidence indicates these forests grew only a few hundred kilometres from the South Pole. The forests were composed of now-extinct *Glossopteris* plants, tall trees with large leaves that formed thick leaf mats in the peat.

By Cretaceous and early Tertiary times, 100–50 million years ago, exceptionally warm 'greenhouse' climates allowed subtropical forests to thrive on Antarctica. On Alexander Island, on the Antarctic Peninsula, 100-million-year-old fossil stumps show that trees grew as dense as in any warm temperate forest today. Identification of the fossil wood and leaves reveals that the trees were ancestors of modern Southern Hemisphere conifers, such as the podocarps and *Araucaria* (monkey puzzle), along with the living-fossil *Ginkgo*. Ferns, liverworts, mosses and some primitive flowering plants grew in the undergrowth.

By about 80 million years ago flowering plants began to dominate polar forests. In Antarctica ancestors of families that live today in South America, Australia and New Zealand, such as the Sterculiaceae, Nothofagaceae (southern beech), Winteraceae (e.g. mountain pepper tree) and Proteaceae (e.g. proteas), became increasingly more common as the climate warmed. Animal fossils show that dinosaurs lived in Antarctic forests during the Cretaceous, and by the Eocene (50 million years ago) primitive mammals ancestral to llamas and sloths lived there, along with 1.83-m (6-ft) tall penguins.

In the northern polar regions conifer forests dominated the land around the Arctic Ocean. Ancestors of modern boreal forest trees, such as spruce, pine, sequoia, juniper and larch, grew

*A leaf mat of southern beech trees (*Nothofagus*) that grew only 500 km (310 miles) from the South Pole about 3 million years ago.*

along with the deciduous conifer *Metasequoia* (dawn redwood). Sycamore, birch, alder and hickory trees also grew there. On the now-barren tundra landscape of Axel Heiberg Island in the Canadian Arctic, layer upon layer of mummified tree stumps and thick beds of dried leaves are the exceptionally preserved remains of 50-million-year-old swamp forests. The forests were home to turtles, warmth-loving crocodiles and primitive hippo-like creatures called *Coryphodon*.

Nowhere on Earth today can forests be found so close to the pole. Trees now can survive only up to 70°N, but in the warm climate of the past forests thrived in high latitudes. In fact, climate models show that polar forests were a positive influence in keeping the high latitudes warm, because their dark colour and low albedo soaked up sunlight (unlike the bright snow surface with high albedo that reflects back solar insolation today). In addition, these forests also had to adapt to the strange polar light regime of long dark winters and summers of continuous sunlight. The trees seem to have coped well by becoming dormant in the dark winter and making the most of the summer sunlight for lush growth.

Ice appears

Evidence of tillites and dropstones of glacial origin in the rock record demonstrates that ice appeared on Antarctica about 40 million years ago, and probably later in the Arctic. Climates cooled, ice sheets built up over land and poured ice out to sea as ice shelves. Although it was once thought that the ice heralded the end of plants on Antarctica, recent discoveries show that vegetation survived even in this glacial landscape. Sandwiched between thick sequences of glacial tillites, mummified twigs of dwarf bushes and leaf mats of the southern beech *Nothofagus* show that this dwarf forest clung to life in a cold tundra environment. These plants grew only 500 km (310 miles) from the South Pole a few million years ago and represent the last vestiges of plant life on Antarctica before the deep freeze set in.

Once ice sheets began to grow on Antarctica about 40 million years ago – probably due to decreasing levels of CO_2 in the atmosphere – Earth's climate changed dramatically. Vast areas of ice over the Antarctic continent reflected back sunlight and cooled the climate further. Cold, dense, salty water sank to the floor of the oceans and spread northwards, changing ocean structures significantly. As tectonic plates moved apart and South America broke away from the Antarctic Peninsula, deep channels allowed ocean currents to flow around Antarctica for the first time, isolating the continent in icy cold waters.

Eventually, about 2 million years ago during the Pleistocene epoch, the Arctic regions cooled enough for ice caps to grow on the land bordering the Arctic Ocean, in North America, Greenland, Scandinavia and Russia. This huge volume of ice was then a major control on our global climates. Ice cores from the ice caps show that these large ice masses were still responsive to change, however, this time driven by orbital cycles. Cycles of cooling and interglacial warming caused ice sheets to wax and wane in cycles of 19,000, 23,000, 41,000 and 100,000 years, in tune with the Earth's position with the Sun (p. 270).

Will there continue to be ice at the poles? As carbon dioxide levels rise and our climate warms, there are already clear signs that the polar ice caps are shrinking. Perhaps climate change will reach a critical threshold that will once again allow forests to grow in the polar regions.

How deep can life live under ice and rock?

The number of subsurface prokaryotes [microorganisms] probably exceeds the numbers found in other components of the biosphere.
W. B. Whitman, D. C. Coleman and W. J. Wiebe, 1998

Until the second half of the 20th century it was assumed that below just a few tens of metres, the subsurface of the planet was essentially devoid of life. However, the biomass of life underground is now estimated to be at least as great, if not greater, than life on the surface of the Earth. In the last two decades increasing numbers of studies have been conducted to discover the nature of the deep subsurface biosphere.

Subsurface environments vary enormously, both physically and chemically, depending on their material and its geologic history, but numerous bore holes have revealed that microorganisms are sustained at many kilometres' depth. Despite their numbers, however, they are thought to occupy less than a millionth of the space available in deep rocks and sediments. Their pervasiveness is limited by the availability of nutrients and water and by other physical and chemical conditions.

Living in the deep subsurface

In general, at depths less than about a kilometre, many of the microorganisms are using organic carbon produced by photosynthesis as an energy source. As temperatures increase with depth, some of this carbon may be activated and made more available to microorganisms, thus fuelling the deep biosphere.

Where carbon availability is ultimately limited or absent, other energy supplies must be found. One possible fuel used by deep subsurface microorganisms is hydrogen, either coupled to sulphate (as in sulphate-reducing microorganisms) or carbon dioxide, as in 'methanogens' (methane-producing microorganisms). Hydrogen can be produced by, for example, the splitting of water by radioactivity or the reaction of water with volcanic rocks in hydrothermal or tectonically active regions below the surface. Such processes offer the possibility of providing an energy source for deep subsurface microorganisms completely independent of surface photosynthesis.

Claims of the discovery of deep, surface-independent, microbial communities have, however, been controversial. Two possible examples include microorganisms living at 2.8 km (1.8 miles) depth in South Africa, which use hydrogen produced from the radioactive splitting of water and sulphate as an energy source, and methane-

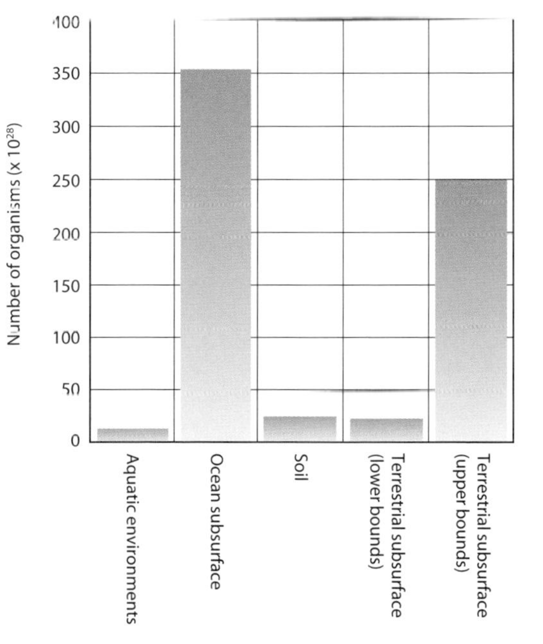

Graph showing estimated numbers of microorganisms to be found in different environments on Earth. The two different bars for the 'terrestrial subsurface' show the possible range.

Cores being recovered from deep ocean sediments. The scientists must wear gas masks because hydrogen sulphide is emitted by the core, as a result of microbial activity.

producing bacteria in Lidy Hot Springs, Idaho, USA, at 200 m (656 ft) depth, living off geothermally produced hydrogen. Microorganisms have been isolated from deep granite rocks in Sweden, and are thought to use deep subsurface gaseous energy sources.

Fractures in deep subsurface rocks can aid in transmitting water (containing nutrients) and energy sources, helping to sustain microbial communities. In contrast, some fractures are also thought to trap water, separating it from other ground water for millions of years. Once depleted of nutrients and energy sources, these habitats would become unsuitable for life, sustaining only a small population of cells – if any at all – illustrating the complexities in understanding the distribution of subsurface life.

Life in deep ice

Remarkably, even the deep subsurface of ice sheets can potentially support microorganisms. Ice recovered from a depth of over 1 km (0.6 miles) below Vostok Station, Antarctica, has yielded diverse bacteria and fungal spores. At 3.59-km (2.2-miles) depth below Vostok, ice cores yielded microorganisms at concentrations of 2,800 to 36,000 cells per millilitre. This ice has been isolated from the atmosphere for approximately 1 million years and the microorganisms probably originate from Lake Vostok, a deep subglacial lake within the ice sheet, which might possibly support an isolated indigenous microbial population. Similarly, the subsurface of permafrost in Siberia has shown to be a stable and long-lived environment for microorganisms, with some showing metabolic activity at –20°C (–4°F).

Such situations highlight the great technical challenges faced in retrieving microorganisms. One problem is potential contamination by external microorganisms from the drilling mud and other drilling procedures. Various tracers can be added to the drill core or drilling mud, including chemical substances and fluorescent beads roughly approximating the size of microorganisms. The absence of these tracers in retrieved material from a core to be studied in microbiological investigations can confirm that it is free of contamination.

Other technical challenges include the problem of retaining the material in the cores at ambient conditions. For instance, new efforts are being made to retrieve cores at the same pressure (up to 250 times atmospheric pressure) that they are found in, in order to isolate microorganisms that might require high pressure for growth.

Scientists examine sections from a core, 2 km (1.25 miles) long, taken from the deep subsurface of the Chesapeake impact crater in Virginia, USA for microbiological analysis. The crater was formed by the collision of an asteroid with the Earth about 35 million years ago.

How deep is the deep subsurface biosphere?
The lowest depth of the deep subsurface biosphere is likely to be set by the Earth's geothermal gradient, assuming that nutrients and water are available. As the geothermal gradient is about 20°C/km depth and the upper temperature limit for life is currently recorded at 121°C (250°F), it follows that the lowest depth of the biosphere would be about 6 km (3.75 miles), although this will vary with differences in rock thermal conductivity as well as in regions where thermal anomalies change the subsurface temperature profile.

Is there life on other planets?
Organisms in the deep subsurface have spurred speculation about the possibility of life in the subsurface of Mars. Protected from surface ultraviolet radiation, shielded from ionizing radiation, and possibly in deep geothermal regions where liquid water is stable, the deep subsurface of Mars may well offer a relatively clement environment for life compared to the surface. Indeed, organisms living independently of photosynthesis in the deep subsurface of Earth have been considered as possible models for extraterrestrial deep subsurface life (see also p. 171).

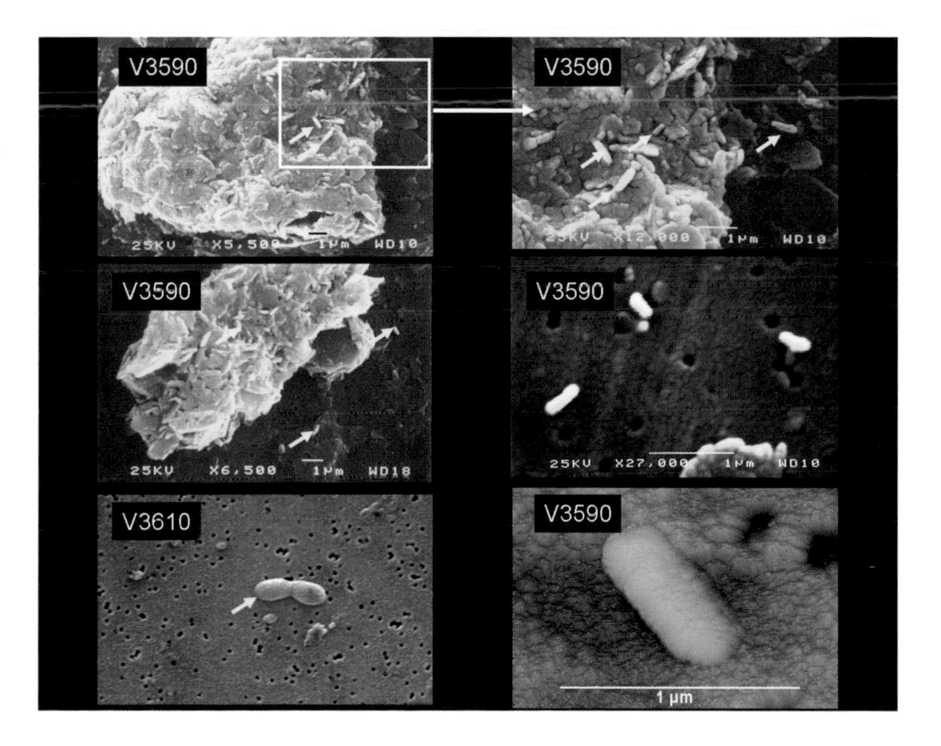

Microorganisms recovered from a depth of 3,590 m (11,780 ft) in the Antarctic ice sheet above the subglacial Lake Vostok.

PETER MAYHEW

Why are islands special?

...[B]y far the most remarkable feature in the natural history of this archipelago ... is, that the different islands to a considerable extent are inhabited by a different set of beings.

CHARLES DARWIN, ON THE GALÁPAGOS ARCHIPELAGO, 1845

Small, discrete and isolated habitats, such as oceanic islands, lakes and mountain tops, differ from other 'mainland' areas in the ecology and evolution of their fauna and flora. As a result, each island has a unique biological history, providing scientists with some of the most telling insights into how ecological and evolutionary forces operate on our planet. In addition, the size, isolation and uniqueness of island populations make them especially vulnerable to changes imposed by humans, and therefore at the forefront of work to conserve Earth's species and communities. But how are islands special in their ecology and evolution, and what have these differences taught us?

Island ecology

One of the simplest things you can ask about the fauna and flora of a place is how many species live there. Many factors influence the species richness of an island, and by studying these factors, biologists have developed their understanding of species richness as a whole. Arguably the most useful finding has been how island area relates to species richness. Predictable relationships are valuable tools to ecologists: the relationship between species richness and area has been used to show that current extinction rates on the planet as a whole are likely to be well above 'background' rates because of the loss in area of natural habitats. These rates are likely to increase with climate change as the area of climatically suitable habitat for species shrinks or moves (p. 278).

If species richness is the most basic property of a flora or fauna, perhaps the most basic property of a species is its population density. Here again, islands turn up surprises. On islands, many species become much more abundant than on equivalent mainland areas. One common reason for this is a lack of the predators and competitors that reduce the size of populations on mainlands. However, the poor species richness of islands also makes them very vulnerable to invasions by introduced species, which find life easier than they would on mainlands. Extinctions can result from such introduced predators, diseases and competitors, but also directly as a result of human exploitation and destruction of habitats. A frightening proportion of island species has gone extinct or is currently threatened with extinction at the hands of humans.

Islands contain many of the world's endemic species – ones that are found nowhere else – and some of these have been isolated for so long that they form distinct families, such as the Kiwis of New Zealand. Others contain representatives of

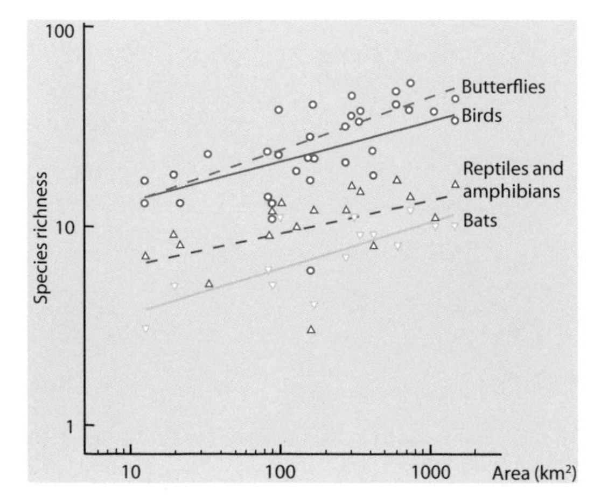

Island area and the species richness of different animal groups on the Lesser Antilles islands in the Caribbean.

taxa that have become extinct on mainlands but survive on islands due to their isolation, for instance the lemurs of Madagascar. Because of this combination of uniqueness and vulnerability, many islands or archipelagos have been designated as conservation priority 'biodiversity hotspots'.

The development of new species on islands

It could be argued that the differences between island and mainland species sparked off modern evolutionary biology: they so impressed Darwin and Wallace that they led directly to the theory of evolution by natural selection. Changes between island and mainland populations can lead to the

Flooded volcano craters in the Galápagos Islands, an archipelago where many endemic species are found on different islands.

Above *A ring-tailed lemur. The lemurs have radiated on Madagascar but gone extinct elsewhere on Earth.*

Right *The Hawaiian silversword alliance has radiated into 28 species occupying a wide range of ecological niches.*

development of new species – speciation (p. 123) – because the isolation of an island prevents inter-breeding with the mainland population, and therefore allows changes to accumulate over time. However, differences between a colonizing species and its mainland ancestors are often just the start of a larger 'adaptive radiation' that occurs on islands when a large number of closely related species evolve from a small number of ancestral species to adapt to the different ecological niches.

Among the most famous adaptive radiations are the 28 species of silversword plants and over 500 species of drosophila fruitflies on Hawaii, and the 1,200 cichlid fishes of the Great Lakes of East Africa. It has even been suggested, and some evidence confirms this, that the whole history of colonization, radiation, extinction and re-colonization on islands may follow a predictable series of events, known as a 'taxon cycle'.

Changes in form

Once a species arrives on an island it may change form in unusual ways, even if it does not give rise to new species. For instance, changes in animal body size often conform to the 'Island Rule', whereby large species become smaller on islands and small species become larger (see also p. 184). While this rule holds within many vertebrate groups, its causes vary. In snakes, for example, diet seems to be important: island populations feeding on birds get bigger, while those feeding on lizards get smaller. In birds, however, changes in body size are associated not with diet but with differences in competition and environmental temperature. The celebrated dwarfism of megafauna on islands – for instance the dwarf elephants of Ice Age Sicily and the small woolly mammoths of Wrangel Island, Siberia – appears to be dependent on the area of the island, with larger islands supporting larger species in a way that is consistent with the number of home ranges that can fit into an area. This suggests a role for competition within a species.

Another possible change in island species is a reduction in dispersal ability, which can be observed in both animals and plants. It manifests

Echium wildpretii, one of many species of this genus endemic to the Canary Islands that have evolved woody stems.

as flightlessness in birds and insects, and reduction of seed structures in plants. The changes may be selected for in order to reduce mortality losses at sea, through loss of costly (and ultimately useless) dispersal behaviour or apparatus, or through increases in fecundity. Flightlessness in island birds, such as the dodo, may have come about because of the paucity of ground predators.

The reproductive biology and growth habits of island plants are special in many ways. The floras of many islands have pale unspecialized flowers, and on other islands wind pollination predominates. Species commonly lose self-pollination, have separate males and females (dioecy), or have male and female flowers that are very different in shape (dimorphism). Another unusual phenomenon is that of herbs growing perennial woody stems like those of shrubs. Many of the above traits are linked evolutionarily and are found in combination. An interesting example of how such traits can evolve together comes from the plant genus *Echium*, which has radiated off the coast of Africa in the Canary, Cape Verde and Madeira archipelagos. The genus is both more woody and more outbred than its mainland relatives. Evidence suggests that the outbreeding habit evolved to counter inbreeding resulting from the low genetic diversity of initial colonists. Woody stems may then have evolved to increase lifespan, maximizing the number of seeds set in a lifetime once the plant is unable to self-pollinate. As with the evolution of animal body size, however, similar outcomes in other groups may have resulted for different reasons.

Where to next?

The ecology and evolution of island life continue to be vigorously studied. Reconstructions of evolutionary trees using molecular evidence unravel the history of island colonization and diversification. Comparisons between mainland and island relatives identify the changes they have undergone, and studies of recent colonizations chart the selective pressures and constraints they experience. The extent of many of the above trends, and their underlying causes, remain to be studied in depth, while at the same time conservationists work to preserve island life for future generations.

PETER R. GRANT

What do we know about Darwin's finches?

Seeing this gradation and diversity of [beak] structure in one small, intimately related group of birds, one might fancy that, from an original paucity of birds in this archipelago, one species has been taken and modified for different ends.
CHARLES DARWIN, 1845

One of the great mysteries of the world has been the origin of species – 'the mystery of mysteries' as the 19th-century philosopher John Herschel put it. How has the world come to be populated by so many species of different organisms, replacing the vast number of others that have become extinct? Charles Darwin answered the question with a comprehensive theory of evolution by natural selection. Galápagos finches played an important role in his theorizing, and in recognition of this they have been called 'Darwin's finches'.

In 1835 Darwin spent five weeks roaming on four islands in the Galápagos archipelago. At the time, his observations on finches seemed to make no obvious impact on the 26-year-old naturalist. Nevertheless, only two years later, after he had returned to England, they helped to guide his thoughts towards a revolutionary theory: that species were not fixed but mutable, that they diversified by gradual change from their ancestors over many generations, and that the mechanism of evolutionary change was natural selection. Genes, the missing ingredient discovered only later, gave substance to the concept of inheritance, and completed the theory.

Since Darwin's time the finches have continued to illuminate the question of how and why species multiply. They are a classic case of adaptive radiation; that is, several species formed in a relatively short time, each adapted to a different ecological way of life.

Ancestral finches

Today, 13 species of finches occupy islands of the Galápagos archipelago lying astride the equator

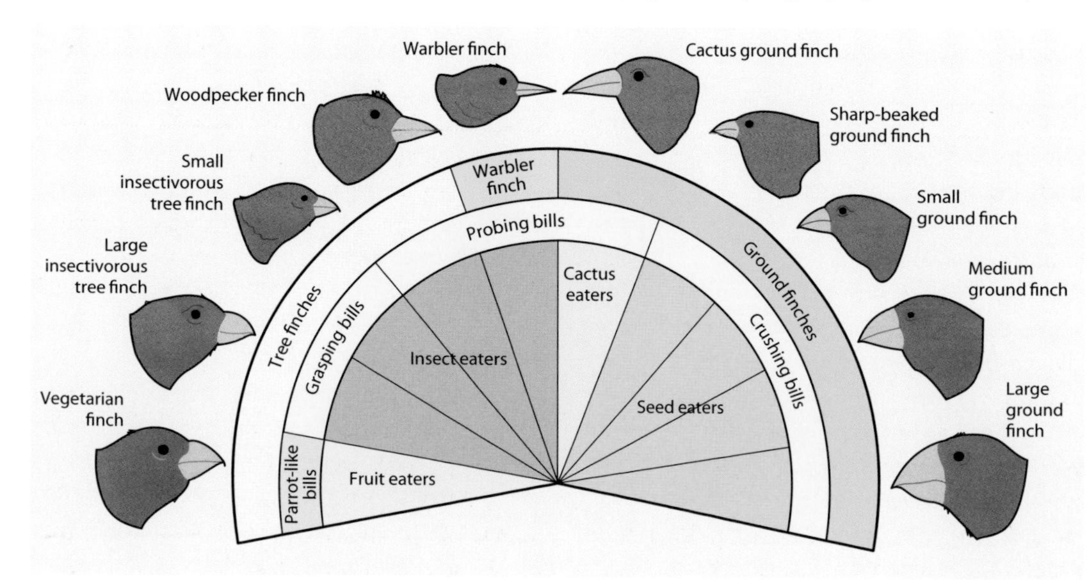

The radiation of some of Darwin's finches, with strikingly varied beak shapes to exploit different food resources.

Above left *The mangrove finch, confined to small patches of mangroves on Isabela Island, is the rarest of Darwin's finches. It is a member of a group of five species of tree finches.*

Above right *Warbler finches, seen here in an illustration by John Gould, evolved their distinctive warbler-like appearance early in the radiation.*

in the Pacific Ocean; a 14th occurs on Cocos Island, 700 km (435 miles) to the northeast. According to DNA evidence, the closest mainland relatives of Darwin's finches are a group of tanagers called seed-eaters. The same evidence suggests that the ancestral species arrived on Galápagos from South or Central America about 2–3 million years ago; Cocos Island was colonized later from the Galápagos. At the time of their arrival, fewer islands existed in the archipelago than now, perhaps only five. They increased in number through volcanic activity, largely centred on a hotspot in the western part of the archipelago.

The Galápagos climate when the finches arrived was hotter and wetter, and the vegetation was probably more like a tropical rainforest than the current mixture of dry-deciduous forest in the lowlands and evergreen forest in the highlands. Thus the finches diversified as their environment changed. Early in the radiation at least four species were formed: the warbler finch, Cocos finch, sharp-beaked ground finch and the vegetarian finch. Later, five species of tree finches and five additional species of ground finches evolved.

The species differ strikingly in the size and shape of their beaks, though in other respects they are rather dull and nondescript. Their distinctive beaks are tools, analogous to pliers, and perform different tasks when used in gathering and dealing with foods of varied types or in different locations. Thus each species exploits food in its own particular niche, using these specialized beaks to feed on insects, snails, fruit, seeds, nectar, pollen and even blood.

Speciation

Almost a century after Darwin had developed a general explanation for the origin of new species, the British ornithologist and biologist David Lack applied it to the finches. The main idea is this: speciation, the evolution of new species (p. 123), begins with the establishment of a new

The beaks of finches are tools, analogous to pliers, for gathering and dealing with foods of different types or in different locations. The various species are equipped with a range of tools for exploiting particular ecological niches.

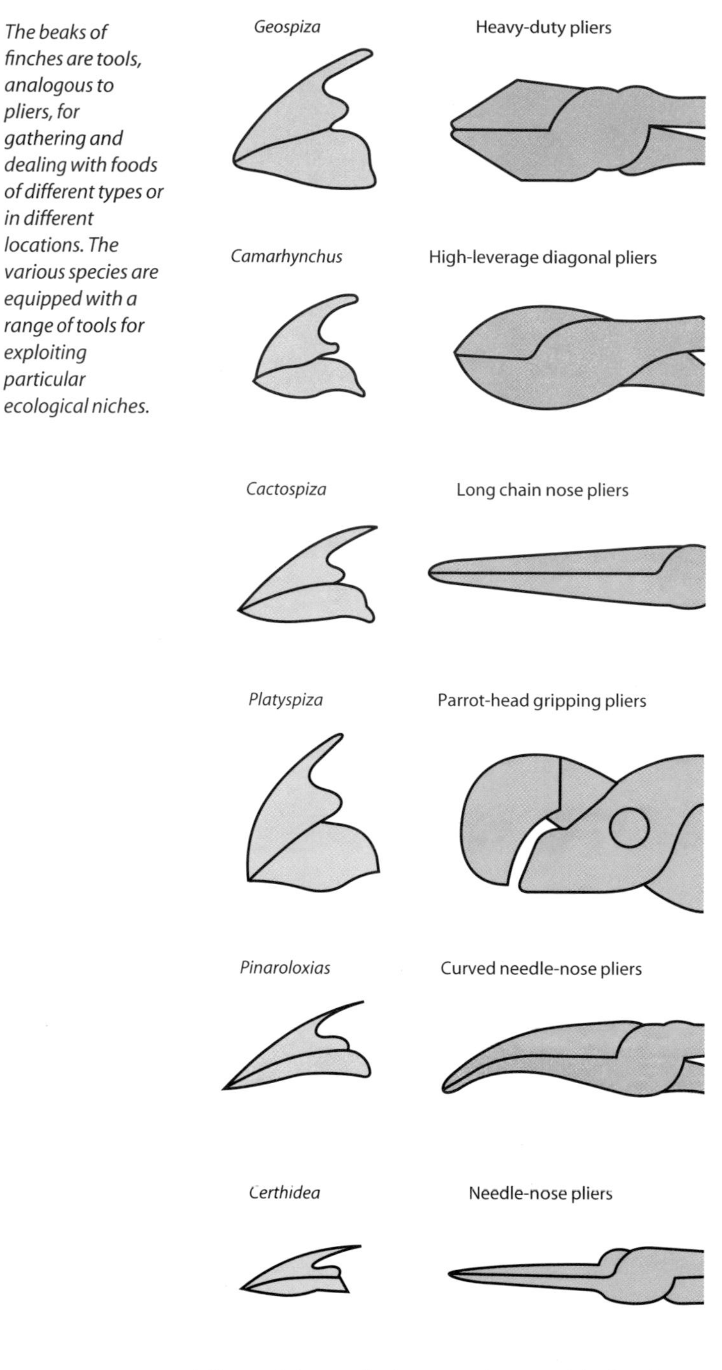

(allopatric) population, continues with the divergence of that population from its parent population, and is completed when members of two diverged populations come together on the same island, living in sympatry, and interbreed rarely if at all. Thus one species gave rise to two, and with the Galápagos finches this happened several times.

The clearest example of allopatric differentiation is shown by the six populations of the sharp-beaked ground finch. Different populations of this species occupy different habitats, and feed on different foods with beaks of varied sizes and shapes. They are locally adapted. The main driving force in the ecological differentiation of populations is thus believed to be natural selection.

Modern research has supported the theory by repeatedly demonstrating the process of natural selection. It has been shown to occur frequently in ground finch populations on the islands of Daphne Major and Genovesa when the environment changes, and to vary according to the particular set of environmental (food supply) conditions. Evolutionary change occurs because the traits subject to selection are heritable.

Establishment of sympatry by two, previously separated populations, is a crucial stage in the cycle of speciation events. The two groups of finches may be similar and compete for food, in which case they are likely to be subject to natural selection that causes their beaks and diets to diverge. This process, called character displacement, results in less competition between them. It has recently been witnessed on Daphne Major Island. By depleting the supply of *Tribulus* fruits during a drought, large ground finches caused an evolutionary shift in the medium ground finch population in the direction of small beak size.

Reproductive isolation

Of equal importance for coexistence is a barrier to interbreeding. Experiments show that differences in both morphology and voice constitute such a barrier, whereas courtship behaviour and plumage do not. Unlike beak traits, which are genetically inherited, song is learned. Learning takes place during a short sensitive period early in life in an imprinting-like process (p. 217). The visual cues of beak size and shape are presumably learned at the same time. Songs, like beak sizes, are presumed to diverge in isolation, therefore the barrier to reproduction originates when populations are geographically isolated on different islands (allopatry) and becomes effective when species live together on the same island (sympatry).

The woodpecker finch is unique in using a twig or a cactus spine as a tool to extract hidden insect larvae.

The barrier to interbreeding is not complete in all cases. Study of finches on Daphne Major Island has shown that they do hybridize rarely, sometimes when young birds misimprint on the song of another species. The fitness of hybrids and backcrosses is dependent on the external environment. Under favourable conditions hybrids and backcrosses of birds with intermediate beak sizes survive and reproduce as well as the parental species, and have no difficulty in acquiring mates or producing offspring. They are relatively unfit under other ecological conditions, but there is no evidence of genetic incompatibilities.

Why Darwin's finches are unique

Darwin's finches are unique in the Galápagos setting in having radiated into at least 13 species, whereas other land birds there have produced none, apart from mockingbirds (4 species). A possible reason is that the ancestral finches may have arrived on the Galápagos first, and by doing so gained the twin advantages of a long time in which to diversify and initial freedom from competitors and predators.

Factors intrinsic to the finches might also have contributed to their success. Flexibility in learning feeding skills is one, as this gives a species enhanced potential for exploiting resources in new or more efficient ways, resulting in unique feeding behaviours. For example, woodpecker finches use a leaf stalk or cactus spine to extract hidden insect larvae, and sharp-beaked ground finches peck at resting sea-birds – boobies – wound them and drink the blood. A second factor is their learned song, which acts as an imperfect barrier to interbreeding and makes them prone to hybridize. Introgressive hybridization (the transfer of genes from one species to another by a fertile hybrid breeding with one of the parental species or the other) may have been an important factor in the radiation, raising the level of genetic variation and facilitating evolution in novel directions.

JOHN LONG

38 The origins of Australia's special wildlife

The animal kingdom as developed in Australia presents us with anomalies and peculiarities perhaps even more remarkable than are exhibited by the plants. But owing to the great difference in the powers of dispersal of the various animal groups, there is less uniformity in the phenomena they present.
ALFRED RUSSEL WALLACE, 1893

Australia is home to an extraordinary fauna of extraordinary animals and birds, such as the kangaroo, koala, wombat, platypus, echidna and the giant flightless emu. In fact 83% of Australia's 268 species of mammals are endemic (that is, found nowhere else), along with 45% of its 740 bird species. Australia also has 794 species of reptiles (89% endemic), the highest number for any country in the world, and these include the second largest lizard, the perentie, and some of the most venomous snakes both on land and in the sea. Australia's only amphibians are all frogs, of which 93% of its 207 species are endemic. Australia is the world's largest island country that most biologists regard as a refugium – a place where species evolved in splendid isolation, thus creating a unique ecosystem. Aside from geographic isolation, what other factors were important in endowing Australia with such a highly endemic, and special, animal fauna?

Origins

In order to understand the origins of today's Australian terrestrial vertebrate fauna we must look back in time to a period when Australia was subsumed within the greater landmass of Gondwana, in the Cretaceous Period, some 120 million years ago. At that time Australia was situated at the bottom of the globe, with the South Pole located not far from southern Australia. Temperatures

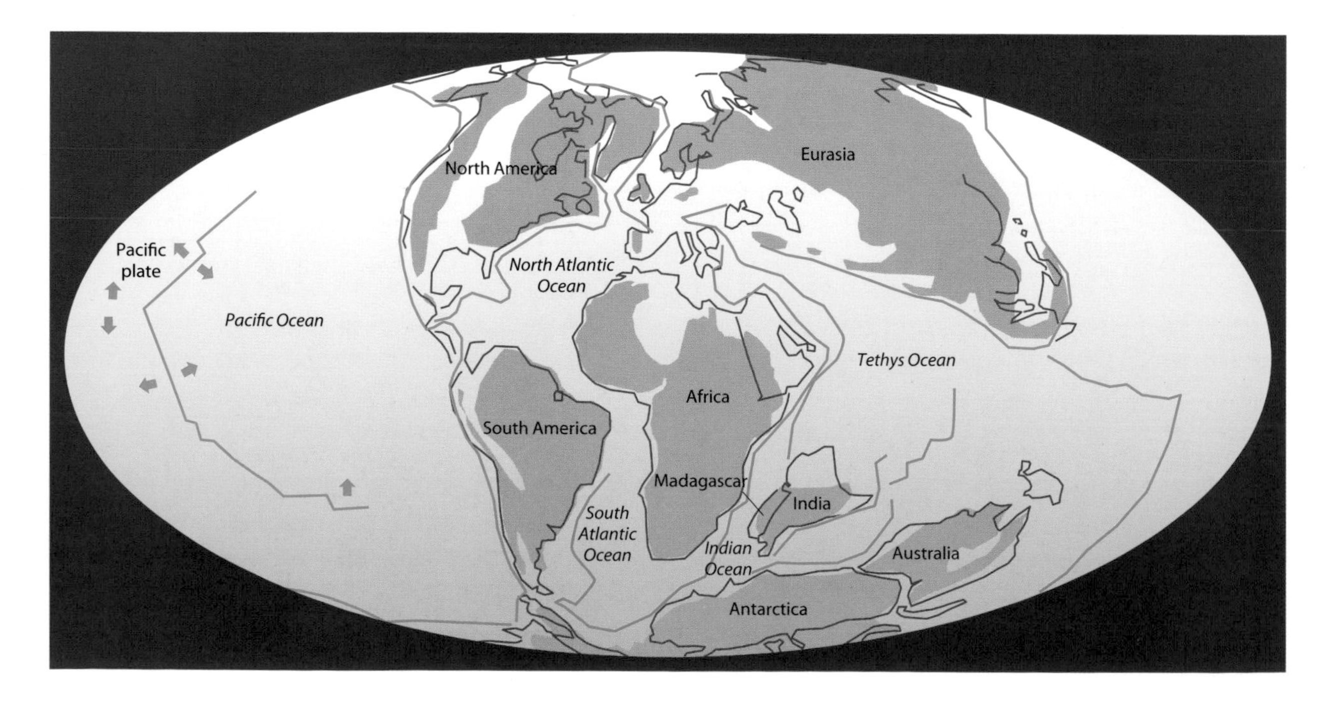

Map of the Cretaceous world at around 94 million years ago. Australia was at a very southerly latitude and still joined to Antarctica as the last remnant of Gondwana.

were cool to cold, but not frozen over as the pole lacked ice caps.

Australia was then home to several groups of dinosaurs. Once thought to be typical of northern hemisphere forms, scientists are now not so sure, as closer examination places some in their own unique groups – such as the plant-eating *Muttaburrasaurus*, 8 m (26 ft) long, and the small armoured *Minmi*, 2 m (6.5 ft) long. And some groups of dinosaurs once thought to be unique to the northern hemisphere have been recorded from much older deposits in Australia, including possible oviraptorosaurs (feathered dinosaurs, with beaked skulls), ornithomimids (*Timimus hermani*; slender dinosaurs, with long back legs) and a possible neoceratopsian (named *Serendipaceratops arthurcclarkei*; the horned dinosaurs).

Mammals first appear in the Australian fossil record in the Early Cretaceous, about 110 million years ago. This early mammal fauna comprises primitive monotremes (egg-laying mammals) allied to the later platypuses (such as *Steropodon*, *Kollikodon* and *Teinolophos*), as well as forms that could be either basal to all the monotremes or closer to the placental line, such as *Ausktribosphenos*. There are no fossil remains of either marsupials or placentals, the dominant living mammal groups in Australia, from this time. Monotremes first occur in the Early Cretaceous of Australia and while they are generally not found elsewhere, one species of Early Cenozoic platypus is known from South America (*Monotrematum sudamericanum*), indicating this group once had a much wider Gondwanan range outside Australia.

Marsupials and placental mammals first appeared about 125 million years ago in China, and from this we may deduce that marsupials at least must have made their way south into Australia by the time it started to split up from Pangaea – the supercontinent linking Gondwana to the Eurasian landmass in the north – perhaps within the Early Cretaceous.

The only record of early Cenozoic mammals in Australia comes from Murgon, Queensland, in the early Eocene (around 53 million years ago), in the form of teeth from a possible placental mammal, a condylarth (*Tingamarra*), a group best known from South America, as well as teeth of a primitive carnivorous marsupial (*Djarthia*) and one of the world's oldest bats (*Australonycteris*).

Reconstruction of Procoptodon goliah, *the largest roo ever, which weighed in at close to 300 kg (660 lb). It is thought to have been an arid region inhabitant.*

The age of marsupials and giant birds

By the late Oligocene-Early Miocene (around 24–23 million years ago) we see the appearance of early members of many of the modern Australian marsupial families – the first primitive kangaroos, koalas, wombats, possums and predatory dasyurids, alongside a range of highly unusual groups that were not destined to succeed, such as ilariids, 'thingodontans' and ektopodontids. Each of these families evolved and diversified through the Miocene in conjunction with the widespread occurrence of wet sclerophyll ('hard-leaved') and temperate rainforest spanning much of central and northern Australia. It is interesting to note the high diversity of some of these groups at this time

Top *The 'mihirung',* Dromornis stirtoni *was the largest bird to walk the Earth.*

Above *A painting of a thylacine from Kakadu highlights the once wide distribution of this mainland predator.*

compared to the modern Australian fauna: for example there were at least eight species of koala in the Miocene (there is only one today).

Kangaroos reached a peak of diversity by the Pleistocene with the sthenurines (short-faced kangaroos) diversifying into a wide range of forms, including the giant *Procoptodon*, the largest of all kangaroos, weighing up to 300 kg (660 lb) and reaching 3 m (almost 10 ft) tall. Tree kangaroos, which today inhabit only tropical rainforest, were quite widespread across the country then, including arid regions like the Nullarbor.

Australia's fossil record of birds is on the whole very poor. Birds are virtually unknown from the Mesozoic, apart from a few bones attributed to an early enantiornithine, and some feathers (which could also be dinosaur feathers), so they tell us nothing about Australian bird origins. Isolated bones dated from about 53 million years ago from Murgon include the world's oldest song bird.

The gargantuan dromornithids or 'thunder birds' (the 'mihirungs' of indigenous legend) appear in the Eocene, known from their three-toed footprints. These were the largest birds ever to walk the face of the Earth. By the Miocene they had reached large enough body mass to cope with the biggest predators of the time, such as the dog-sized marsupial lions (*Wakaleo*) or large thylacines (*Thylacinus potens*). The massive *Dromornis stirtoni* stood nearly 3 m (10 ft) tall, and with an estimated body mass of 500–650 kg (1,100–1,430 lb) it outweighed any of the New Zealand moas, Madagascan elephant birds or Ter-

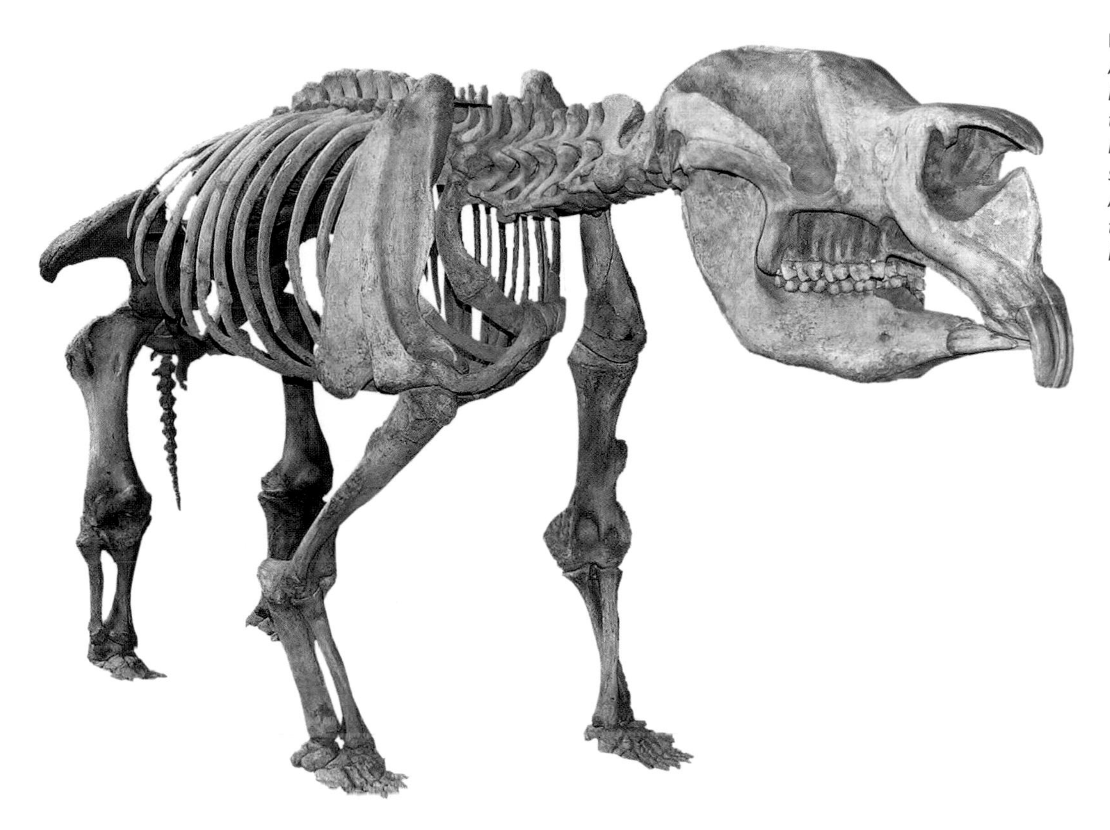

Diprotodon was Australia's largest land mammal and the world's largest marsupial. Fossil sites in central Australia indicate they lived in large herds.

tiary 'terror birds' like *Diatryma*. Dromornithids evolved from ancestral duck-like ancestors (anseriforms) and represent a fine example of the evolution of a group in isolation to gargantuan proportions. The Pleistocene dromornithids survived up to the final megafaunal extinction event, about 45,000 years ago, and were cohabitors with modern Australian emus.

Emus date back as far as the late Miocene. Molecular phylogenies indicate that they are closely related to the Kiwis of New Zealand, suggesting a common ancestor before Australia separated from New Zealand around 80 million years ago in the Late Cretaceous. Most of the modern families of endemic Australian birds appeared by the early Miocene, around 20 million years ago. Australia's gift to the world are crows and ravens, the earliest corvids originating in Australia before global migration.

Extinction and the ice ages

During the Pleistocene (1.8 million years–11,400 BC), best known for its series of global ice ages, the Australian terrestrial vertebrate fauna reached a peak of size, and this assemblage is generally referred to as the Australian megafauna (that is creatures that have a body mass of over 40 kg/88 lb). The largest marsupial ever was *Diprotodon optatum*, found throughout Australia over the past 2 million years. A wombat-like grazer, it reached 2.5 m (8.2 ft) in length and weighed up to 2.7 tons. *Zygomaturus* was a slightly smaller swamp-dwelling form, possibly occupying a tapir-like niche.

Australia's largest predator was the 6-m (20-ft) long, 800-kg (1,760-lb) varanid lizard, *Megalania prisca*, which roamed eastern Australia. The top mammal predator was *Thylacoleo carnifex*, a robust marsupial bearing enlarged secateur-like shearing premolars. It was up to 1.6 m (5.25 ft) long and weighed around 120 kg (265 lb). In addition to the giant kangaroos already mentioned, there was a guild of medium-sized predatory kangaroos, like *Propleopus*, with serrated premolars and stabbing incisors which they used to kill their prey. The giant wombat, *Phascolonus*, reached up

Emus are living examples of surviving megafauna. Like many Australian creatures they have an outstanding ability to survive in a harsh, unpredictable landscape.

to 1.6 m (5.25 ft) long and weighed in at about 250 kg (550 lb), while gracile wombats like *Warendja* probably climbed trees in search of food.

The Australian megafauna was extinct by around 44,000 years ago. There has been much debate about what caused this event, but an increasing body of evidence points to the arrival of humans in the country around 50,000 years ago as marking the beginning of the decline. Humans could have impacted on megafauna in two ways: by direct hunting and by the practice of fire-stick farming which irrevocably altered the vegetative landscape of the country. Climate change is another potential culprit, but research indicates that the demise of the megafauna coincided with a period of relatively mild climate, being between ice age extremes, and that anyway the megafauna was by this time well adapted to the ice age arid cycles of the Pleistocene.

Today the modern Australian fauna retains two megafaunal species of mammal, the red and grey kangaroos, and one bird, the emu, that survived the Pleistocene extinctions. The extant Australian fauna is characterized by animals that are readily adapted to the harsh range of temperatures and unreliable rainfall of the vast majority of the Australian continent, as well as its overall low biomass (implying low food resources).

Marsupials have the ability to abort young without incurring a high energy loss to the mother, or in some cases can hold their developing foetus in a suspended stage until times of abundant food allow its continued development leading to birth. In other cases, such as the koala, Australian animals have also reduced their brain size in evolving to the current species, in order to save energy and survive in such a low nutrient landscape.

CRISPIN LITTLE

Breathing sulphur: life on the abyssal black smokers

We were alone. Where, I could not say, hardly imagine. All was black, and such a dense black that, after some minutes, my eyes had not been able to discern even the faintest glimmer.
JULES VERNE, 1870

Hydrothermal vent communities were first discovered in 1977 near the Galápagos islands. Since then they have radically altered our view of life in the darkness and cold of the deep sea, in part because the primary energy source for the vent ecosystems is not the Sun, as for most of life on Earth, but sulphur.

Diving to the abyss

Some of the best-studied hydrothermal vents are those on the East Pacific Rise (EPR), an ocean spreading centre (p. 61), close to the equator at 9° North, off the west coast of central America. At this point the vents are at a depth of 2,500 m (8,200 ft), where the ambient conditions are challenging – pressures are a crushing 250 bars and temperatures a chilly 2°C (35.6°F). Even to get to these vents is an expensive and logistically complex undertaking, requiring a large oceanographic research vessel and a submersible capable of diving to abyssal depths, such as the famous Human Occupied Vehicle *Alvin*.

It takes two days' steaming due south from Mexico to get to the spot, and once on site there is nothing to see but open ocean. The only indication that you have arrived comes from the echo-sounder showing that the sea bed has risen almost a kilometre from the Pacific abyssal plain to the top of the EPR. It then takes two hours to reach the bottom in a submersible, a journey that

Above Alvinella pompejana *is one of the most thermotolerant animals known and builds its tubes on the side of the hottest vent chimneys.*

Left *The robotic arm of the deep-sea submersible* Alvin *being used to measure the temperature of vent fluid gushing out of a black smoker chimney.* Alvinella *tubes surround the vent orifice.*

begins for the first few minutes in sunlight, but is thereafter in total darkness. Once on the seafloor, a 100-m (328-ft) wide chasm, called the axial summit trough, runs along the centre of the ridge. Here, the submersible lights reveal an expanse of almost completely black, glassy lava, seemingly devoid of life.

Animals and smokers

Numerous pure white crabs and squat lobsters are the first signs that active hydrothermal venting is close. Then an amazing sight comes into view: huge fields of tightly packed vent mussels and white giant vent clams up to 20 cm (8 in) long, as well as dense clusters of *Riftia*, a 2-m (6.5-ft) long vestimentiferan tube worm. Swimming and crawling around the bivalves and worm tubes are numerous other strange creatures – fish, crabs and shrimps, and, on closer inspection, a great diversity of small snails.

All this life is centred on the hydrothermal vent chimneys, rising up metres into the water column and some pouring out black smoker fluid. The submersible pilot operates an external mechanical arm to place a temperature probe into the top of one of the active chimneys, recording a value of around 370°C (698°F), hot enough to melt lead (and *Alvin*'s perspex viewing ports). There are even animals living here: the 'Pompeii' worm, *Alvinella*, constructs tubes directly on to the outside of the chimneys. The most thermotolerant animal that has yet been found, it regularly endures temperatures of around 60°C (140°F).

Chemosynthesis

When the hydrothermal vents were first discovered it was not clear why there was such an incredible abundance of life around them and not on the adjacent seafloor. The energy source was obviously not organic carbon floating down from the surface waters as marine snow, as this would not explain the clustering of anomalously large animals around the vents. It had to be something associated with the hydrothermal fluid itself.

Research in the subsequent 30 years has confirmed that this is indeed the case, and that the vent ecosystem is powered by geochemical sources, most importantly hydrogen sulphide (H_2S) from the vent fluid itself. This, and other reduced compounds, is combined with oxygen from seawater by prokaryotic microorganisms (bacteria and archaea) in the primary production

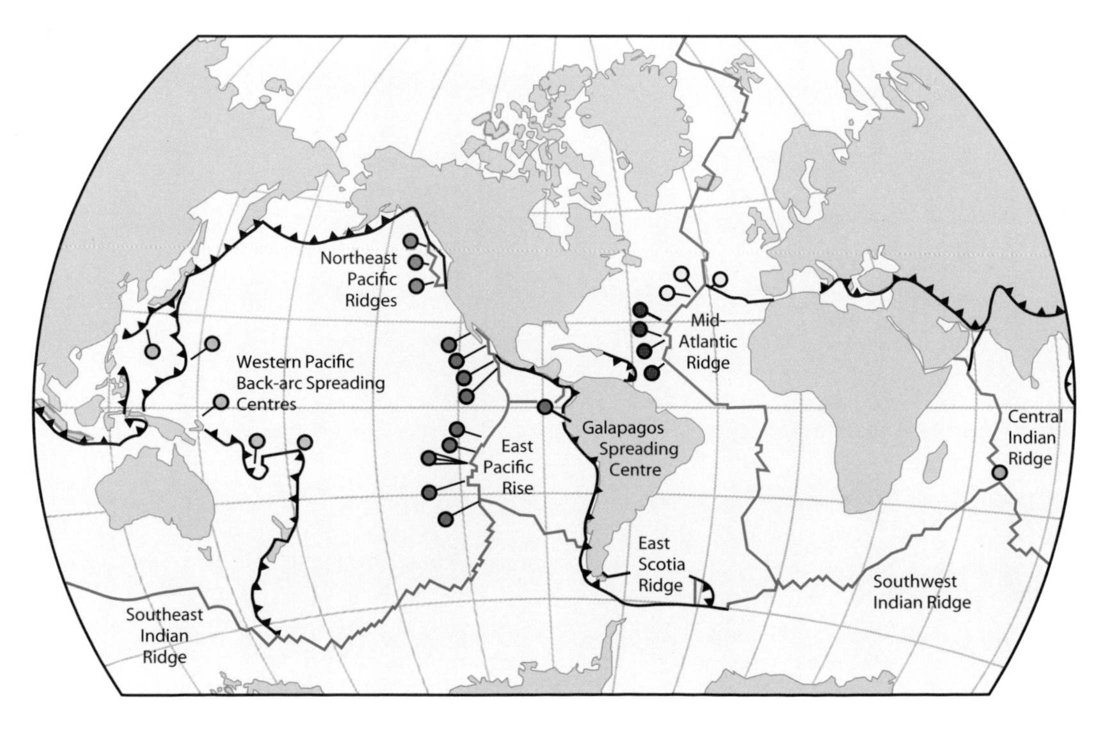

Map of the world's spreading ridges and subduction zones. The coloured circles show the hydrothermal vent sites that have been discovered so far. The colours indicate the six currently recognized vent community biogeographic realms.

step, a process called chemosynthesis. All rock and animal surfaces are coated with the prokaryotes, sometimes in such abundance that they form mats thick enough to be visible to the naked eye, which are grazed by many of the vent animals, in particular the snails and crustaceans.

Symbiosis

A strange and important relationship exists between the chemosynthetic prokaryotes and the vent animals. It takes the form of an internal symbiosis (endosymbiosis), with the largest and most dominant vent animals having bacterial endosymbionts that provide them with most, and in some cases all, of their energy requirements.

The most dependent animal-bacterial relationship is found in the vestimentiferan tubeworms, such as *Riftia*, as the adults have no digestive tracts, instead having a large organ called the trophosome that takes up a large proportion of their body. This is packed with millions of sulphide oxidizing bacteria, which the worms must provide with both oxygen (from seawater) and hydrogen sulphide (from vent fluid). These compounds are taken up into the blood stream across the highly vascular plume (the red structure that sticks out of the tube) and bind separately on to modified haemoglobin molecules, which prevent the compounds from combining.

The blood carries the haemoglobin molecules to the trophosome where the hydrogen sulphide and oxygen are taken up by the bacterial symbionts for oxidation, releasing the energy needed to build carbohydrates for growth. Excess symbiont cells and/or their metabolic by-products are then absorbed by the worm in return. The worm thus receives nutrition, while the advantage of this arrangement for the symbionts is that they are provided with optimal growing conditions without competition from other prokaryotes.

Molluscs and shrimps

A similar association with endosymbiotic sulphide oxidizing bacteria is found in the vent clams (*Calyptogena*) and mussels (*Bathymodiolus*), but in these cases the symbionts are in cells called bacteriocytes located in the gill tissue. *Calyptogena* is entirely dependent on its bacteria for nutrition, but *Bathymodiolus* still retains some filter-feeding capability. Some species of *Bathymodiolus* have both sulphide oxidizing and methanotrophic bacteria in their gills, thus enabling them to make use of methane in vent fluid as an additional energy source. This dual symbiosis is also found in two species of large provannid gastropods found at vents in the West Pacific. The Mid-Atlantic Ridge

The plume of the giant tubeworm, Riftia, poking out of the white tube, is a highly vascular gas exchange organ. It can rapidly be withdrawn if a predator, such as a crab, approaches.

A hydrothermal vent scale worm found living among giant tubeworms and mussels. A diverse fauna of polynoid worms inhabits hydrothermal vents, some thriving in hot water near the tips of black smokers. Scale worms are some of the many smaller animals that make up the bulk of the diversity at vent sites.

vent communities are dominated by enormous swarms of the vent shrimp *Rimicaris exoculata*, which has several extraordinary adaptations to the vent environment.

The interior surfaces of the flaring lower carapaces of adult *Rimicaris* host colonies of filamentous sulphide-oxidizing bacteria. These are cropped regularly and used as food by the shrimps, a relationship that has been described as 'exosymbiosis'. To feed their bacteria farms the shrimps must swim in the hot vent fluid, rich in hydrogen sulphide. *Rimicaris* can sense infrared emissions from the vent fluid using highly modified visual organs containing specialized pigments that lie underneath a transparent area of the dorsal carapace.

Geographical variations

Although the ridges formed where plates are spreading apart in the East Pacific, the Indian and Atlantic Oceans are all connected, there are distinct differences in their vent faunas, for reasons that we do not yet understand. For example, the Mid-Atlantic Ridge vent communities have no vestimentiferan tube worms and are instead dominated by *Rimicaris* and vent mussels. The isolated West Pacific vents are host to small tube worms, large provannid gastropods with symbionts, mussels, but no *Rimicaris*. Indian Ocean vent communities have a mixture of West Pacific and Atlantic vent animals with *Rimicaris* swarms, large provannids, mussels, but no vestimentiferans. There are certain to be other differences in yet-to-be discovered vent faunas along unstudied ridges, particularly at high latitudes, where oceanographic research is logistically difficult. The next 30 years will surely prove as exciting as the first 30 for hydrothermal vent research.

On the Mid-Atlantic Ridge vent shrimps replace the alvinellid and vestimentiferan tubeworms of the east Pacific as the most abundant animals at the vent chimneys.

PENELOPE BOSTON

Extremophiles and life on other planets

Imagination will often carry us to worlds that never were.
But without it we go nowhere.
CARL SAGAN, 1980

Looking for life on other planets has been a cherished dream since we first began to think about venturing into space to find other worlds. But our original conception of Martian monsters and Moon maidens has been replaced by the realization that we have 'aliens' right here on Earth in environments so extreme that we could not possibly endure them unaided. In the late 1960s, scientific attention was focused on the microorganisms growing colourfully in the extremely hot waters of geisers and thermal pools in Yellowstone National Park (p. 23). Then, in the late 1970s, biological communities were discovered around deep sea hydrothermal vents (p. 167), prompting the radical notion that deep beneath the ocean surface a chemical energy source (hydrogen sulphide) could run a living community without energy directly from the sun.

This revolutionary discovery opened the imaginations of microbiologists and other biologists to a wide array of other habitat possibilities. Around the same time, the first extraterrestrial life detection experiments flew to Mars aboard the two Viking Landers. The connection of extreme life forms here on Earth and our need to plan the search for possible lifeforms in radically different planetary environments formed a combination that galvanized us into a new era of biology.

Astrobiology (the study of life beyond Earth), geomicrobiology (life in the Earth's rock environments) and extremophile biology all emerged from this synthesis. At the same time, technology advanced rapidly to enable us to see organisms with powerful electron microscopes, and to use DNA to reveal the full microbial diversity in nature previously only glimpsed in the small percentage of microbes (much less than 1%) that we can actually grow in the laboratory.

What is an extremophile?

'Extremophiles' are organisms that occur in any extreme environment, for instance in acid conditions at one extreme and alkaline at the other. Many of them can endure saturated salt solutions, temperatures above the boiling point of water, dry cold dehydration way below freezing point, and even enormous doses of ionizing radiation and intense ultraviolet radiation. But the underlying idea of extreme life is a slippery concept. We use ourselves and our own preferences as a notion of what is extreme or normal, but does the term 'extreme environment' have any real meaning?

For our type of life chemistry – carbon-based biomolecules dissolved in water – there are real

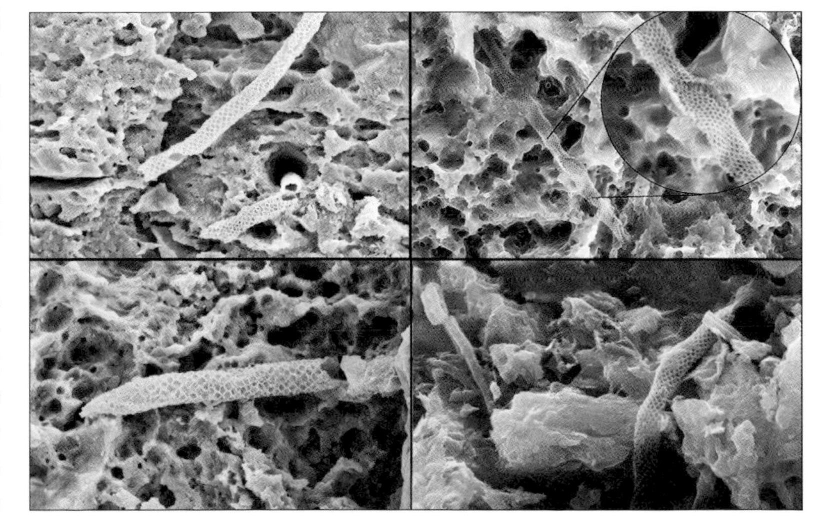

Scanning electron microscope images of a 'bacterial garden' growing on bare lava walls in a cave in the Cape Verde islands. Each measures around 1 to 2 microns in diameter (a micron is one-millionth of a metre).

Scientists at work studying extremophiles in a sulphuric acid saturated cave in Mexico.

limits. The freezing and boiling points of water, acid or alkaline extremes and their effects on the structure and function of biomolecules, and the presence or absence of oxygen all provide the settings within which Earth life must operate. High-energy radiation can destroy biomolecules, and toxic metals or other compounds can equally ruin the chemistry of life. However, in spite of the extremes of physical and chemical conditions in different environments, life is very resourceful and has had a great deal of time to develop clever means to bend the rules, and ooze, squirm or wiggle its way around these apparently rigid physical and chemical boundaries.

For example, thick microbial mats live in caves dripping with sulphuric acid. Others thrive in intensely alkaline surface ponds. We can scoop up handfuls of tiny, colourful photosynthetic bacteria happily living in saturated salt solutions. Bacteria that can endure huge doses of ionizing radiation were accidentally discovered while experimenting with radiation sterilization of food in the 1950s and they are now found living in nuclear power plant cooling ponds.

In the rock environments on Earth's surface and in its caves and subsurface fractures we find organisms deriving their life energy from apparently unappetizing inorganic materials such as manganese, iron and sulphur compounds, or methane and hydrogen gases (p. 151). These surroundings may seem 'extreme' to us, but for the native microbes of these environments, they are ideal.

Extremophiles on Earth and the search for life in the Universe

In the mid-1970s, when the first missions went to Mars to look for life, most of the organisms that we now call 'extremophiles' had not been found. If they had, the missions might have been designed radically differently. The study of extremophiles on Earth is broadening our understanding of what is possible and what is not. When we go to other worlds in future, perhaps we will find an entire planet where the inhabitants evolved under conditions similar to those on Earth that we consider 'extreme'.

The variety of habitats on planets increased first with the discovery of potential liquid oceans under the icy crusts of some of Jupiter's moons (Europa and Enceladus). In the past few years, astronomers have found hundreds of planets around stars beyond our own Sun (exoplanets) and more are being found every month. Many of these planets are radically different from any in our own solar system. Some are the size of Jupiter, but orbiting as close to their star as our planet Mercury; others are in orbit around suns whose radiation output greatly exceeds that of our Sun. Because life is a planetary phenomenon, the vast expansion of our understanding of what a planet can be is also broadening our view of what life might be.

How many of these exotic extra-solar planets harbour life we cannot yet tell, but they will be at the forefront of astrobiology in the future. Our galaxy is beginning to open up to us in terms of life, and extremophiles here on Earth are the key to that understanding.

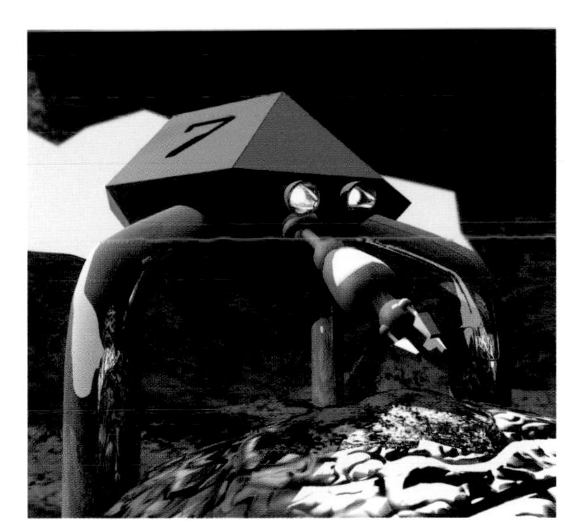

Above *The Atacama Desert in Chile is one of the driest places on Earth – some parts receive rain only once in a few decades. Rosa Bonacorsi takes data in the Atacama next to rocks which have desert varnish organism communities living on them.*

Left *An imaginary 'microbot' on a future extraterrestrial life detection mission investigating a colourful microbial mat on another planet.*

Extremophiles and extraterrestrial intelligence?

The presence of extremophiles on Earth is preparing us to look for microbial life on other planets, and perhaps even multicellular organisms. But we will probably find that intelligent life is much rarer. On Earth, it appears to have taken several billion years from the apparent inception of life until we arose as an intelligent species. We have only been in existence for a relatively short period of time, and manifesting a complex technology for even less.

SETI (the search for extraterrestrial intelligence) has to rely on the detection of another intelligent lifeform that is actively broadcasting in the radio band frequencies. As if this were not long enough odds of making successful contact, we know that if an intelligent species arose in a radically different environment from ours, we might not even be able to recognize it or communicate with it.

If our own planetary development pattern is typical, then the discovery of the existence of extremophile microorganisms on another planet may not say much about whether that planet is or will become intelligence-bearing. However, our current understanding of intelligence is even more limited than our understanding of life itself. There may be many more radically different strategies for development of intelligence than we can imagine at this time. The deeper lesson that we have learned from our discovery and study of extremophile microorganisms is that nature can produce results far different from those that our limited imaginations expect.

Plants & Animals

The diversity of life never ceases to amaze. And yet biologists are still very uncertain about how many species there are on the Earth. Some 1.8 million species have been described and named, but that is probably only a small fraction of the numbers out there. Every year, thousands of new insect species are discovered, and there seems to be no end to their diversity. This has suggested that we may share the Earth with anything from 10 to 100 million other species. Even new species of mammals and birds are still occasionally identified.

Many collaborative studies have been set up to document all the Earth's species, but it may be an endless task. It has been suggested instead that we should focus efforts on a sampling approach – concentrating on certain areas, especially those where life is most diverse, and seeking to understand their biodiversity in detail.

Size is a key biological attribute. The size of a plant or an animal determines where it can live, how large its populations can be, what food supplies it requires, and its vulnerability to predation. For most of the history of the Earth, all species were microscopic, and indeed much of life today is invisible to us. Microscopic species can exist in vast numbers and, although huge numbers die every day, they can breed remarkably fast. So, it is a mystery in some ways why larger, multicelled plants and animals evolved at all.

Even the animals and plants that we are most familiar with vary remarkably in size. It is easy to make a case that small is best: cockroaches are not large, but they are very tough and breed fast, and they are nearly impossible to eradicate. Large

African elephants are the largest land animals living today. Biologists are fascinated by limits that seem to exist on size and what determines them.

Chameleons have an amazing ability to change colour to match their surroundings. Research has now shown how this process works, and yet the speed and precision of the colour matching are still a mystery.

animals, such as elephants and whales, exist in small population sizes and are vulnerable to extinction from food shortage or hunting. Of course, in any competitive interaction, the larger animal usually wins, so perhaps bigger is better after all.

Biologists are still investigating why organisms vary so much in size. It is likely that there is no rule: evolution seeks opportunities, and the world has space for small and large organisms. Perhaps surprisingly, the largest living organism is not an elephant, a whale or a giant redwood tree, but is in fact a fungus that covers 15 ha (37 acres) and is estimated to weigh over 100 tons.

There are many methods of animal locomotion. Moving on legs, flying through the air or swimming through water are all specialized skills, and engineers often marvel at the efficiency of animal design. No inventor has yet replicated the beauty and skill of a flying bird – and perhaps never will. Experimental studies can demonstrate the biomechanical limits of plants and animals, but also just how good they are at what they do, such as the ability of some mammals to dive to great depths for long periods, withstanding cold and high pressures, all without access to air.

It is often said that engineers can learn from animals. This is sometimes true, but not always. Leonardo da Vinci famously produced drawings of endless mechanical contrivances based on his observations of anatomy and locomotion. Biologists, though, often take a more analytical view: insects and birds flap their wings, but flapping is a laborious and inefficient way to move through the air. A rotating wing, such as on a helicopter, is a much more efficient design for a single locomotory element (aeroplanes use fixed wings for lift, and must have propellers or a jet engine to drive them forwards). So why has no animal ever invented a rotor? Evolution is constrained by the materials available, and cannot leap to an entirely new design as an engineer can. Birds and insects evolved from ancestors with legs, and the wings had to be modified legs of some sort.

When studying animals it is easy to be anthropocentric, and this is particularly true of vision and colour. Biologists may speculate and experiment on animal camouflage and colours used for warning or display, but if the animals see the world differently, we may be missing the point. The discovery that birds see ultraviolet light, hidden to us, revealed a new world of colour patterns in plants. Dogs have been studied most, and it turns out that, as had long been suspected, they do not see detail as well as we do (so the moving images on a television are probably a blur to them), but they can see much better than we can in twilight conditions. New evidence also shows that they do see colour, though in a limited way, contrary to a long-held myth.

MICHAEL J. BENTON

Estimating present-day global biodiversity

The innumerable species, genera, and families of organic beings, with which this world is peopled, have all descended, each within its own class or group, from common parents, and have all been modified in the course of descent.
CHARLES DARWIN, 1859

In Darwin's day, naturalists thought there were perhaps 10,000 or 100,000 species on the Earth. And it was this vast diversity of life (or biodiversity as it has become known, using E. O. Wilson's felicitous term from 1992) that stimulated Darwin to seek a natural, rather than a supernatural, explanation. As a young man, circumnavigating the globe on board the *Beagle*, Darwin had observed that there were far more species on Earth than were required: islands had their own species of birds, flowering plants or turtles, similar to those on neighbouring islands, but also different and incapable of interbreeding. Species could clearly arise relatively fast (p. 123), and maintain their distinctness, and there seemed to be no limit to this diversity.

Biologists today still seek to estimate the scope of biodiversity – these estimates range over at least an order of magnitude, from perhaps 2–3 million species at the lower end, to 30–100 million at the upper end. The lower estimates represent summaries of the number of species that have actually been documented. For example, Robert May has calculated that some 1.7–1.8 million species of modern microbes, plants and animals have been named so far by taxonomists, and that figure must be a minimum estimate of current biodiversity.

A new species of clouded leopard discovered in Borneo in 2006. Genetic analysis showed the island species is distinct from its mainland relative.

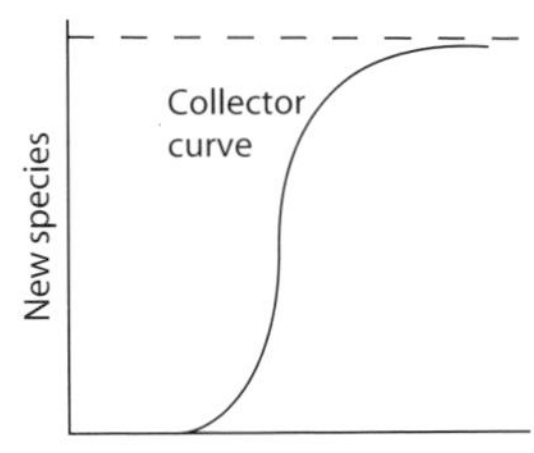

A collector curve for well-known groups such as birds and mammals: apparently nearly all species that exist have now been named and described, so the rate of discovery levels off after initially rising steeply.

New species are named all the time. Each year, on average, systematists add one or two new species of mammals and three new species of birds to the lists (based on estimates of taxonomic activity from 1900 to 1975), as well as dozens of newly named microbes, fungi, plants and marine animals, and some 7,250 new species of insects. Can we use our knowledge of the rate of naming new species to predict how many will eventually be named and so estimate the actual diversity of life on Earth?

The collector curve

If the additions are the same each year, the pattern of increase would be linear. However, the pattern may be more complex. It could be argued, for example, that the discovery of new species would follow a collector curve, i.e. a pattern of rapid rise when new species were being discovered all the time, followed by a slowing-down of the rate as the number of described species approaches the true figure for diversity.

For birds or mammals it is likely that the levelling-off phase, the asymptote, was reached in about 1900, and that the current slow rate of discovery of new species means that virtually all species of birds and mammals have in fact been discovered. This evidence could lead to a modest estimate of perhaps 3 million species for true current biodiversity. However, a little reflection suggests that this comfortable impression is probably very far from the truth.

It is misleading to extrapolate numbers from the groups of living organisms that are most fully documented, such as birds and mammals. For insects, microbes, fungi, parasites, meiofauna (creatures that live in sands and soils), deep-sea organisms and many others, taxonomists find new species wherever they look. In a random soil sample, a taxonomist of bacteria can discover many hundreds of previously unrecognized species; and in a sample of mud from the deep ocean floor a marine biologist may find dozens of undescribed species.

For such poorly understood groups, the rates of discovery and description depend only on the time and effort expended by taxonomists. This

BUG BOMBS AND SPECIES RICHNESS

In the 1980s, in a famous sampling experiment, the beetle expert Terry Erwin decided to sample all the bugs from a single species of tropical tree, and use this figure as a basis for extrapolating an estimate of the global biodiversity of tropical beetles. He sampled the entire arthropod fauna from the canopy of a number of trees of the species *Luehea seemannii* from Central and South America by setting off 'bug bombs' under the selected trees. These devices pump powerful insecticide into the tree, and Erwin then collected and classified all the dead arthropods that fell to the ground.

Erwin estimated that 163 species of beetles live exclusively in the canopies of *Luehea seemannii*. There are about 50,000 tropical tree species around the world, and if the number of endemic beetle species in *Luehea seemannii* is typical, this implies a total of 8.15 million canopy-dwelling tropical beetle species in all (163 x 50,000). This figure excludes forms that live in more than one tree species. Beetles typically represent about 40% of all arthropod species, and this leads to an estimate of about 20 million tropical canopy-living arthropod species. In tropical areas, there are typically twice as many arthropods in the canopy as on the ground, and this gave an estimate of 30 million species of tropical arthropods worldwide.

Erwin's work was challenged in 2002, when it was shown that tropical beetles were not so fussy, and that many lived happily on a number of closely related trees. So, in Erwin's calculation, the figure of 163 exclusive canopy dwellers, is cut to about one-tenth, and the global biodiversity of beetles downsized from 8.15 million to less than a million. The total global diversity of insects is then estimated at 3–6 million.

Right above *A bug bomb is set off beneath a tropical tree. Fine netting slung beneath will catch all the insects and other creatures, allowing a detailed census to be taken.*

Right below *One of the beetle species collected by Terry Erwin in his campaign to estimate the global biodiversity of tropical beetles. This one is* Ellipticus dorbignyi, *from the 'Pleasing Fungus Beetle' family.*

Tropical reefs are famously diverse. Here, in the Red Sea, hundreds of species of corals, sponges, shellfish, arthropods and fish live in busy harmony – a rich fauna because of the warm water and abundant food.

means that it is impossible to predict when the curve will show an asymptote, and consequently final totals for many groups cannot yet be estimated. Regrettably, then, estimates of actual global biodiversity based on current records can only work for such heavily studied groups as birds and mammals, but not for those vast sectors of life where taxonomic work has barely begun.

Synonymy and sampling

There is also another factor that can mean that estimates of the number of species are actually too high. It is suggested that perhaps 20% of 'new' names are given to species that have already been named – the so-called 'synonymy problem'. There are many examples of a taxonomist publishing a new name for an exotic specimen – and finding out only later that the same species had already been named by someone else, perhaps in another institute, in another country, even in another language. Strict revisions of taxonomic work continue all the time, and they identify an average synonymy rate of 20%.

Biologists have tried another approach to estimate biodiversity. For highly species-rich groups, it might be more appropriate to carry out a detailed survey, find the maximum diversity, and then extrapolate worldwide. A well-known experiment concerns tropical rainforest beetles (see Box). The estimate came as a considerable surprise when it was published: 30 million species of tropical arthropods must imply a global diversity of all life in the region of 50 million. Some biologists even talked of figures of 100 million or more.

Work published in 2002, however, suggested that the figure of 30 million species of tropical arthropods was an overestimate, and was probably closer to 5 million. Global biodiversity then falls from 50–100 million species to 10–15 million species. This is still an astonishing total, and taxonomists have a long way to go before they have catalogued it all. It has commonly, and rightly, been said that we may be killing off species without our knowledge – until they are identified and named, we don't know when and where they existed, nor do we notice if they cease to exist.

PETER MAYHEW

Why are insects so diverse?

On examination, this problem turns out to be unexpectedly complicated.
G. G. SIMPSON, 1949

The insects comprise a group of six-legged arthropods with three major body parts: head, thorax and abdomen. Any natural history museum worth its salt will have a display cabinet full of insect specimens of many shapes, sizes and colours, testifying to their staggering diversity. Insects make up over half (about 920,000) of all species described to date. Explanations for this diversity must address two broad issues. The first is rates of both speciation (p. 123) and extinction, since the diversity of insects today, as with all 'natural groups' of organisms, derives from a single ancestral species through the opposing processes of speciation and extinction. The second issue is which features of insects have contributed to these rates.

Biologists have three major sources of evidence on insect speciation and extinction rates: first, the fossil record; second, the relationships between living species as depicted by evolutionary trees; and third, our historical records of the decline and extinction of modern species.

Evidence currently suggests that rates of speciation in insects have generally been quite modest, and certainly comparable with those of many other less-diverse groups. The fossil record shows that each insect family (groups of closely related species) gave rise to only about one-tenth of a

A collection of beetles laid out and labelled for the Institute of Biodiversity, Costa Rica, showing their impressive variety.

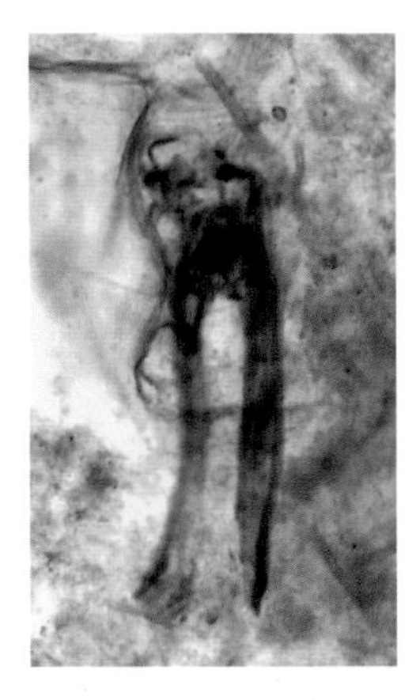

The oldest insect fossil remains: jaws of Rhyniognatha hirstii, *found in Rhynie, Scotland.*

new family per geological stage. In mammals the value is about triple that. However, the insects are a lot older than the mammals, and age can have a huge influence on diversity. The earliest known insect has been found in Scotland and is around 400 million years old. A group of organisms speciating at a modest but constant rate over that length of time will soon generate large numbers of species, because groups of species, like populations, will grow exponentially if unchecked. Indeed, there is little evidence that the numbers of insect groups have been held in check over time: the numbers of fossil insect families approximately doubled in the last 65 million years.

Given modest speciation rates, insects should have modest extinction rates, and the evidence generally supports this. Of the insect families that coexisted with the last of the dinosaurs, over 95% are still with us today, compared with only about a quarter of terrestrial vertebrate families. Most of the insect species known from the last glacial period in Britain are also still alive. Current extinction rates in insects are rather poorly known because most are not well recorded, but they seem to vary considerably.

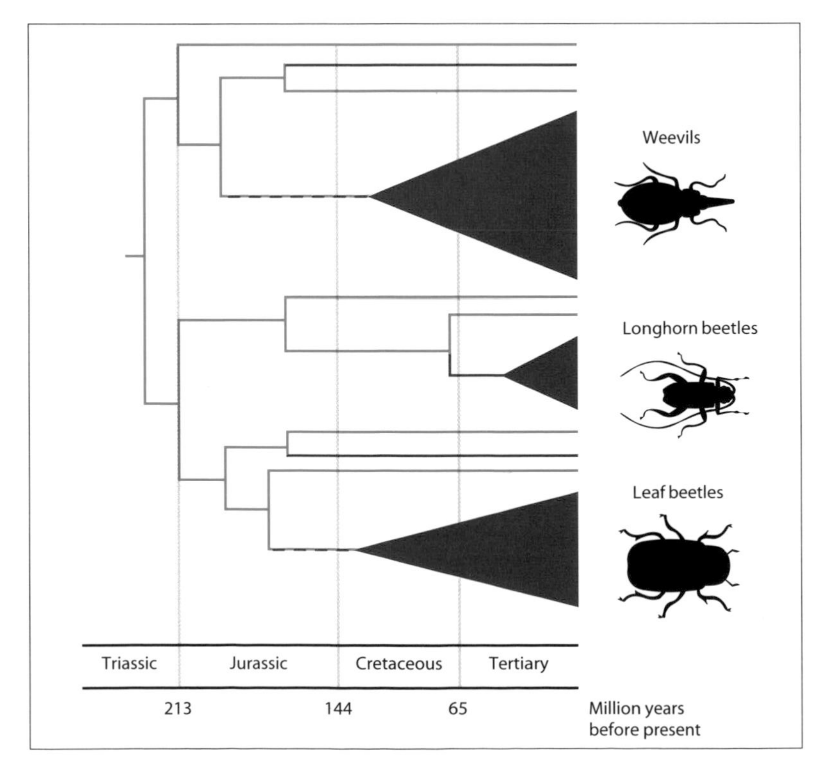

The radiation of plant-feeding beetles. The red branches feed on flowering plants, while their closest relatives in green do not. The 'red' lineages are much more species rich.

Factors in speciation and extinction rates

What determines insect speciation and extinction rates? Arguably, two of the most influential characteristics have been the development of wings, especially folding wings, and herbivory (feeding on plants), in particular feeding on flowering plants. In both these cases, important evidence comes from comparing the species richness of groups with the innovation, as against the species richness of their closest relatives without.

Only three groups of animals have wings (birds and bats are the other two), but all three contain substantially (and, collectively, unexpectedly) more species than their closest relatives. However, the most primitive winged insects (dragonflies and mayflies) are relatively species poor; very diverse groups only arose later, after folding wings, and possibly other innovations, developed. Interestingly, a significant predictor of current extinction risk in many insect groups is their dispersal ability, which is often linked to flight. So it is likely that the development of wings has played a major part in the low extinction rates of insects, allowing them to colonize new habitats and recover quickly from disturbances.

In contrast to flight, the ability to feed from plants has evolved on numerous occasions within the insects. Most groups that evolved herbivory have become more species rich than their closest relatives, but the most impressive examples perhaps come from some of the beetle groups that feed on flowering plants. Just three such beetle groups contain over 100,000 described species. It is likely that herbivory enhanced or sustained diversification rates by opening up a diversity of niches to exploit.

Other intriguing findings are less easy to interpret. Across animal species as a whole, there are many more small-bodied than large-bodied species, leading us to expect that small body size might explain rates of diversification in insects. However, the statistical evidence currently, and mysteriously, does not support this. So is body size genuinely not influential, or are we not conducting the right analyses? Another curious fact is that the oldest surviving insect orders, the bristletails and silverfish, are no more species rich than their

closest non-insect relatives: only later subgroups of insects became really diverse. It would therefore, ironically, be more correct to refer not to the diversity of 'insects' at all, but rather to some, as yet not confidently identified, subdivision.

Where next?

Current studies of insect diversity have to work with (often very) incomplete data on fossil and current diversity, on evolutionary relationships, and current population status. Improvement in these areas will increase our confidence in providing answers. Current outstanding questions of importance include the effect of body size on diversification; which of the many ultimate factors have been the most influential, and how they depend on each other; and whether and how they have exerted their influence on speciation and extinction rates. From all these sources we can hope for exciting future developments in our understanding of this so-evident and so-intriguing mystery of the natural world.

Above *A dragonfly, one of the primitive groups of insects that cannot fold their wings, and that are relatively species poor.*

Left *A true bug (*Dysdercus albofasciatus*), a member of the group of insects that can fold their wings, and have become very species rich.*

43 Why are some organisms small and some large?

The most obvious differences between different animals are differences of size, but for some reason biologists have paid singularly little attention to them.
J. B. S. HALDANE, 1927

Opposite *A blue whale, the largest animal ever to have lived on the planet. Why are all whale species big?*

Haldane was a genius who had great insights; here is the rest of the above passage: 'In a large textbook before me I find no indication that the eagle is larger than the sparrow, or the hippopotamus bigger than the hare, though some grudging admissions are made in the case of the mouse and the whale. But yet it is easy to show that a hare could not be as large as a hippopotamus, or a whale as small as a herring. For every type of animal there is a most convenient size, and a large change of size inevitably carries with it a change of form.'

All 80-odd species of whale are big – the blue whale is the largest animal that has ever lived. So *why* are none as small as a herring? Why are plants on land enormous – trees for example – but those in the sea mostly microscopic? And it seems that close relatives of ours who went extinct only 12,000 years ago, *Homo floresiensis*, evolved a dramatically smaller size when they colonized an island in Indonesia – why? Before we answer these questions, we have to look at the big picture.

Why are any organisms big?

For the first six-sevenths of the history of life on Earth, all organisms were microbial. Large – visible – organisms only evolved around 600 million years ago. Even today, most of life consists of organisms too small to be seen with the naked eye, such as bacteria. In fact most animals too are invisible – numerically, four out of five animals is a microscopic nematode worm. So we could ask first: why are *any* organisms large?

Around a billion years ago, the oxygen concentration in the oceans rose to high levels. This allowed the evolution of multicellular organisms that rely, at least in part, on the diffusion of oxygen through tissues, which, in turn, requires a high oxygen pressure. The stage was now set for the ecological predator-prey relationship to drive an evolutionary arms race (p. 30). If you want to be a predator it helps to be bigger than your prey and to avoid predation it pays to be bigger as well.

So, hypothetically, but plausibly, predator-prey arms races drove the initial evolution of large body sizes. This cannot go on indefinitely though, and other factors come into play. Elephants enjoy immunity from natural predation at their current size, which is fortunate, as they are probably reaching the upper limit of their body size. Why? As bodies get bigger the diameter of the legs needed to support them also gets bigger (see also p. 190). An elephant twice as tall but with all its proportions the same would weigh eight times as much (height x length x breadth), but the surface area of a cross-section of a leg would only be four times as great and so each square inch of leg would be supporting twice as much weight as before. The bones, however, could not take this pressure, and the legs would need to be proportionally wider. An elephant's legs are already very wide with respect to its body compared to, say, a gazelle. Make them proportionally even wider and soon we no longer have a viable animal.

The question we have just addressed is in fact 'Why does size matter?', but we have not answered the question of why, in an evolutionary sense, elephants 'want' to be big in the first place. A bird species might 'want' to be big in order to catch and kill smaller prey and fly off with them.

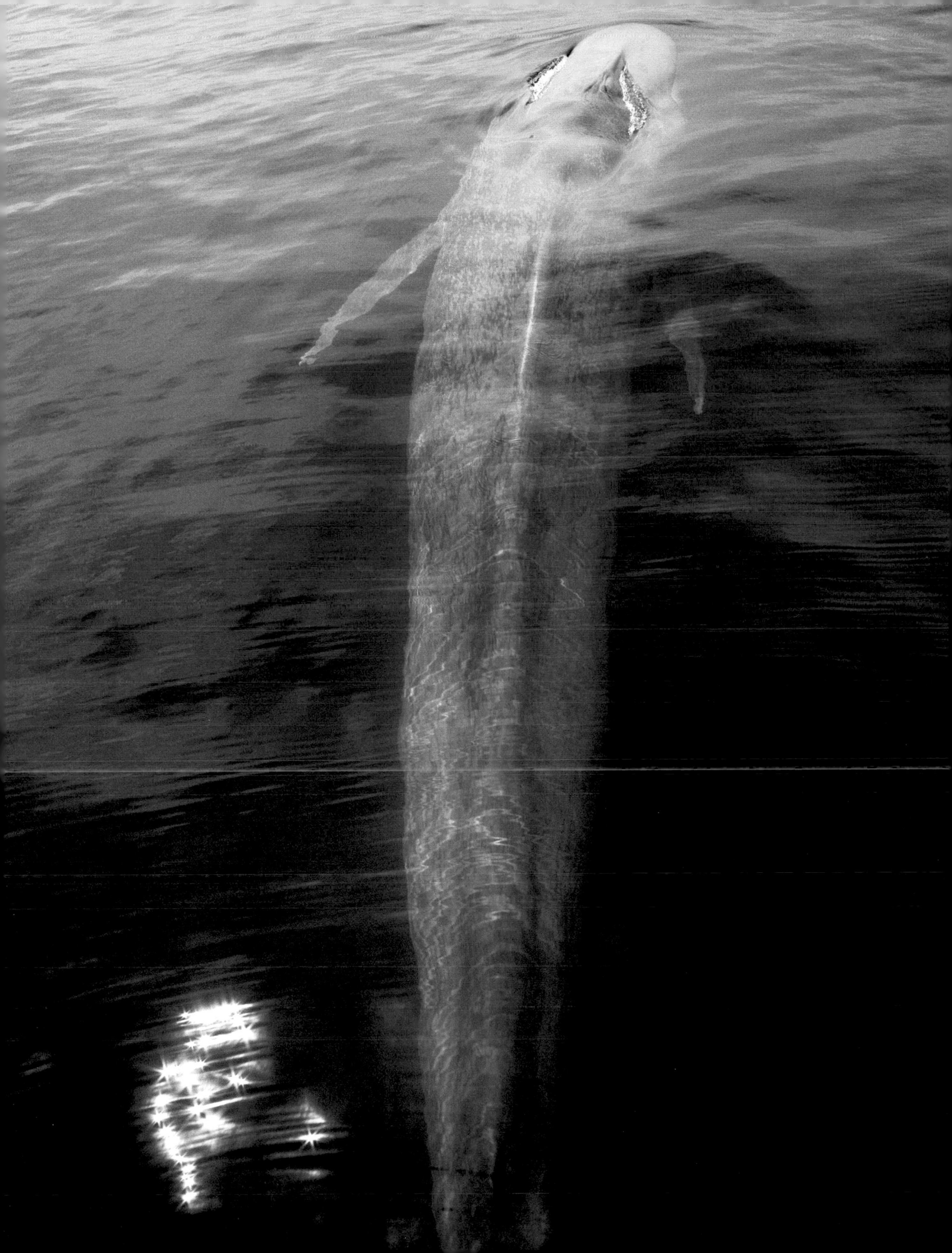

But, with increasing size, muscle strength does not increase as fast as muscle volume – weight – so big birds like eagles rely mainly on soaring rather than powered flight, and a bird big enough to fly off with an elephant is impossible. So size *matters* to the mode of flight – soaring versus flapping – but the *reason* eagles are big is to kill smaller prey and fly off with them.

Spherical cows

A joke about the other-worldliness of mathematicians has one starting a conversation thus: 'Consider a spherical cow…'. It is in fact very informative to consider organisms as spheres, as this allows us to use basic geometry to understand some important biology. As a sphere gets bigger, its volume increases more rapidly than its surface area: the volume is proportional to r^3 (r = radius) and the surface area to r^2. So, if you double the radius of a sphere the surface area increases four times and the volume increases eight times.

Elephants are probably reaching the upper limits of their body size – their legs are already very wide in proportion to their body size when compared with antelopes, and could not get much wider.

A warm-blooded animal produces heat in the volume of its body and loses it through its surface. So a bigger animal can maintain its body temperature on proportionally less food than a smaller one. A mouse eats about a quarter of its body weight every day, whereas we – most of us – eat much less. Mice could not eat enough food to keep warm in an extremely cold environment, which is why the Arctic is dominated by large animals like bears. And this is also why whales are big – the oceans are cold. A whale the size of a herring would freeze to death. If you want to be a warm-blooded animal in the sea you need to be big. Tuna, which are warm blooded, are large, as are warm-blooded sharks.

Another arms race occurred between plants competing for light. The light that one plant absorbs is not available to its neighbours, so plants do not want to be shaded. One way around this is to become a tree and grow taller than your neighbours who then 'want' to grow taller than

Eagles, such as this bald eagle, are big enough to catch and carry off prey, but their large size means that they rely mainly on soaring, rather than powered flight.

you and so on. So why are there no 'trees' in the ocean? It would make sense for a plant to have its leaves on the surface to catch the light and to dangle roots in the deeper layers where most of the nutrients are. But there are two problems. One is structural – the ocean waters are turbulent and exert enormous forces that trees could not withstand. The other is that a tree on land germinates in a suitable place and stays there – a tree in the ocean would end up, most probably, in an unsuitable location.

Mysterious shrinkage

There are many things about size that we understand, but there are others we do not. 'Island dwarfism' is a widely observed phenomenon that is not well understood at all. When animals colonize islands, large animals evolve to be smaller and small ones evolve to be bigger (see also p. 154). Now-extinct elephants on Sicily were the size of an average dog. This happened to us too: after we colonized the island of Flores in Indonesia it seems we evolved to be 1 m (just over 3 ft) tall, with brains the size of a grapefruit. The remains of Flores Man – who lived well into the time that *Homo sapiens* had colonized Asia and Australia – have been found alongside tools and the skeletons of giant rats and the pygmy form of the elephant *Stegodon*.

Why does this occur? Perhaps it is because predation is much less of a problem for species on islands. In this case, if a large size has evolved for predator defence, then the removal of the threat will shrink the species. Similarly, if a very small size also evolved as a predator avoidance strategy, then small species will get bigger.

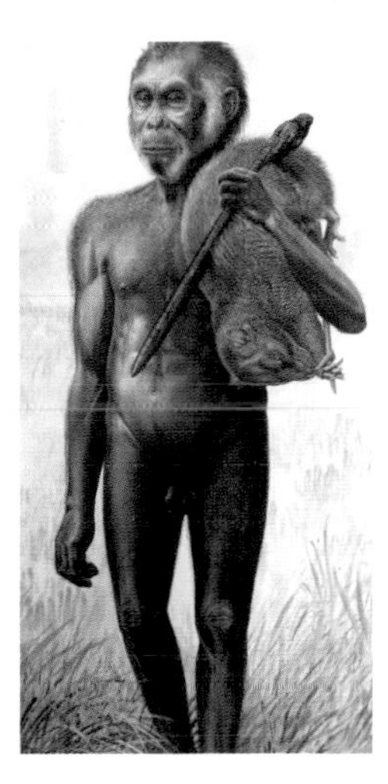

Reconstruction of Homo floresiensis*: remains of this human species from Flores show that it was 1 m (just over 3 ft tall), possibly as a result of 'island dwarfism'.*

Small is beautiful

Let us end where we began – with microbes. One advantage of being microscopic is that it enormously expands the space you can inhabit. The big creatures we are familiar with inhabit, essentially, a two-dimensional world on the surface of the Earth. Microbes, on the other hand, are found absolutely everywhere we have looked, including deep in the ocean sediments and even solid rock (p. 151). They are teeming on you and inside you, in the follicles of your eye-lashes, all over your skin and in your gut. Being tiny opens up entire worlds for creatures to inhabit.

JOHANN BRUHN & JEANNE MIHAIL

What is the largest living organism?

Do I contradict myself?
Very well then I contradict myself,
(I am large, I contain multitudes.)
WALT WHITMAN, 1891

The three authors of a letter to *Nature* in 1992 did not anticipate the variety of reactions that would be generated by it. They had simply described one of the largest and oldest living organisms – a subterranean fungus nearly invisible from above ground. But what captures the human imagination more than extremes of size and longevity, if not their achievement in a unique fashion by unexpected life forms? The lesson: always remain open to the discovery of that which exists beyond initial appearances.

Discovering the 'humongous fungus'

In 1986, three years after 2,000 pine seedlings were planted in an experimental clear-cut in Michigan's Upper Peninsula, one of the present authors (Bruhn) observed the initial, anticipated, seedling mortality. Caused by the lethal 'honey mushroom', *Armillaria ostoyae*, the mortality was apparent as scattered dead seedlings. Curious about the spatial arrangement of the *Armillaria* fungus population that caused the mortality, Bruhn then used a 'vegetative compatibility' test to ascertain the pattern of *Armillaria* in the forest floor. The results indicated a population of non-overlapping *A. ostoyae* individuals that was too large to have originated since planting. But more interesting was the fact that *A. ostoyae* shared the plantation with two contiguous individuals of a related fungus, *Armillaria gallica* (previously called *A. bulbosa*), which is not so lethal and so is less obviously detectable.

While the two *A. gallica* individuals didn't overlap, they intermingled with the *A. ostoyae* individuals and extended beyond the plantation boundaries into the surrounding forest. To clarify this spatial pattern, Bruhn invited James B. Anderson of the University of Toronto and Myron Smith of Carleton University, Ottowa, to conduct genetic analyses of mapped *Armillaria* cultures

Right *Rhizomorphs of* Armillaria gallica, *used for exploratory growth towards nearby food sources.*

Far right *A group of* A. gallica *mushrooms fruiting on an oak log. They are connected by a supporting network of rhizomorphs beneath the bark.*

Aspens within the 'Pando' clone, Fishlake National Forest, southern Utah. All the trees visible in fact are sprouts from the rootstock of a single aspen, which covers 43 ha (106 acres). Could this be the largest living organism in the world?

from the plantation and the surrounding forest. The results confirmed Bruhn's findings. The scientists unambiguously distinguished the largest *A. gallica* individual from its neighbours, estimating that it spanned at least 15 ha (37 acres), weighed more than 100 tons and had remained genetically stable for over 1,500 years. It was this that was the subject of the three scientists' letter to *Nature*.

How can *Armillaria* reach such enormous sizes? Species of *Armillaria* fungus decay wood by means of microscopic filaments called mycelia, leaving little behind except carbon dioxide and water. The energy derived from wood decay fuels further decay as well as the production of visible, above-ground mushrooms and the exploratory growth of root-like rhizomorphs. It is these rhizomorphs that permit *Armillaria* to traverse the forest floor in search of fresh woody substrates to feed on. In effect *Armillaria* individuals are modular in design with indeterminate growth.

Big trees and whales, and bigger armillarias

Of course, assessments of size depend on what exactly you are measuring. Smith, Bruhn and Anderson in their piece in *Nature* noted that giant sequoias can exceed 1,000 tons; they also noted that the total living mass of their largest *A. gallica* individual was similar to that of an adult blue whale (exceeding 100 tons). In response, other scientists headed by Michael C. Grant of the University of Colorado nominated an aspen clone in Utah, USA, as the largest living organism. Comprising approximately 47,000 stems of root sprout origin and covering 43 ha (106 acres), this 'tree' weighed over 5,000 tons.

Considering the global distribution of *Armillaria* species, larger individuals doubtless exist. In May 1992, an *A. ostoyae* individual covering 600 ha (1,480 acres) near Mt Adams in Washington, USA, was nominated as the largest living organism. But then in 2003, a group of scientists headed by B. A. Ferguson reported another *A. ostoyae* individual, this time covering 880 ha (2,175 acres) in Oregon, with an estimated age of 2,400 years.

Because *Armillaria ostoyae* generally kills its hosts it is much easier to detect than *A. gallica*, yet the latter is the more prolific producer of rhizomorphs. In fact, it is unlikely that we will ever know how large an *A. gallica* individual may become. We were simply fortunate to encounter a very large individual in the course of our studies.

Fragments, organisms and individuals

The mass of an *Armillaria* individual fluctuates over time in response to substrate availability. An *Armillaria* individual may become fragmented during periods of starvation, yet reunite later in times of plenty. If so, then is continuity a defining feature of an individual?

Other life forms (for instance aphids and mould fungi) reproduce asexually, producing myriad genetically identical organisms – together these might also be considered an individual. To paraphrase Stephen J. Gould, single mould colonies, aphids and *Armillaria* mycelia may be considered as organisms, though each comprises genes and organelles at a most basic level of organization, while at the same time representing individuals within species at a higher level of organization. Again we must look beyond initial appearances!

45 Engineering limits to the body size of land animals

It would be impossible to build up the bony structures of men, horses, or other animals … if these animals were to be increased enormously in height; for this increase in height can be accomplished only by employing a material which is harder and stronger than usual, or by enlarging the size of the bones, thus changing their shape until the form and appearance of the animals suggest a monstrosity.

GALILEO GALILEI, 1638

Large size brings problems. In the film, the giant gorilla King Kong looks immensely strong, but in real life he would collapse under his own weight. Measuring 30 ft (9 m) high, he is five times the size of a real gorilla – that's five times in each of the three dimensions, and so five cubed (125) times the volume of the real ape – he would weigh about 25 tons. This weight has to be supported by legs that are five times as wide as the legs of real gorillas, and five times as thick from front to back: they have five squared (25) times the cross-sectional area, and are thus 25 times as strong. Loaded with 125 times the weight of a real gorilla, those legs would certainly break (see also p. 184).

King Kong may look extremely strong, but if he were as big as he was shown in the film his legs would be too weak to support him.

That would not be King Kong's only problem. The chemical processes of metabolism that keep us alive produce heat. Mammals and birds use some of that heat to keep their bodies warmer than their surroundings, but the metabolism has to continue at a certain rate whether the heat is needed or not. As a rough general rule, a mammal that is 125 times as heavy as another has to metabolize about 40 times as fast. King Kong would have to lose excess heat from his skin, which has only 25 times the area of a real ape's skin. The heat loss would be hampered by his fur, five times as thick as a real gorilla's. King Kong would overheat in a hot climate.

Dinosaurs

King Kong is imaginary but dinosaurs were real and must have overcome the problems of very large size. The largest living land animals today are African elephants, which can weigh about 6 tons, but many of the dinosaurs were much heavier. We cannot measure their weights because their muscles and internal organs have decayed. All we can do is make models or drawings of the animals as we think they would have appeared in life, and use these reconstructions to estimate the volume of the living animals. This leaves a lot of uncertainty since we do not know how stout or skinny they were – and various estimates may be half or twice as much as each other depending on whether the animal was judged to have a stocky or a slender build.

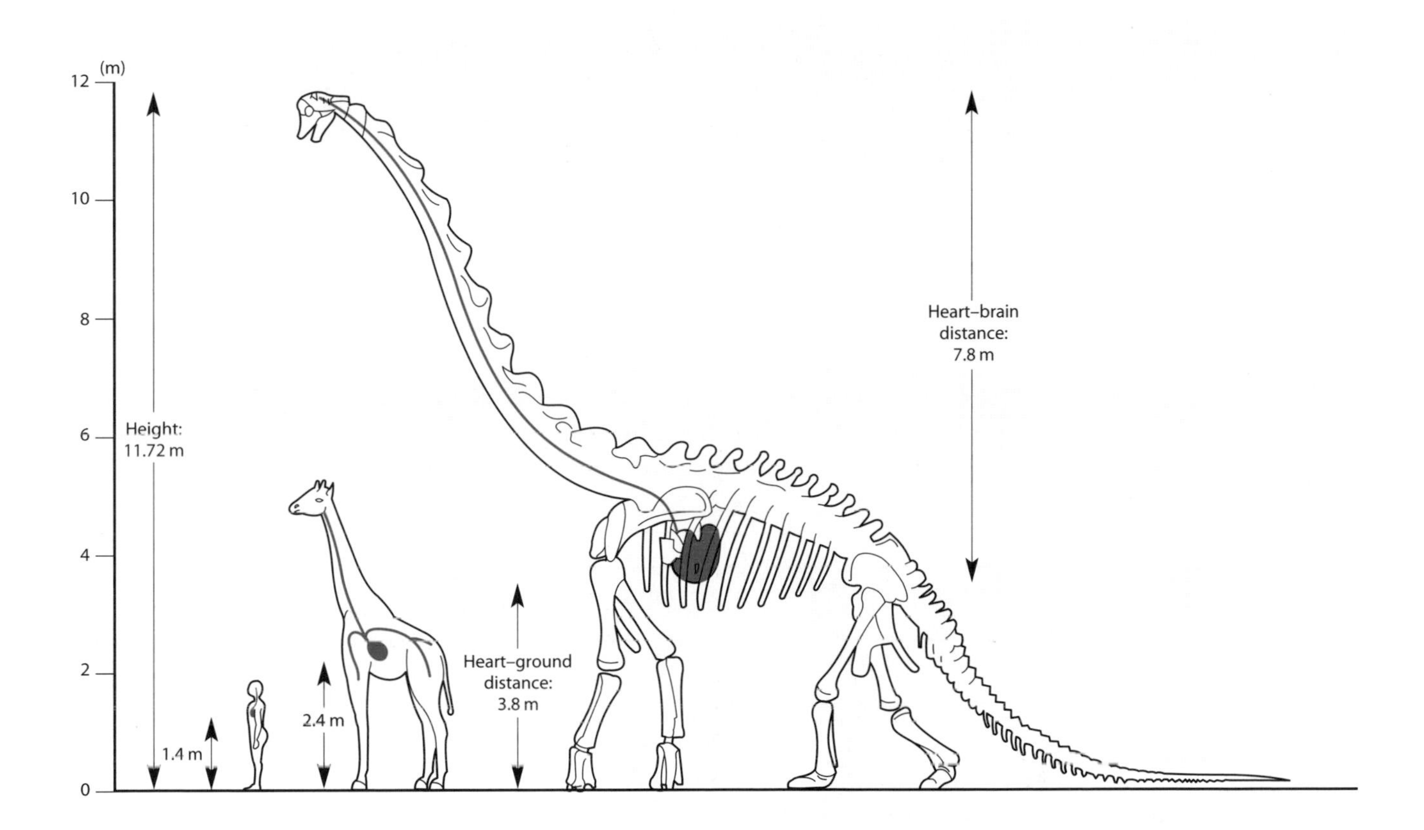

The largest dinosaurs belong to a group known as the sauropods. Among them is *Brachiosaurus*, for which we have a reasonably complete skeleton and which seems to have weighed about 50 tons (see also p. 38). A few even larger bones show that some sauropods must have been heavier, perhaps 70 or 80 tons in the case of *Argentinosaurus*.

It used to be thought that large sauropods could not have walked on land, but would have had to wade in water deep enough for buoyancy to support most of their weight, for instance moving along the bottom of lakes. However, many footprints of large sauropods have been found in rock which must once have been mud, and some are so sharply defined that it seems inconceivable that they could have been formed under water.

The dimensions of dinosaur leg bones have been used to calculate their strength. For instance, measurements on the leg bones of *Apatosaurus* (a 35-ton sauropod; see p. 14) led to

Above *The heart of the dinosaur* Brachiosaurus *must have pumped blood at a high enough pressure to drive some of it up to the brain. The heart-brain distance was far greater than in humans or even giraffes, so exceptional blood pressure must have been needed.*

Left *The femur of the dinosaur* Antarctosaurus, *a contender for the title of biggest-ever land animal.*

the conclusion that it could have run around as briskly as an elephant (a 3-ton Indian elephant has been filmed running at 6.8 m per second, or 3 mph). It has been calculated that even a 200-ton sauropod could have had legs strong enough for walking on land. So why could sauropods be much bigger than King Kong? First, because they walked on four legs instead of two; secondly, their bones were relatively thicker than those of a scaled-up gorilla would be; and finally, they seem to have walked with their legs very straight, more like elephants than apes.

The potential danger of a large animal overheating depends on whether it is 'warm-blooded', like mammals and birds, or 'cold-blooded', like modern reptiles. Warm-blooded animals metabolize five or ten times faster than cold-blooded ones of the same size, even in warm environments. There is much controversy about whether any or all of the dinosaurs were warm-blooded (see p. 35), but rough calculations cast doubt on whether a warm-blooded *Brachiosaurus* would have been able to avoid overheating in a hot climate.

Brachiosaurus faced another problem associated with its large size. It seems that it kept its neck erect, thereby raising its brain almost 8 m (26 ft) above its heart. From this it has been estimated that its heart would have had to pump at four times the blood pressure of most mammals, and twice the blood pressure of even a giraffe. It is hard to imagine how its heart could have been strong enough to achieve this. A suggestion that the blood system in the neck may have worked like a siphon does not solve the problem, because the neck veins would collapse and block the flow.

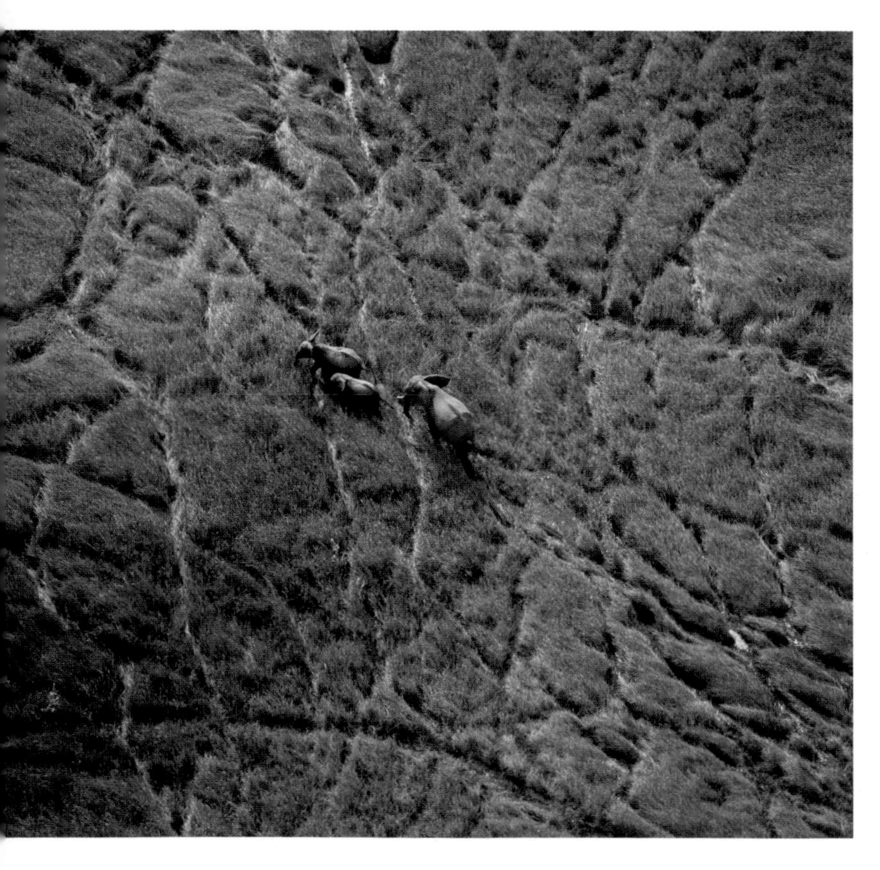

African elephants, here in the Okavango Delta, Botswana, are the heaviest living land animals, but some dinosaurs were more than 10 times as heavy. Larger animals need a bigger area of territory to supply them with food.

Population sizes

So far we have discussed only physical limits to body size – limits that depend on the strengths of bones and muscles, and on the dispersal of heat. There is another, ecological limit that is even harder to quantify. Larger animals need bigger areas to supply their food: a given area of grassland can support far fewer buffaloes than voles. But to have a good chance of long-term survival, a population of animals needs a minimum number of individuals, generally reckoned to be some tens of thousands.

In 1979, Africa supported around 1.3 million elephants weighing around 3 tons each. It can be estimated that the same resources could support 150,000 50-ton dinosaurs, even if they were warm-blooded. So even the biggest dinosaurs may have had populations comfortably over the limit for viability.

The limits of possibility

The largest land animals known to have existed were dinosaurs estimated to have weighed around 80 tons. Even if we were certain that these were the largest land animals ever, we should not conclude that larger ones were impossible. The maximum size of animals depends less on what is possible, than on what is competitive in the struggle for existence. The reasons given here, however, suggest that the largest dinosaurs were not very far below the limit of possibility.

Running, hopping, skipping: animal locomotion

46

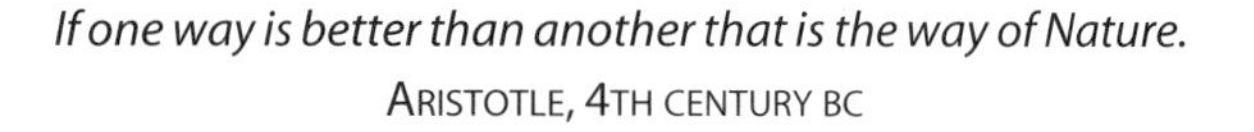

If one way is better than another that is the way of Nature.
ARISTOTLE, 4TH CENTURY BC

Locomotion in animals fulfils many roles and has many requirements. Hunting, escape and migration have very different demands for acceleration, manoeuvrability and efficiency, all of which pose fascinating mechanical challenges. Here, we consider three bipedal gaits, and their mechanical implications.

Running

Running is a gait familiar to humans, and adopted by a range of bipeds (many birds, and probably a good variety of other dinosaurs). It is characterized by an even timing between left and right footfalls, and a ballistic phase during each step when neither foot is in contact with the ground. Biomechanicists conventionally consider running as a form of mass-spring system, with the legs both acting with spring-like characteristics and also containing truly elastic elements (typically long, stretchy tendons).

The spring-like mechanics are apparent from fluctuations in energy form in the different

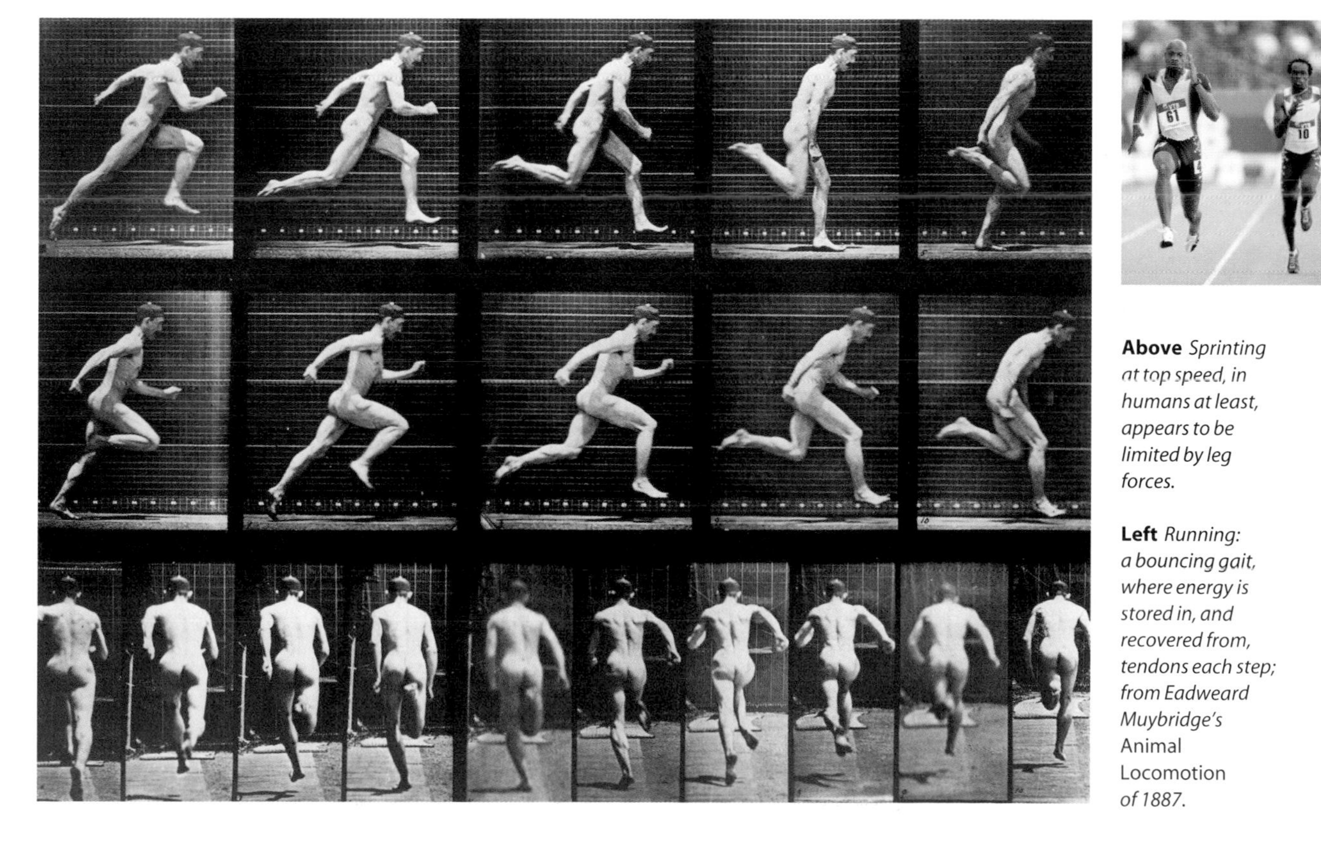

Above *Sprinting at top speed, in humans at least, appears to be limited by leg forces.*

Left *Running: a bouncing gait, where energy is stored in, and recovered from, tendons each step; from Eadweard Muybridge's* Animal Locomotion *of 1887.*

phases. During each ballistic phase kinetic energy (KE, due to speed) interchanges freely with potential energy (PE, energy due to height) – that is simply the action of gravity. While the foot is in contact with the ground, and the leg is loaded, the situation is quite different: the leg compresses, slowing the body (both horizontally and vertically) as it continues to fall; both kinetic and potential energy reduce to a minimum at mid-stance. Evidence that this loss of energy is actually stored, and largely recovered, comes from two observations.

The first is metabolic: the rate at which oxygen is consumed during running can be studied by measuring changes in gas composition between inhaled and exhaled air. When compared with the rate at which mechanical energy (KE+PE) has to be replaced, the rate at which oxygen is consumed appears to be far too low. This means that either muscles are very much more efficient at converting chemical energy into mechanical energy than we think, or elastic mechanisms are involved, so the full energy requirement to power each step does not have to be provided afresh.

The second observation is even more direct. Stiffness and elasticity of the long, stretchy tendons typical in good runners (think of your own Achilles tendon, or the chewy bits in a turkey drumstick) can be calculated by measuring force and extension on samples stretched in the lab. Such material testing indicates that approximately 93% of the energy put into a tendon can be recovered. The degree to which tendons actually stretch during running can also be measured using a variety of techniques, including ultrasound, and so the energy stored in the major tendons can be calculated. Combined with the metabolic studies, we can conclude that not only do tendons stretch and recoil, but this process of energy recovery allows runners to bounce along expending a small proportion of the energy otherwise required.

Hopping

Two-legged hopping is best known in kangaroos and wallabies, but is also used by other mammals and small birds, and is often considered as just a special case of running. Elastic mechanisms, especially for the larger hoppers, are certainly important, just as for running, and the long, elastic tendons are obvious to the naked eye. However, from certain viewpoints, hopping should be inefficient. By having both feet in contact with the ground at the same time, the loads transmitted to the body, including the relatively inelastic guts and brain, are doubled compared with running: hoppers have departed from wheel-like efficiency mechanisms of spreading the load through time as far as it is possible to go.

Despite this, metabolic studies of kangaroos and wallabies hopping on treadmills show that hopping can be strangely efficient. Above a certain, relatively slow, speed, the rate of oxygen consumption remains approximately constant with increasing speed. This suggests that the *efficiency* of hopping, the metabolic cost per unit distance, gets better with increasing speed – the opposite of what happens to runners – and this unexpected result has been attributed to some

aspect of elasticity. Why this is not available to runners as well as hoppers remains controversial.

Interestingly, although hopping and running are generally considered mechanically similar (both can be considered 'bouncing' gaits), specialist hoppers and runners cannot switch between the two gaits. This suggests that the spring-like mechanical set-up for hoppers and runners is fundamentally different. One obvious characteristic of kangaroos and wallabies is their long back, neck and tail, all of which contain considerable tendinous structures. Perhaps these might allow for a mechanism of storing and returning large amounts of energy that is fundamentally different from runners.

Skipping

Skipping gaits are adopted by bipeds that are for some reason 'overbuilt', that is, ones that have disproportionately strong legs, but that wish to locomote at above walking speed. This is clearly the case for both situations in which skipping is natural in humans. Children of skipping age are disproportionately sturdy. To maintain elastic similarity (the same ability to store energy in tendons) they 'should' have proportionally more slender legs than adults, whereas they are actually proportionally chunkier. The second natural use of skipping among humans is when astronauts 'walk' on the Moon. Because the Moon's gravity is too low to overcome the centrifugal effects of vaulting over relatively stiff limbs (i.e. walking) at anything over a desperately slow pace – 40% of

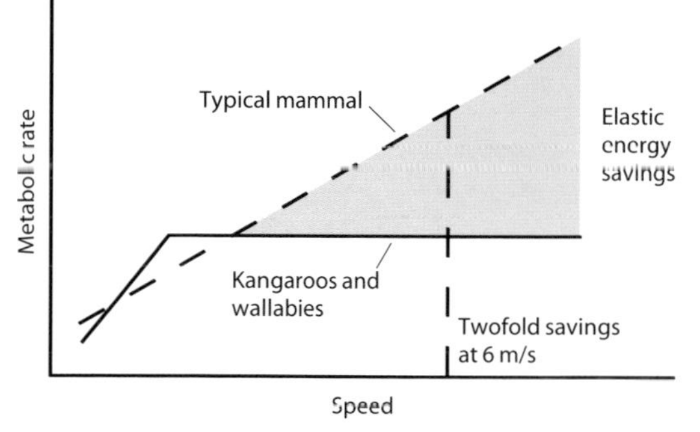

Above *Hopping kangaroos: perhaps the most extreme, elastic bouncing animals.*

Left *Kangaroos and wallabies find hopping at higher speeds no more tiring than at low, becoming considerably more efficient than typical mammals.*

WHAT STOPS US RUNNING FASTER?

The difference between fast and very fast running, whether looking at an individual or comparing average to elite sprinters, is most apparent in the forces experienced by the legs: high sprint speeds require high leg forces. If the forces along the legs are increased by requiring a sprinter to go round a tight, but banked, bend, foot timing adjusts to spread the load across a longer stance; this, however, also leads to a slowing of the athlete. Indoor 200 m sprint results can be successfully predicted from outdoor results by taking into account the increased forces due to running around bends.

Surprisingly, the same does not apply to galloping greyhounds. When racing, these dogs hit speeds around 64 km per hour (40 miles per hour), and they are able to maintain these speeds when sprinting around the bends on the track. What is more, centrifugal forces effectively increase the greyhound's weight by 71% on the bend, and yet are not adequately compensated for by increasing the time each foot spends on the ground. As a result, the legs experience an average force on the bend of 165% of that on the straight. It appears therefore that maximum speed in greyhounds is not force-limited, and it remains a mystery why they cannot run faster along the straight.

Greyhounds appear unconstrained by limb forces, and are able to sprint at high speeds around tight bends.

Earth-bound walking pace – gaits involving flight phases are naturally adopted by astronauts. Just as in the case of skipping children, astronauts are 'overbuilt' for their environment. With body weight only 16% of that on Earth, astronaut legs 'should' be incredibly more gracile on the Moon, with both tendons and muscles scaled to have 16% of their earthly cross-sectional area.

Skipping is also seen in *Propithecus*, a type of lemur from Madagascar. These are specialized 'leapers', with a remarkable capacity for propelling themselves from branches using powerful hind legs. Locomotion across level ground is achieved by a form of skipping – once again, bipeds possessing too-powerful legs resort to skipping.

The observation that skipping in primates is associated with 'overbuilt' limbs suggests a potential explanation for the mechanical benefits of skipping. It is reasonable to suppose that, under certain conditions, the net effectiveness of having one leg loaded adequately – whether in terms of elastic storage and recoil, or due to the loading regime on the leg muscles – and the other leg simply doing its best under unfavourable conditions, might exceed that of using both legs loaded below optimal performance.

Left *Astronauts on the Moon naturally adopt a variety of skipping gaits.*

Opposite Propithecus *– a specialized leaping lemur species that skips at high speeds when on level ground.*

JIM USHERWOOD

Flying and walking: learning from nature

What has been will be again, what has been done will be done again;
there is nothing new under the sun.
ECCLESIASTES 1:9

It is often assumed that engineers have learnt marvellous lessons through observing animal aquatic, aerial and terrestrial locomotion, and that these studies might allow us to understand how to build faster or more efficient vehicles. The sketches of Leonardo da Vinci provide an excellent example of inspiration for flying machines based on his observations of birds and bats. Unfortunately, few of da Vinci's designs for flying machines have proved satisfactory or even viable. Indeed, the big engineering successes in vehicle design – carts with wheels and aeroplanes with propellers – have only been achieved when the biological equivalents – running legs and flapping wings – have been ignored.

A page from the notebooks of Leonardo da Vinci, showing some of his sketches of bird flight – inspiration from biology is not a new idea.

The catalogue of failures for attempts at flapping aeroplanes is long, comic and often tragic. Surprisingly, two key considerations in engineering, those of constraint and compromise, are often overlooked when engineers consider biology. Here, two case studies are presented. The first is an example of how the constraints and compromises of animals make biomimetics unwise. The second is a case where, once biology-like constraints have been applied, engineers and evolution are converging on similar solutions.

Efficiency in hovering: mistaken biomimetics

The impressive performance of insects and, to a lesser extent, birds and bats, in hovering and slow, carefully controlled flight is striking, and has inspired a recent funding boom from military sources to develop flapping 'Micro Air Vehicles'. These MAVs are intended as hand-portable spy drones capable of flying in a slow, controlled manner suitable for investigating the insides of buildings. Ideally, they should be quiet, efficient and capable of carrying a payload of cameras, chemical detectors etc. The failure of, or at least the immense difficulty encountered by, engineers in meeting such requirements using insect-like flapping wings provides a 21st-century repeat of the lessons learnt by trial and error a century ago by the developers of aeroplanes.

Hovering and slow flight with flapping wings is terribly inefficient, even in specialist animal fliers that have had millions of years of selective pressure to evolve. Modern studies of slow insect and bird flight show it requires huge powers in aerodynamic terms when compared with helicopter-like flight; for instance, a slow-flying pigeon requires 3–4 times the power of an equivalent helicopter. Hovering insects, birds and bats use exceedingly high (35–45°) angles of attack, resulting in terrible lift:drag ratios.

Why, then do animals do this? Because they are *constrained* to use reciprocating aerofoils (instead of propellers), and so have to *compromise* between aerodynamic and inertial power efficiency. Unless a flapping flyer has access to perfectly elastic springs, new energy must be given to the wings each time they accelerate – an inertial power cost totally avoided by propellers once they are up to speed. Flapping animals are therefore compromising aerodynamic efficiency by using high angles of attack; this results in a lower wingbeat frequency for a given lift (reduced inertial power), albeit at the cost of a lower aerodynamic efficiency. Unless the micro air vehicle engineers have access to perfectly elastic

Above
A 'dragonfly' microdrone in a research laboratory. This micro air vehicle being developed by a French company is just 6 cm (2.4 in) wide and has wafer-thin silicon wings which flap when powered by electricity.

Left *A hovering hummingbird hawkmoth: very effective, but aerodynamically very inefficient.*

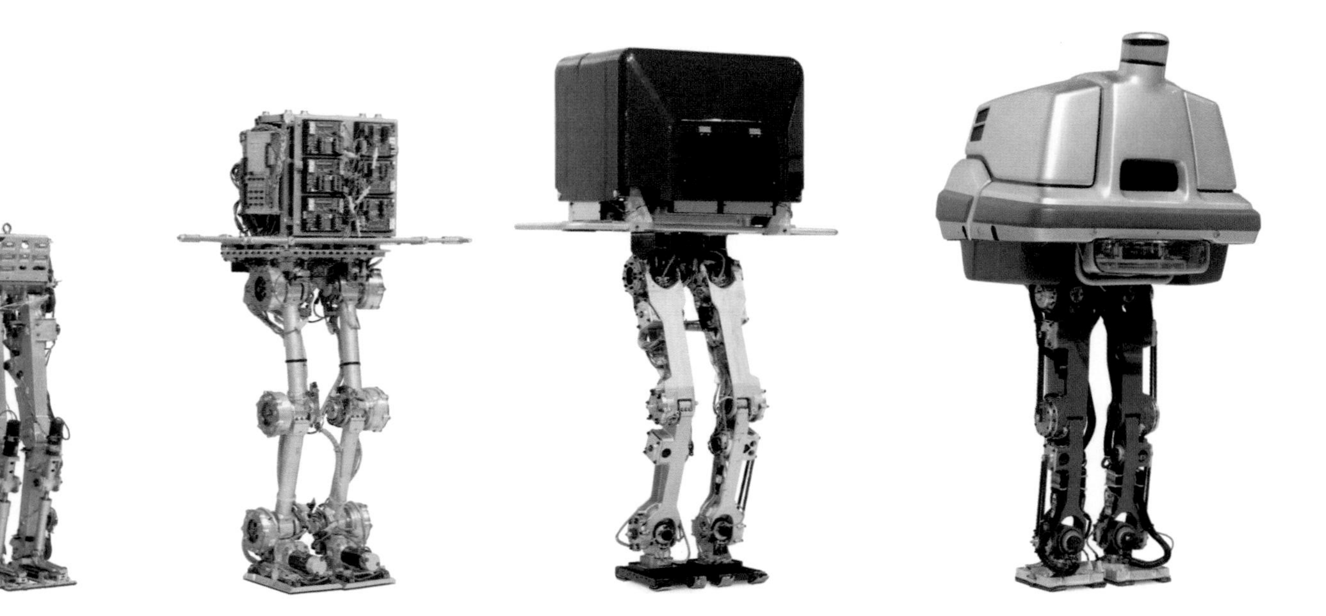

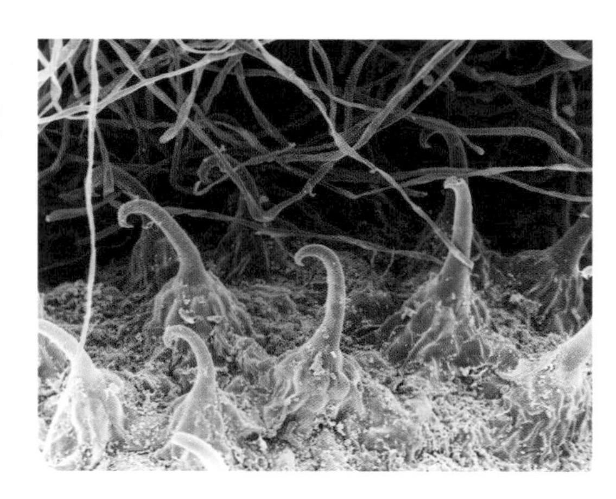

Right *A scanning electron microscope image of the hooked fruit of cleavers or goosegrass – the inspiration for Velcro. This is a rare example of direct, widely adopted biomimetics.*

springs, hovering mechanical flappers will also experience high inertial costs, and designs based on simple helicopters would be far more effective.

Biomimetics, bioinspiration and bioconvergence

Engineering science is exceedingly useful in understanding animal form and function; however, the converse, the use of biology as a direct model for engineering design, is considerably less clear-cut. In fact, the case of Velcro, which was famously invented after observation of the hooks of seed burrs, stands out as a near-unique example of direct biomimetics with wide and successful application. That is not to say that it is impossible for useful engineering lessons to be learnt from biology. A recently coined term, 'bio-inspiration', is a good description of a reasonable middle ground, in which the limitations experienced by biological systems are actively considered, but successful ideas – a prime example currently being investigated is gecko-like adhesion – can be explored within engineering.

True cases of biomimetics tend to be accidental: the engineers may not be setting out with the intention of copying biological solutions, but find that their best solutions happen to be closely related to those observable in nature. Such a phenomenon might be termed 'bioconvergence', and is neatly exemplified by the development of powered two-legged robots.

Powering walking: bioconvergence?

Biology has invented bipedal walking a number of times: not only have humans evolved bipedality, but so did the dinosaurs, which include the birds today. Engineers developing bipedal walking robots have, understandably, often followed the basic human plan. Such robots as Honda's Asimo are extremely impressive machines, capable of dancing, climbing stairs and even playing football. However, beyond the anthropoid morphology – arms, legs, a familiar number and direction of joints – the physics behind such machines are quite different from those of biologi-

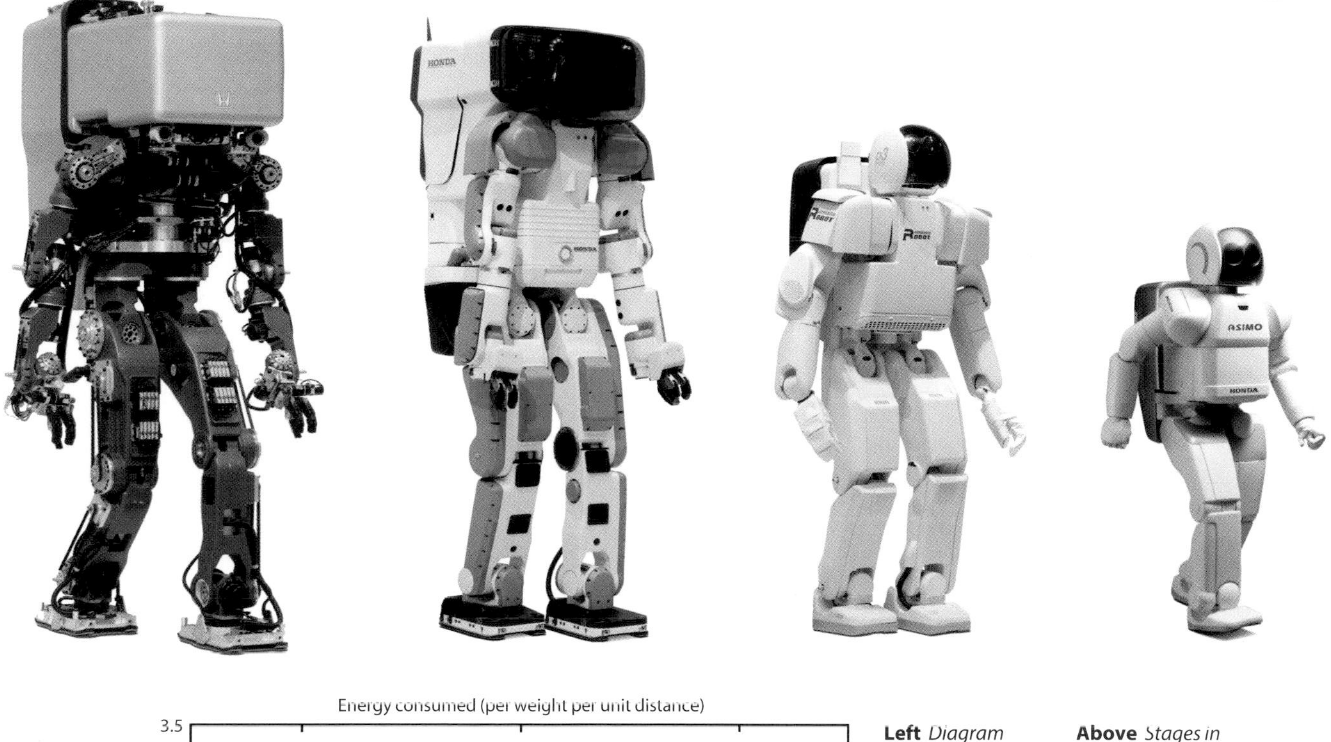

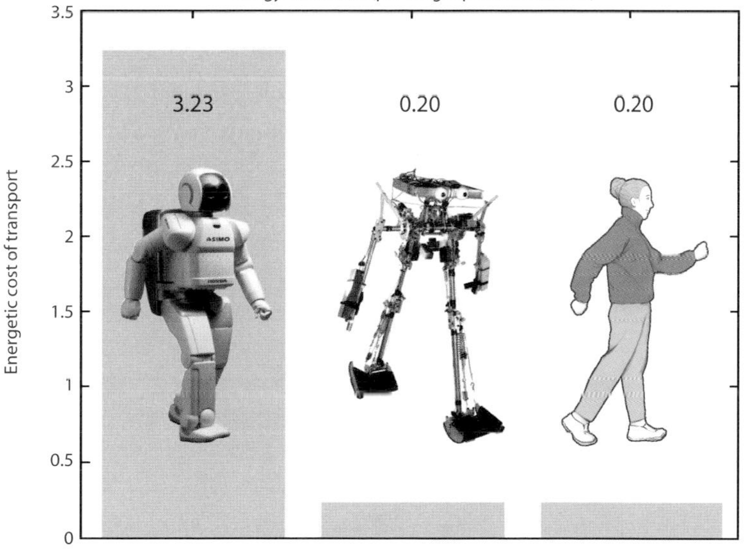

Left *Diagram comparing the energy efficiency of Asimo, the Cornell biped and a human when walking. Walking by 'vaulting' over stiff limbs allows relatively efficient locomotion.*

Above *Stages in the development of Asimo, a walking, dancing, football playing robot created by Honda, which does not use human-like walking mechanics. Very impressive, but not designed for efficiency in steady, level walking.*

cal bipeds. Asimo and friends walk with a high degree of active control about each joint, and their bodies do not vault up and down with each step.

Recently, however, developments inspired by stiff-legged tinker-toy walkers, which have entertained children and engineers alike for decades, have been taken one step further. It has been convincingly demonstrated that if a robot is allowed to walk with relatively stiff limbs, and gravity is allowed to play its part, resulting in an 'inverted pendulum' motion as the body vaults over the leg, the power requirements are drastically reduced compared with the highly complex Asimo-like robots. So, more efficiently powered robots are now rapidly converging on the design principles that can be observed in human walking. Whether it was biology, good physics or watching children's toys that inspired this new generation of engineered walkers is open to debate. Probably all of them, and it is very exciting to be on that boundary between engineering, physics and biology where inspiration can come from so many sources.

JULIAN PARTRIDGE

How do dogs see the world?

Properly trained, a man can be dog's best friend.
COREY FORD, 1952

Vision in domesticated dogs has many characteristics in common with ancestral wolves, but also shows the effects of artificial selection and breeding by human owners.

Domesticated from wolves some 15,000 years ago, dogs have become valued companions for humans, assuming roles ranging from hunting and herding to acting as guides for the blind. Despite this long association, and the curiosity of most owners, misunderstandings abound about the visual world of dogs. Although the eye of a dog is similar to ours (p. 104), dog vision differs in several key ways.

Acuity

Acuity is the ability to see detail in objects and depends firstly on the quality – especially focus – of the image that the lens and cornea form on the retina at the back of the eye. In dogs, just as in humans, this image suffers from varying degrees of astigmatism and spherical and chromatic aberrations, as well as short- or long-sightedness. A surprising number of dogs are short-sighted (myopic), with different breeds varying in the prevalence of this focusing error. One study found that more than half of all tested German Shepherd dogs and 64% of Rottweilers were myopic (one by as much as –3 dioptres, so it would focus objects around 30 cm/12 in away). Such high levels of myopia might explain some inappropriate behaviour; certainly many dogs would benefit from glasses!

Myopia in dogs tends to get worse with age (unlike humans where the reverse is more common), as does a dog's ability to accommodate (i.e. to change focus for objects at different distances). Like humans, dogs focus by changing the shape of their lens, but the dog lens is stiffer than ours, giving it less ability to accommodate. For instance, humans can, at least as children, focus on objects from the far distance to as close as 7 cm (2.75 in) from their eyes, but dogs without myopia can only accommodate as close as 30 to 50 cm (12–20 in) and may resort to smell when locating objects literally under their noses.

Optics aside, another important limitation to visual acuity in dogs is the organization of the retina. In the dog retina, just as in ours, the outputs of several photoreceptors converge on a single

Eye shine in wolves. Light is reflected from a mirror-like layer behind the light receptor cells, thereby increasing visual sensitivity, which is particularly beneficial when hunting in semi-darkness.

ganglion cell, these being the cells that connect to the brain via the optic nerve. High levels of convergence under-sample the optic image, reducing acuity. Dog optic nerves typically have fewer than 200,000 nerve fibres compared with the human ones which have about 1.2 million, despite the fact that the eyes can be of similar size. On this basis alone, acuity in a dog can be predicted to be less than in humans.

Dog retinas also differ from humans' because they lack a fovea, an area of very high cone photoreceptor and ganglion cell densities in the central part of our retina (the part you are using to read this text, in fact). Instead, a dog retina has a horizontally orientated, oval-shaped region in which photoreceptors and ganglion cells are at somewhat elevated densities compared with the surrounding areas. Dogs thus have a horizontal region within their visual field in which they see with higher acuity than other regions. Such a horizontal band of high acuity, a 'visual streak', is known to be associated with animals that live in open habitats and need to scan the horizon for prey or predators.

Wolves have a more defined visual streak with higher densities of ganglion cells than most dogs, suggesting that they have higher acuity. Indeed, domestication appears to have reduced acuity, and may have altered the way the dog's brain processes visual information. The net effect is that human visual acuity is some four to nine times better than that of a dog, even that of a sighthound or retriever. In practice, a dog will need to be as close as 6 m (20 ft) from an object before it will see detail visible to us at 24 m (78 ft) or more.

Brightness sensitivity

If dogs lack the visual acuity of humans they certainly outperform us in having higher sensitivity; something that becomes apparent when walking a dog at twilight. Dogs need as little as one-sixth of the light needed by humans in order to see.

Above *Hidden by tall grass and twigs, a Siberian tiger reveals the effectiveness of its camouflage, produced by high contrast elements.*

Right *A male Satin bower bird decorating its bower with blue objects to attract and impress females for mating.*

At the core of understanding animal coloration is a knowledge of behaviour, ecology, evolutionary biology and vision. Contemporary research on animal coloration is thus a truly multidisciplinary enterprise.

Framework

Simply put, an animal's body is a trade-off between camouflage, for protection from predators, and gaudiness, for displaying and impressing mates. With this framework, much can be understood about the conundrum faced by animals, and their multiple, beguiling solutions. Paralleling this distinction is that between sexual selection and natural selection, both of which were identified and named by Darwin.

Sexual selection (p. 233) refers to mating advantage – specifically the advantages that individuals of the same sex and species have in sexual reproduction. Natural selection primarily concerns survival advantages. Sexual selection is powerful – it can be used to explain the evolution of gaudy, extravagant traits that would seem to confer survival disadvantages. Classic examples include the peacock's tail, a bower bird's bower, ornamentation in birds of paradise, and noisy song production.

Sexually selected traits tend to be costly, not only in terms of the increased risk of detection and predation, but also in the time and energy required to construct the traits. Surely the male peacock would be safer from predation if it did not carry such a vast tail, and the male bower bird would spend its time better seeking food rather than building a display structure? The trade-off in this tension between sexual and natural selection is clearly in favour of the former: the male birds gain such advantages, in being able to mate with more than one female, that the risks of detection or starvation are outweighed.

The diversity of sexually selected colour displays in birds, fishes, reptiles and insects is staggering. Classic experiments involve scientists manipulating the trait in some way, and seeing if it affects mates preferred, or number of offspring sired. One recent finding is that iridescent colours – those that change colour with the angle of viewing – are often sexually selected. Iridescent patches are effectively 'flashing lights' and are inherently conspicuousness. Colours rich in ultraviolet (UV) reflectance may also tend to be sexually selected. An explanation can be found in the fact that many animals are sensitive to UV wavelengths, but most leaves (and other vegetation) reflect little UV, and thus UV/blue rich traits are inherently conspicuous against these backgrounds.

Warning coloration and mimicry

Warning coloration is the exception to the general rule that protection from predation is conferred by cryptic coloration, with gaudiness only for mating advantage. Here, instead, gaudiness indicates the unprofitability or distastefulness of the prey, and the predator gains by learning the association. In this way, certain insects such as ladybirds and tiger moths are unpalatable to birds, and they sport warning colours, typically reds and yellows. Some poisonous frogs and spiders also have such colours for the same reason.

Iridescent chest feathers of a male European starling in breeding plumage: the intense reflections which vary with the angle of viewing are thought to be used by females to assess mates, and also by males to assess rivals.

Potential prey may evolve warning coloration to become either distinct from harmless or profitable prey, or similar to distasteful or unprofitable prey. When distasteful prey resemble one another, it is called Müllerian mimicry, and when prey resemble each other but one is distasteful and the other not, this is Batesian mimicry. For example, the coral snakes of North America are venomous, and carry stripes in the classic warning colours of red, yellow and black. The harmless red milk snake looks almost identical, and clearly gains protection by looking like its dangerous neighbour.

A venomous coral snake displays its warning coloration (above); the tropical milk snake (below) is a harmless mimic of of the coral snake.

*A katydid, or bush cricket (*Championica montana*), demonstrating exquisite masquerade camouflage against a moss background.*

Camouflage and compromises

Camouflage can be assisted by countershading, where the lower surfaces of the animal are lighter than the upper, helping to cancel out the effects of shadows cast by the animal's body. Thus, at least in side view, the existence of the body is more difficult to determine. Examples of countershading are found in mammals, fish, sharks and birds.

While the concept of camouflage might seem straightforward, there are many interesting twists. Camouflage by crypsis occurs when the animal comes to resemble a random sample of its visual background, though it is often difficult to determine accurately the relevant backgrounds. Another type of camouflage is termed 'masquerade'. Here the prey is detected as distinct from the background, but is not recognized as edible; examples are when insects resemble a leaf or twig. 'Disruptive coloration', involves, counterintuitively, an animal having high visual contrast elements in its colour patterns. These are thought to distract the predator's attention, and/or break up the body outline, and so make detection difficult. While the above types of camouflage have long been identified, their role in explaining the evolution of the coloration in any particular species is usually untested.

Just as fingerprints and the irises of the eye are individually unique in humans, it has recently been recognized that many elements of an animal's coloration are individually unique. In this way, individual animals can be identified by their colour pattern. Other current work focuses on

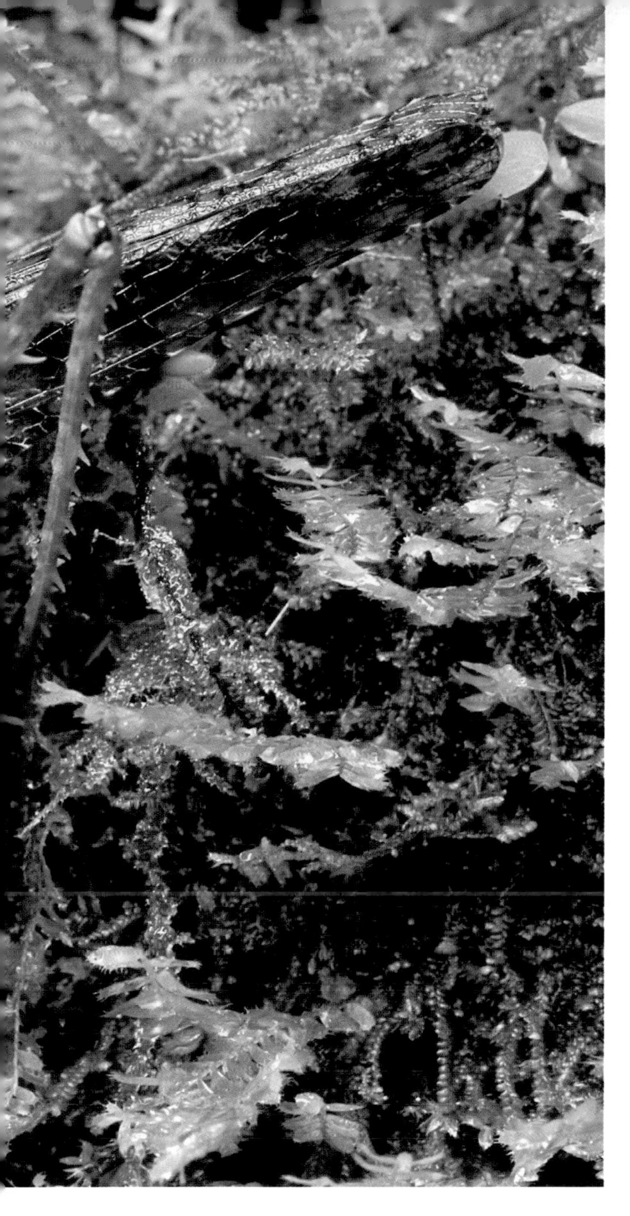

the role of shadows in camouflage. Shadows from leaves and the dappled light in a forest canopy are highly coloured environments that are difficult for visual systems. Just how predators and prey exploit these natural filters remains elusive.

Revealing displays

Many studies now show that coloration (and other features of a display) can be indicative of the condition and health of an animal. To potential mates, condition might convey information about an animal's ability to procure material resources (for the mate, or the offspring), or it can indicate that the animal has 'good genes', which would be beneficial to the offspring. Thus individuals can make informed decisions about the benefits of alternative mating partners by virtue of coloration and display. Consequences range from refusing to mate to post-copulatory sperm expulsion.

There is evidence for self-reinforcing coevolution of traits (usually in males) and preference (usually in females), leading to the evolution of traits that are not indicative of condition. These can be termed arbitrary, aesthetic or Fisherian traits, the eminent biologist Ronald Fisher having originally hypothesized the genetic process leading to their evolution.

Another major insight into animal coloration is the increasing awareness that animals do not see colours as humans do (see p. 202). Many birds and insects (and some fish and reptiles) are sensitive to ultraviolet wavelengths (to which humans are blind), and have an extra class of cone cells in their retina. They therefore probably perceive an extra dimension to colour vision.

Thus, despite all the work since Victorian times, it is clear that there are still many mysterious issues in the origin and maintenance of animal coloration.

A male zebra finch photographed in normal colour (above), and with ultraviolet wavelengths only (below). In zebra finches, and numerous other bird species, ultraviolet wavelengths are used by females in their assessment of males as potential mates.

SASCHA K. HOOKER

How do animals adapt to the deep?

From space, the planet is blue. From space, the planet is the territory, not of humans, but of the whale.
HEATHCOTE WILLIAMS, 1988

Opposite *An anglerfish with luminescent lure. Current estimates of the number of new species yet to be discovered in the deep ocean range from 10 to 30 million.*

Below Alvin, *a research submersible, can dive to 4,500 m (14,760 ft); in 1960 two men on the* Trieste, *a navy bathyscaphe, reached 11 km (6.8 miles) deep in the Mariana Trench.*

The ocean provides 99% of the space by volume of the Earth's potential habitat, but little is known about the animals that live there, especially at great depth. The deep sea is a high-pressure, cold, dark and low-oxygen environment that appears relatively inhospitable. Despite this, many species have adapted to meet the difficulties of life at depth and thrive. One of the greatest challenges animals face is increased pressure, and this is most pronounced for air-breathing animals that negotiate regular changes in pressure as they dive to feed. It is this aspect that also makes the study of deep-sea animals far from easy.

Scientists use sampling equipment to collect deep-sea animals, but unless a pressurized container is used the reduction in pressure as animals are brought to the surface often kills them. Observations can be made using manned submersibles, but most have a maximum depth of only a few hundred metres. Beyond this depth, submersibles need a thick titanium pressure sphere so that the passengers do not suffer from the enormous pressure. Alternatively, remotely operated vehicles can be used, and these are required to access the deepest ocean depths.

For the study of air-breathing diving animals, technological developments in animal-attached tags allow data on depth and movements to be collected. Microphones and digital video recorders can also be attached, allowing scientists to hear and see what the animal has experienced while underwater.

The ocean depths

In the deep ocean, sunlight penetrates only the upper few hundred metres. This lack of light limits primary production (photosynthesis), and bacterial respiration occurring at depths between 500 and 800 m (1,640 and 2,625 ft) causes the water layer to be very scarce in dissolved oxygen. Like light, temperature decreases with depth to around 0–5°C (–18 to –15°F) at depths greater than 1,000 m (3,280 ft).

From the atmospheric pressure that we are accustomed to (1 atmosphere, or atm), pressure increases by 1 atm for every 10 m (33 ft) descended in the ocean. So at a depth of 1,000 m (3,280 ft) the pressure is over 100 times that at the surface. Pressure has little effect on liquids or solids, but has a great effect on gases. A popular way to demonstrate this is to attach a styrofoam

Below left *Many marine mammals, birds and reptiles regularly travel to great depths to find food. Elephant seals, sperm whales and beaked whales are exceptional divers.*

Below right *Marine mammals show several adaptations to maximize their oxygen stores, reduce their metabolic rate and thus their oxygen consumption, and to minimize problems caused by pressure-induced increases in blood gases.*

cup to deep-ocean sampling equipment. Such cups return to the surface thimble-sized, as the huge pressures have crushed the air in them, forcing structural change. Air spaces in animals similarly become compressed at depth, potentially causing damage to any surrounding rigid structures.

Deep-sea species have been found at all depths, with fish and amphipods (small, shrimp-like crustaceans) even as deep as 8,400 m and 9,800 m (27,560 ft and 32,150 ft) respectively. Many of these show a variety of adaptations to the harsh conditions, with highly specialized species associated with unique environments such as sulphur vents (p. 167). Adaptations to low-light include large eyes and bioluminescent lures. Many deep-sea species also have less rigid lipids in their cell membranes, since the high-pressure, low-temperature environment makes fat more rigid. The cold, low-oxygen conditions cause many deep-sea species to be sluggish, and hot-blooded divers such as marine birds and mammals have adapted to exploit this and catch such slow-moving prey. However, in order to do so, they themselves must cope both without access to air while underwater and also with the effects of changing pressure on their bodies.

Air-breathing divers

The use of animal-attached tags has confirmed the astounding abilities of numerous marine vertebrates that dive to great depths. Many of our insights into diving physiology, however, have come through the use of captive and trained animals, which can be easier to study than animals in the wild. In some cases, researchers can create a captive environment around wild behaviour, such as the transportation of Antarctic Weddell seals to a solitary remote man-made ice-hole, thus restricting them to this 'laboratory' hole to obtain their air supply.

Marine mammal adaptations to the physical effects of pressure include their compliant ribcage, which allows their lungs to collapse fully. Similarly, many species lack facial sinuses and have extensive vasculature in the middle ear and other sinuses, which are thought to fill with blood at depth, so preventing pressure squeeze. Higher

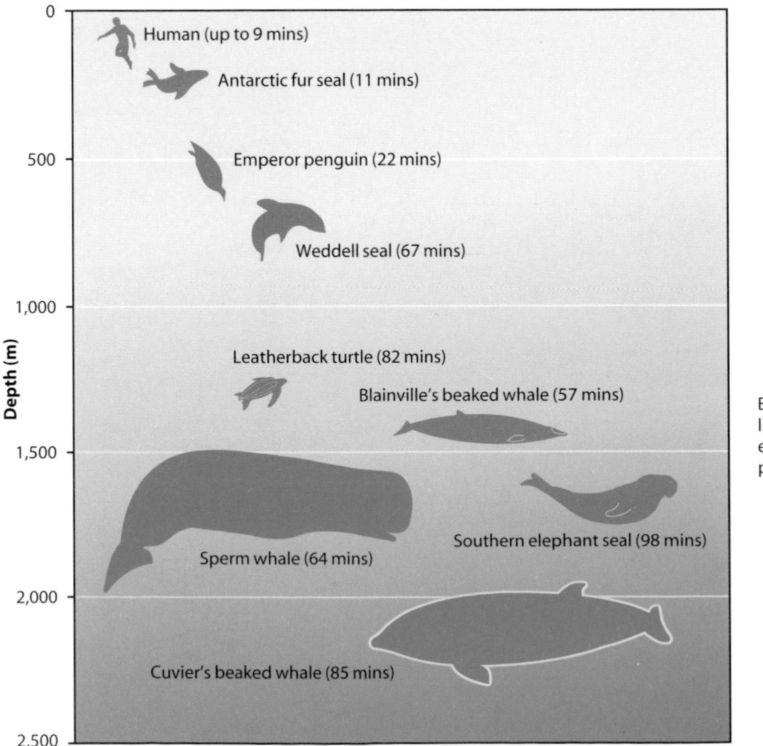

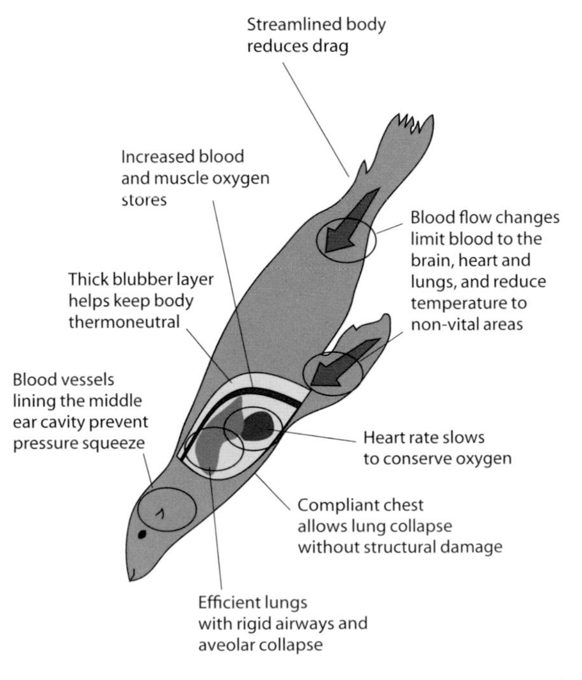

An Antarctic fur seal diving. Recent studies have shown that Antarctic fur seals exhale on the way back to the surface from depth, a behaviour which may help prevent shallow-water blackout.

pressure also increases the concentration of gases diffusing from the lungs into the blood, causing blood gases to become elevated. Once the lungs are collapsed, however, this does not occur.

Increased nitrogen levels in the blood and tissues can cause decompression sickness ('the bends'), when bubbles form and grow on ascent, leading to severe pain and even death. This can be a problem for scuba divers breathing compressed gas at depth, but has also been found among free-diving Japanese pearl fishermen who perform repeated dives to 30-m (100-ft) depths, leading to a cumulative increase in levels of blood nitrogen. Elevation of blood oxygen at depth can also mislead the body into 'thinking' it has more oxygen than it really does. On ascent, the re-expansion of the lungs can reverse the diffusion gradient and pull oxygen from the blood back to the lungs, potentially leading to shallow-water blackout and drowning.

Exhalation prior to diving helps to avoid these problems by causing lung collapse at shallower depths. However, this means that the lungs cannot be used as an oxygen store. Instead, diving mammals such as elephant seals and grey seals rely on elevated oxygen stores in their blood and muscle. In addition, heart-rate is slowed, circulation is restricted to all but crucial organs (heart and brain), and temperature is reduced – all of which helps reduce metabolic rate and thus oxygen consumption.

As is often the case, however, nothing is ever clear cut. Recently, cameras and other equipment carried by Antarctic fur seals have shown that they exhale during the last half of the ascent of all dives. This species therefore inhales prior to diving, and thus presumably requires this extra lungful despite the additional risk entailed. Exhaling may minimize gas transfer between lungs and blood on ascent, and so avoid the reduction in blood oxygen that could cause shallow-water blackout. This is not yet certain, however, and, as with much research into the deep ocean, generates more questions than answers.

Animal Behaviour

Animals can seem to us to be friendly, frightening, intelligent or stupid. It is easy to think of animal behaviour with respect to humans – the loyal dog, the aloof cat – and yet it is important to understand how animals behave in their own terms. Behaviour is a mixture of instinct and learning, and it is broadly true that mammals and birds exhibit more evidence of an ability to learn than, say, bees and worms.

The evolution of behaviour is harder to determine. Why for example do many ants and bees give up their chance to mate and produce young? Clearly in this way they further the success of their colony by helping the queen produce and care for her offspring, but superficially at least it seems to go against evolutionary theory of the survival of the fittest. Why indeed are animals sometimes nice to each other? It is understandable, in an evolutionary sense, for an animal to be nice to a close relative – to warn it of danger, or provide it with food – because evolutionary self-interest is involved and the animal is helping preserve shared genes. But why would an animal put itself in danger by warning a whole herd or flock of the imminent arrival of a predator?

Many aspects of animal behaviour became clear when Darwin distinguished sexual selection from natural selection. The giant antlers of deer and the showy tail of the male peacock are more to do with attracting a mate (sexual selection) than survival in the wild (natural selection): the risk of being killed by a predator is outweighed by the evolutionary benefits of siring many offspring. It is important to note that there are critics of sexual selection theory, at least in the uniform

Migrating geese can find their way to their destination over vast distances. They may make mental maps of their journeys, and parents use their memory of the route to lead their young.

Meerkats on the lookout for others in the group. A fascinating area of research on animal behaviour is to seek to understand how apparently altruistic behaviour could have evolved.

model generally assumed: some species, such as seahorses and pipefish, have gaudy, showy females, while the males dutifully look after the eggs in special brood pouches.

Animals communicate in many ways. One of the most prevalent – pheromones – is literally invisible to us. Other 'hidden' signals, such as the waggle dance of the honey bee, are now much better known. The astonishing ability of animals to find their way to a precise spot over long distances, even when they have never previously been there, is another mystery. It often has more to do with their innate ability to construct a map of their surroundings, rather than some hidden navigational extra sense, though this is still a very active area of research.

How human are animals? We must beware of reading human emotions into animals, but many mammals do seem to show fear, happiness and sadness. Charles Darwin realized this, and presented extensive evidence in his book, *The Expression of the Emotions in Man and Animals*, published in 1872. In this he extended his study of modern humans to the apes, and showed that chimpanzees and gorillas have many seemingly human-like responses. Scientists have now discovered that spindle cells, once thought only to exist in the brains of humans and other great apes, are also to be found in different types of whales, in a part of the brain linked to empathy and intuition.

Are animals conscious? Of course animals sleep and are awake, or conscious, but the mystery lies in how far they are conscious of themselves, of the world, of what happened yesterday. Experiments have shown that apes and monkeys at least can recognize themselves in a mirror. Dogs and other animals learn words in human languages – the names we give them, words for food and orders to lie, stop, fetch, and so on. These aspects of animal behaviour can be uncomfortable for scientists trained in certain fields because there is a risk that the observer might be deluded. As in all areas of human endeavour, and none more so than the study of animal behaviour, the observer must observe him- or herself before drawing conclusions.

Instinct and learning in animal behaviour

I will not attempt any definition of instinct, but everyone understands what is meant when it is said that instinct impels the cuckoo to migrate and to lay her eggs in other birds' nests.
CHARLES DARWIN, 1859

Instinct is the innate knowledge that guides behaviour. Its basis, as Darwin might have guessed, lies in the wiring of particular units of the nervous system – in circuits that have in many instances been mapped cell by cell. Learning, on the other hand, is generally thought of as quite different – the alternative to instinct. Although exactly how these two processes work remains an intriguing mystery, researchers now agree that they are tightly interconnected and mutually dependent.

Innate recognition and responses

The first reasonably complete behavioural study of one of these aspects was an illuminating 1930s investigation of egg recovery by geese. The ethologists Konrad Lorenz and Niko Tinbergen (who later shared the Nobel Prize with Karl von Frisch, see p. 248) observed that a goose egg sometimes rolls out of the nest when the parents rotate the clutch (an unlearned ritual performed several times a day to keep the embryo from sticking to the inside of the shell). After the parent settles

Left *Egg rolling behaviour in the study by Lorenz and Tinbergen: when nesting geese spot an egg outside the nest, they rise, put their beak on the other side, and roll the egg back between their legs. This highly stereotyped behaviour is typical of ground-nesting birds.*

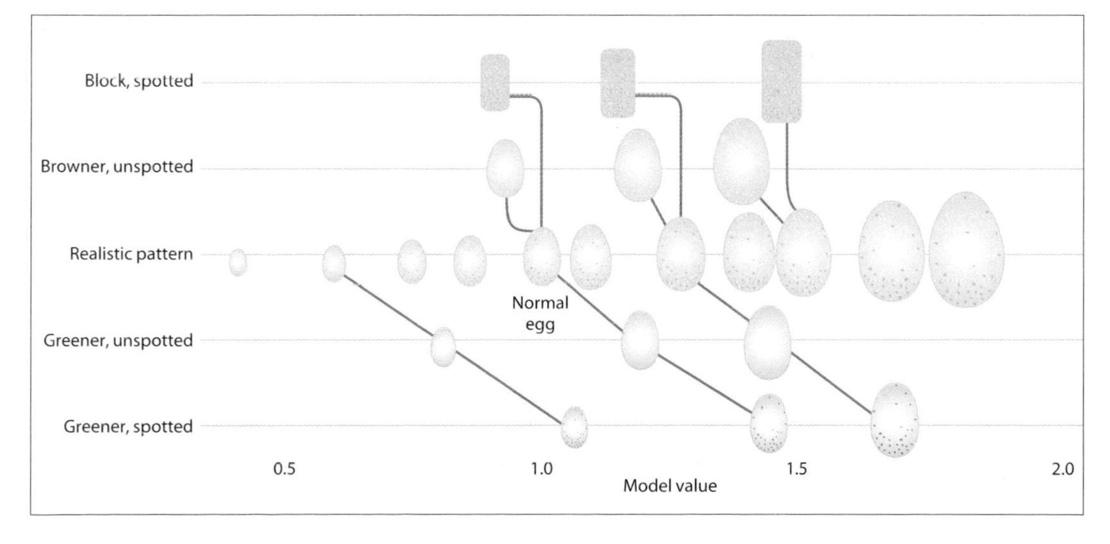

Below left *Egg titration: objects further to the right in this diagram do a better job of stimulating egg recovery. Size, shape, speckledness and colour all play a role in triggering this behaviour.*

Imprinting: goslings learn to recognize their parents during a brief time after hatching and can even be imprinted on people – here Konrad Lorenz. They are unable to learn after this critical period.

back on the nest, it fixates on the errant egg, reaches out with its neck, places its bill on the far side of the egg, and then rolls it gently back between its legs into the nest.

The behaviour is stereotyped: all geese recover eggs the same way; none ever tries pushing, or using a wing or foot. Moreover, once the goose has put its bill beyond the target, the egg can be removed without interrupting the full expression of the rolling behaviour. That geese will roll non-existent eggs means that the behaviour is a unit – what Lorenz and Tinbergen called a fixed-action pattern.

But what triggers this elaborate innate response? Tinbergen and his colleagues discovered that gulls (which have a recovery behaviour almost identical to geese) depend on a set of schematic cues – what are now known as sign stimuli. So sketchy is the bird's idea of an egg that it will roll in grapefruit and beer bottles. Tests on choice of objects show that the key stimuli involved include an unbroken outline, ovalness, size, speckling and greenness. Each of these features adds independently to the stimulating value of the 'egg'.

Programmed learning

Lorenz rediscovered a curious kind of learning now called imprinting: young animals in some species learn to recognize their parents based on a few sign stimuli – movement away from the nest and an exodus call in geese. The learning has a critical period beginning at a fairly standard time after birth, and ending as little as a few hours or days later. Often the learning is irreversible. In the absence of more naturalistic cues, ducklings can be imprinted on balloons and toy trains – or people.

Many other kinds of learning in the wild seem almost as carefully guided by innate processes. In songbirds, juveniles pick out their own species' song from the many sounds around them by responding to cues embedded in the melody. During a fairly extended critical period they memorize the song, and then (after a good deal of experimentation in learning how to sing it) reproduce it the next spring. Birds learning to recognize dangerous animals innately cue off the alarm calls of members of their own species, memorizing the object of the mobbing behaviour and attacking the same kind of animal in the future. So mindless

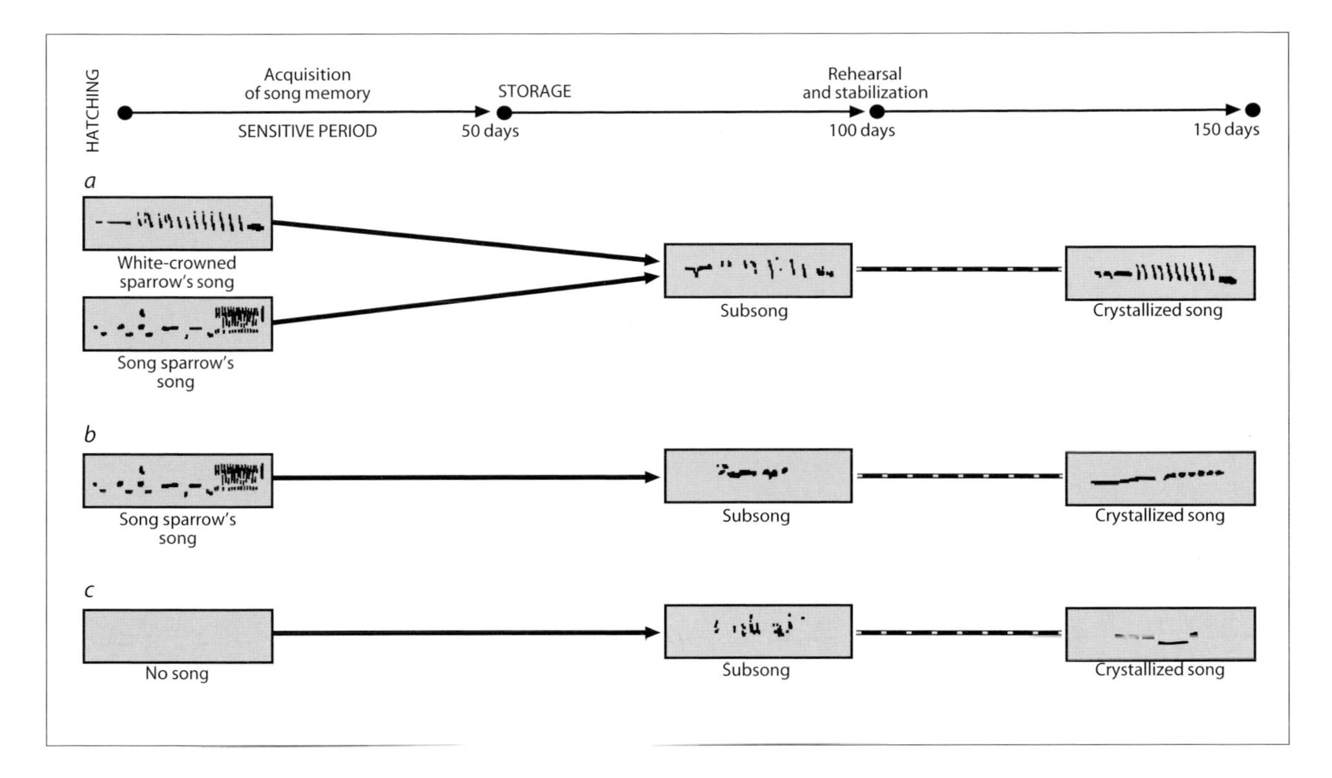

is this learning that birds can be taught to hate laundry bottles.

Learning by rats or pigeons in laboratories was originally thought to be free of innate guidance, but in recent decades much evidence has accumulated to suggest otherwise. For instance, pigeons will learn to associate colours with a food reward, but not sounds. Since seeds are silent, pigeons which ignore these cues altogether learn faster and leave more offspring compared to those who attempt to find acoustic correlations. It's not that pigeons cannot learn auditory stimuli: as a cue for predicting danger, sounds are readily memorized.

Even the experimentation that accompanies trial-and-error learning – the most common way animals in the wild learn new behaviour – seems to benefit from innate guidance. In the lab, pigeons only experiment with the beak when the problem to be solved involves food; for escaping danger, the birds will not learn to peck, but will pick up treadle-hopping behaviour (something they cannot be taught to do to obtain food). Since the beak is the main tool for obtaining food while running away on the legs is important in danger avoidance, these biases create animals that discover and learn the correct responses faster and more reliably.

Latent learning, concepts and insight

It was once thought that animals were limited to learning specific cue and behavioural correlations in the presence of immediate reward or punishment. In fact, animals of many species can learn in the absence of reinforcement and then apply that knowledge to an unrelated situation later. Moreover, a great deal of behaviour is exploratory: for instance, mice spontaneously exercise in running wheels for no reward; indeed, they prefer difficult wheels – rectangular ones, or wheels with internal barriers that must be vaulted over. The same mice will spontaneously explore and learn a thousand-choice-point maze with no reinforcement, a feat far beyond the capacity of our species.

Species-specific learning implies innate preparation and drives. Studies of chimps show that they are able to solve novel problems only if they have had previous experience playing with the tools that will later be helpful. The deep-rooted need to explore and 'fool around' is critical to their subsequent success.

Song learning: given a choice between various songs during their critical period, young white-crown sparrows pick out their own species' song from another's and memorize it. Much later they learn how to sing it themselves. Chicks exposed to no song, or only to songs of other species, later produce only a simple innate melody.

Above *New Caledonian crows forage in the wild with two types of tools they manufacture from twigs and leaves. In the lab, some of these birds are able to discover that a novel tool can be made by bending wire. This hooked wire is ideal for recovering food-laden buckets from plastic tubes.*

Until relatively recently animals were assumed to be only capable of fairly specific, case-by-case learning. More recent studies have shown creatures ranging from honey bees through fish to birds able to learn concepts like 'tree' or 'asymmetry'. Being able to generalize from specific instances to create conceptual classes might enable individuals to make more sense of their world. Some of this concept formation, however, is species specific: pigeons readily learn the concept of relative size, which humans find difficult (because we search for a more complex pattern), while the concept of equality is trivial for us but nearly impossible for the birds. Like learning in general, concept formation may be some mix of innate expectation and learning.

Examples of insight – solving a problem in the mind rather than by trial and error – are rare but convincing. Ravens raised in a flight cage away from adults from birth can solve quite unnatural problems requiring them to recover meat hung from branches on string. Each invents a separate solution. New Caledonian crows can discover that bending a straight wire into a hook creates a tool for removing a bucket of food from a tube. Border collies can infer the names of objects the way children do when they learn language. Parrots can be taught to understand rudimentary English and demonstrate the ability to count, identify common features and form concepts (p. 251).

And yet the hand of instinct is seen in each example. The crows, for instance, cannot learn to bend the hook at the outset; they must discover anew each time that the straight wire will not work. Perhaps not only learning, but all the higher-level cognitive functions of humans and other animals will prove to be species-specific gifts conferred upon us by our genes.

Right *Common octopus in a learning test in a laboratory tank. The octopus has a highly developed brain, and is thought to be the most intelligent of the invertebrates. Tests such as this have shown that octopuses have a good memory, can recognize colours and shapes, and can think through puzzles.*

MARK BROWN

Is an ant colony a superorganism?

Go to the ant, thou sluggard; consider her ways, and be wise: Which having no guide, overseer, or ruler, provideth her meat in the summer, and gathereth her food in the harvest.
PROVERBS 6: 6–8

Ants fascinate us. They have played a starring role in our stories, from folk-tales to films, throughout human history. One reason ants have such a hold on our imaginations is their apparently complex, coordinated and cooperative behaviour. While many have made the analogy between these insect societies and human societies, and drawn lessons from them (including the author of Proverbs), some ant biologists have suggested that an ant colony is more like an individual organism – a superorganism. Is an ant colony really a superorganism, rather than a collection of discrete individuals? And does thinking of an ant colony as a superorganism help us understand how ants behave?

What is an ant colony?

There are over 12,000 species of ants (in contrast, there are just under 5,500 species of mammals), and there are many different types of ant colony. The simplest has a single queen (who has mated with one male) and numerous workers. The queen is the mother of all the workers, and her job is simply reproduction. The workers, who are all female, forage for food, build, maintain and defend the nest, feed and groom the queen, and rear their siblings, the queen's offspring. The ant colony is a sorority, and males are only produced when the colony requires new sexual (or winged) ants. The new sexuals, queens and males, leave the nest, mate and found new colonies. Ant

Ants work together to rear their mother's offspring, here a queen fire ant with her daughter workers and brood.

*A bivouac nest of the New World army ant (*Eciton burchellii*). Army ants cling together by their claws to protect their mothers and sisters.*

colonies like this can, depending on the species, live from a few years to decades, and contain tens to hundreds of thousands of workers. But this is only one of many kinds of ant colony. Ant colonies can have one or several queens, and a queen can be mated with up to 20 or more males (although most queens mate with just one). Also, a colony can live in one or several nests, or even no nest at all. The army ants of Central America are nomadic, creating temporary bivouacs for themselves out of their own, living bodies.

Why do ants live in colonies?

Charles Darwin suggested that all life has evolved through the action of natural selection on individual organisms. The animal behaviour we see today is a result of this process, and so animals behave in ways that increase the number of offspring they produce. Why, then, do worker ants live in colonies, sacrifice their own reproduction and work to help their mother queen reproduce? Darwin wrote that this 'at first appeared to me insuperable, and actually fatal to my theory'. Over a century later, another English biologist, W. D. Hamilton (see also p. 225), solved this problem when he pointed out that because of the way ants are related to each other (see Box), workers gain more reproductive benefit from rearing sisters than rearing their own daughters. That is, workers pass on more copies of their genes through their sisters than they would through their own offspring. So, female workers live together with their mother for their own reproductive benefit.

Ant colony as superorganism?

The division of work in an ant colony – the queen reproduces and the workers do everything else – can be compared to the division of our bodies into reproductive tissue and the somatic (body) tissues that support it. Just as we use our bodies and behaviour to acquire food, defend ourselves from enemies and generally act to increase our chances of reproducing, the behaviour of ant workers appears aimed to increase reproduction by the queen. In the early 1900s, William Morton Wheeler (an American ant biologist) developed the idea that ant colonies were superorganisms. As a superorganism, the colony becomes the indi-

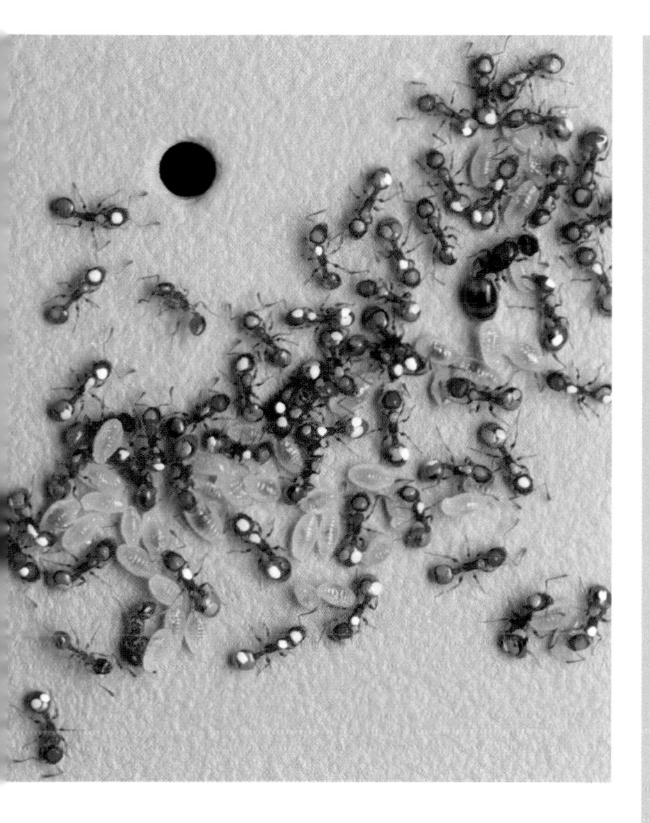

Above left *A colony of the ant* Temnothorax curvispinosus, *including a queen, workers and brood. The workers have been given unique paint marks to facilitate tracking their behaviour during collective tasks.*

ANT RELATEDNESS

In humans, males and females have two sets of chromosomes – they are diploid. Mothers and fathers each copy half their genes to new offspring, so they are each related to their offspring by 0.5. On average, brothers and sisters share 0.25 of their maternal genes and 0.25 of their paternal genes, and so half of their genes with each other. For a female, the genetic benefit of either producing daughters or having sisters is equal – they both share 50% of her genes.

In ants, males have only one set of chromosomes – they are haploid. Ants are known as a haplo-diploid species. In a haplo-diploid ant family, a queen produces sons on her own – male ants have no father. The queen copies half of her genes into each son, and so is related to them by 0.5, but each son receives all his genes from his mother, so is related to her by 1.0. This is an example of asymmetric relatedness. Queens and daughters are related to each other by 0.5, but sisters are related to each other by 0.75 because they share all of their father's genes (0.5) plus 25% of their mother's genes. Consequently, the genetic benefit of producing sisters (0.75) is higher than that of producing their own daughters (0.5).

Diploid organisms (e.g., humans)

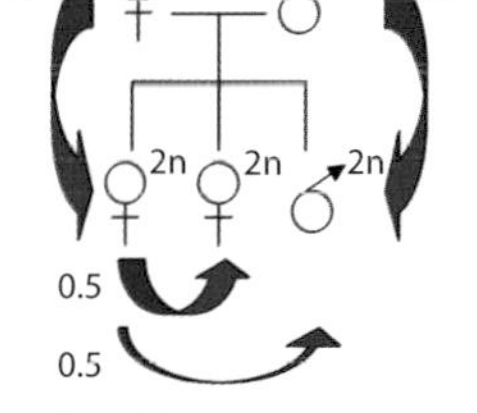

Haplo-diploid animals (e.g., ants)

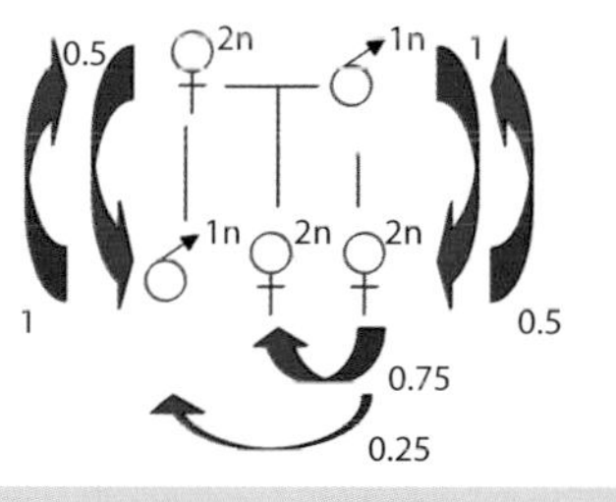

This diagram shows how male and female ants are related to each other, and how this compares to a system we are more familiar with – ourselves.

vidual upon which natural selection acts, and ant behaviour can only be understood by looking at how it benefits the colony as a whole.

This has often been seen by biologists as a rather shocking idea – natural selection acts on individuals, not groups – and consequently, prior to Hamilton's explanation of why ants live in colonies, the idea of the ant colony as superorganism fell out of fashion. However, like fashion, it has recently come back into favour. But why?

Ant behaviour is not directed by the queen, but comes about through the decisions of individual ants when they encounter biological, chemical or physical cues. Together, ants produce complex nest structures, collect food from their environment and rear new members of the colony. The key to the re-emergence of superorganism-thinking is the relationship between these cooperative behaviours and a colony's production of new workers, queens and males. Cooperative behaviour that leads to the production of more offspring by the colony will be favoured by natural selection, irrespective of whether it is functional at the level of the individual ant. For example, in colonies of honey-pot ants some individuals act as larders, storing food in their grossly swollen abdomens to the point that they can no longer walk.

So, in terms of cooperative behaviour, the colony is a unit and is being selected just like an individual non-social animal would be. To all intents and purposes, the colony is indeed a superorganism.

When is an ant colony not a superorganism?

But is that the whole story? Not quite. Ant colonies are not utopias of cooperation. Ironically, the same relatedness asymmetry that causes ants to work together to produce more sisters also results in conflict in the colony. In essence, workers and queens disagree about the relative proportions of

The glistening bodies of honey-pot ants contain food for the whole colony. Sugary fluid is stored in their gut, so much so that their abdomens are enormously swollen and they can no longer walk.

new queens and males that a colony produces, and even who should produce the males.

Because an ant queen has equal reproductive benefit from producing daughters and sons – they each carry half of her genes – a queen should produce the same number of each sex. In contrast, workers pass on more of their genes through rearing sisters than brothers. This leads to a tug-of-war between queens and workers over a 50:50 versus a 75:25 sex ratio. Studies suggest that queens usually lose – after all, there is only one queen pulling and many workers – so colonies tend to produce the workers' preferred sex ratio.

A further twist is that in some ant species, workers can lay haploid eggs that develop into males. Because female ants share more genes with their own sons, and then their nephews, and then their brothers, workers should compete with their mother, and with each other, to lay male eggs. Such conflict may reduce the efficiency, and thus the reproductive output of the colony as a whole. In contrast, if the colony is a cooperative superorganism, workers should give up the chance to have sons in order to rear their mother's offspring. Evidence suggests that the superorganism triumphs over anarchy, with selection for greater productivity of the colony outweighing selection on individual workers to have their own sons.

Further, more complex conflicts emerge in colonies with multiple queens, or multiply-mated queens, or multiple nests. While this is a focus of much current research, as yet we do not understand how all of these conflicts are resolved. Pan out – an ant colony moves across the terrain from its nest, like the tentacles of a sea anemone bringing in food. Focus in – individual ants work together, but also compete with each other and their mother for reproductive supremacy. Ultimately, whether an ant colony is a superorganism depends upon what you want to understand about ant behaviour. Ant colonies are thus a fascinating mixture of cooperation and conflict and thinking of them as superorganisms can help us understand some, but not all, of this behaviour.

RAGHAVENDRA GADAKGAR

Why are animals nice to each other?

Natural selection, it should never be forgotten, can act solely through and for the advantage of each being.
CHARLES DARWIN, 1859

It may sound strange, even perhaps malicious, to label niceness as a mystery. But that's just what it is for evolutionary biologists, who like to label anything that they cannot easily explain through Darwin's theory of natural selection as a mystery. Natural selection, graphically described by Darwin's phrase 'the preservation of favoured races in the struggle for life', prepares us to expect competitive selfishness rather than cooperation and altruism. After all, how can an individual that pays a cost in helping another be expected to win the race to survive and reproduce?

And yet we find many examples of animals doing just that. Honey bee workers kill themselves in the process of stinging predators that might destroy their nest. Helpers at the nest of the bee-eater postpone rearing their own offspring and spend time and energy in assisting their parents to raise an additional brood. A ground squirrel risks attracting the attention of the predator to itself by giving an alarm call to warn its neighbours. Why aren't such individuals eliminated by virtue of lowering their chances of survival and reproduction?

Ground squirrels, which give alarm calls warning others of danger while putting themselves at a slightly greater risk of predation, provide an opportunity to investigate how such altruistic behaviours spread through natural selection.

Many birds, such as the bee-eaters seen here, show cooperative breeding, where older siblings who are reproductively mature postpone breeding and remain in their natal nests to assist their parents in raising an additional brood.

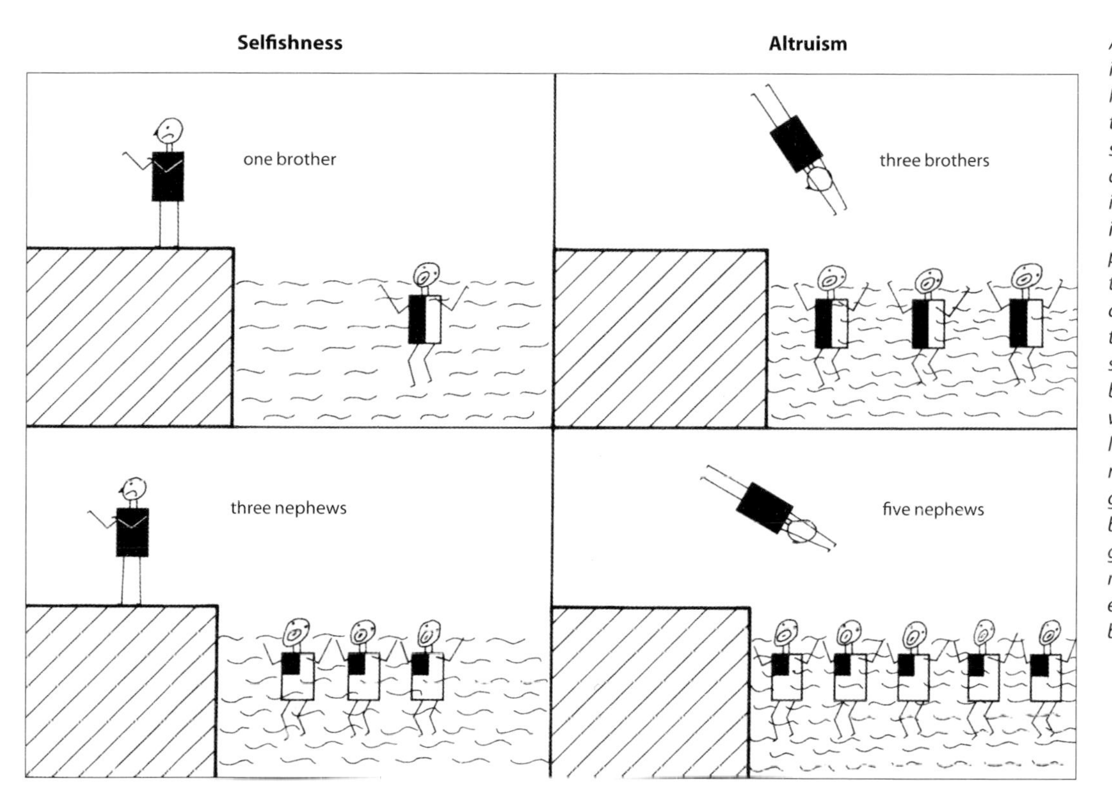

A cartoon illustrating J. B. S. Haldane's idea:. the shaded sections of the drowning individuals indicate the proportions of their genes which are also present in the altruist standing on the bank. The altruist is willing to risk his life when the number of his genes expected to be rescued is greater than the number in his body expected to be lost by his drowning.

A modern evolutionary theory

Not surprisingly, humans have displayed an absorbing fascination for cases of cooperation in the animal world, long before the evolutionary puzzle associated with them became evident. Precise evolutionary thinking on this matter can be traced back to J. B. S. Haldane, who realized that risking one's life to save drowning relatives can indeed be favoured by natural selection, provided that more copies of genes that give rise to such behaviour are recovered in the saved relatives than are lost in the risk taker. W. D. Hamilton (see also p. 222) formalized essentially the same idea in what has since come to be known as Hamilton's Rule, which states that an altruistic gene will spread in a population when the benefit to the recipient, devalued by the coefficient of relatedness between altruist and recipient, is greater than the cost incurred by the altruist.

Thus the alarm-calling behaviour of the ground squirrel is no longer a mystery if the probability of saving individuals carrying genes for alarm-calling is greater than the probability of losing one copy of such a gene due to the death of the caller. Similarly, the helping behaviour of the bee-eater can be explained if its assistance at its parents' nest results in the rearing of more additional siblings than the number of offspring it might have produced instead of helping. Not only does this theory provide a logical explanation of why cooperation evolves more easily among kin, it also shows why close kinship is not always essential. If the benefit is very much greater than the cost, even low genetic relatedness will suffice.

Testing the theory

The theory is elegant indeed, but the hard part is to show that animals behave as if they obey Hamilton's Rule. Here, most observers have chosen the easy option of assuming that the cost and benefit terms are equal, and of testing the simpler prediction that altruism is more often directed towards close relatives than it is to distant relatives or non-relatives. This simpler prediction is sometimes, but not always upheld. Thus an excessive and often exclusive focus on measurement of relatedness, and the neglect of the cost and benefit terms in empirical studies, has sometimes given the false

Ropalidia marginata *is a tropical paper wasp abundant in South India. Each nest may contain from one to around a hundred adult female wasps, but only one – the queen – is fertile, while the others assist in raising her offspring. Because the queen is no different in size or shape from the sterile helpers, such a society is considered 'primitive'.*

impression that Hamilton's Rule is inadequate to explain altruism. Where the cost and benefit terms have been measured, Hamilton's Rule has indeed provided a powerful tool to understand altruism.

Studies on the white-fronted bee-eater in Kenya have shown that not only the presence of helpers at the nest, but also the bizarre behaviour of a father harassing his sons to return and act as helpers, is consistent with the predictions of Hamilton's Rule. Computation of the costs, benefits and relatedness involved in different strategies shows that by harassing sons and bringing them back to help rear additional offspring, a father gains a substantial fitness advantage. In contrast, sons reap about the same fitness benefit whether they resist their father's harassment and carry on with their own family life or whether they succumb and return to act as helpers.

In our study of the primitively social wasp *Ropalidia marginata* in South India, we have used Hamilton's Rule to compute the costs and benefits for different wasps of remaining in their mother's nests as sterile helpers versus leaving to found their own new colonies and reproduce. It turns out that for some individuals staying back is a more profitable strategy, while for others leaving brings more fitness. We have succeeded in predicting correctly the fraction of the population that should opt for a sterile helper role in a social setting rather than a reproductive role in a solitary setting.

Hamilton's Rule is, however, inadequate when cooperation is directed towards non-relatives. The theory of reciprocal altruism (which is based on the idea that favours are returned after a time lag) provides a powerful explanation for cases of cooperation among non-relatives. There is also a relatively untested but very promising radical new idea that altruism may simply be a handicap that the most successful individuals can afford to take on without paying the same cost that unsuccessful individuals would have to pay.

While more needs to be done on the theoretical front, empirical studies measuring all three terms – cost, benefit and relatedness – are now what is mostly required to clinch our understanding of the evolution of altruism. But I would hazard a guess that we are poised to demystify the evolution of niceness in the natural world.

How do animals navigate?

Doth the hawk fly by Thy wisdom and stretch her wings toward the south?
BOOK OF JOB 39:26

Each autumn, millions of monarch butterflies stream south from the northeastern United States en route to one of a few isolated mountaintops in Mexico. None has ever seen the destination: all of them were born in the summer at least a thousand miles north of their wintering grounds. At the same time hundreds of millions of juvenile birds are also migrating south for the first time, each to a species-specific range. Salmon, sea turtles and dolphins make equally epic journeys. How do they know where to go?

Cognitive maps and local route finding

Some animals do not migrate, but nevertheless navigate within their familiar home ranges. One approach for local navigation is to memorize all the landmarks in the region. The animal consults the resulting internal map when it needs to set course for another goal – its nest for instance. Although this sounds simple, map-based navigation is considered a higher cognitive accomplishment.

Whether an animal is navigating locally or migrating across the globe, it needs a compass to determine direction. A honey bee knowing that she is 200 m (660 ft) south of the hive must be able to figure out which direction is north before she sets off home. For most animals a sense of location is also important. The same bee with a sense of direction must also know where she is in relation to the hive if she is to use her compass to set a course back. An internal map can provide location, as can dead reckoning – judging where you are by keeping track of the distance and direction of each leg of your outward journey.

The compass sense

The most common compass used by navigating and migrating animals is the sun. But the sun is a problematical guide: its direction changes from east to west over the course of the day, and not at a steady rate (moving most quickly in azimuth near noon). Moreover, this pattern changes with the seasons and with latitude, so no simple rule or formula can turn the sight of the sun into a direction. And, of course, the sun is below the horizon for much of the day, or may be obscured by a cloud; for aquatic animals, the bending of light by water makes the problem even more difficult.

The answer to the problem of sun movement may be simply to relearn it every so often. When the sun is hidden behind a cloud or under the horizon, but blue sky is visible elsewhere, birds

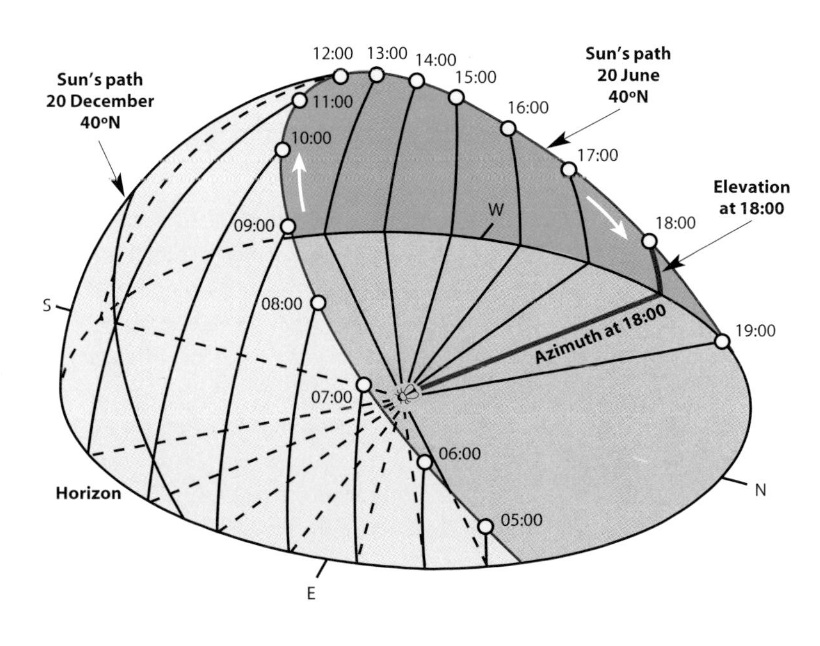

The sun's arc is shown hour by hour for the summer and winter solstices at 40° north, roughly the latitude of New York and Madrid. The arc varies with latitude: further south the path through the sky is more vertical and the variation between summer and winter is less; further north the path is lower and the variation much more extreme.

Right *The sun is surrounded by a regular pattern of polarized-light 'vectors' created by scattering in the atmosphere. The sun always lies on an arc passing perpendicularly through a polarization vector in the sky (the distortion in this two-dimensional fish-eye diagram obscures this relationship). The degree of polarization, indicated by the thickness of the polarization vectors, depends on the angular distance from the sun.*

Above right *Some species of European warblers (red line) migrate south by steering southwest until meeting the magnetic dip angle for southern Spain (or for 40 days, whichever is first), and then southeast until the dip angle is roughly horizontal (or another 30 days). Warblers of the same species starting further east (blue line) fly southeast and then southwest. In either case the route avoids the Alps and the Mediterranean.*

Opposite *A migrating herd of porcupine caribou (*Rangifer taranbus*) crossing a narrow stream in the Arctic National Wildlife Refuge, Alaska.*

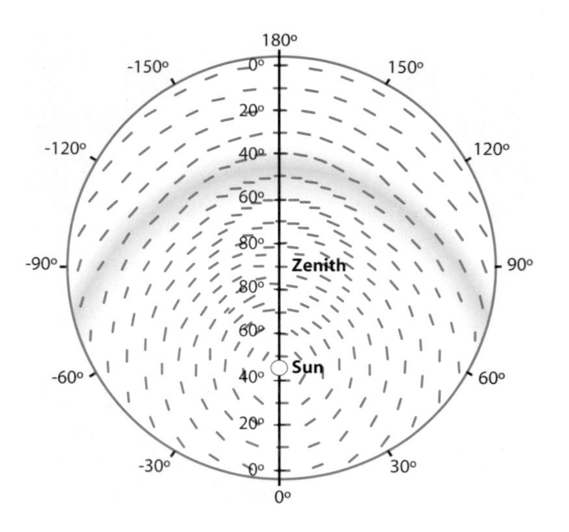

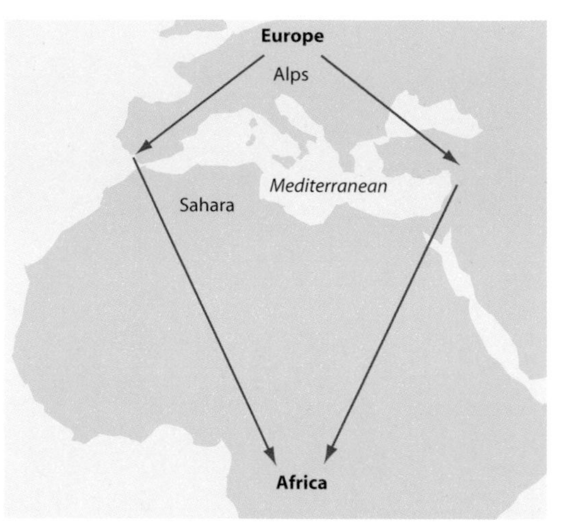

and insects use a back-up system: the sun-centred pattern of polarized light in the sky. This pattern, invisible to our eyes, is clearest in the ultraviolet (a wavelength also invisible to us). Many insects and birds use polarization to infer the sun's location.

Overcast skies and nighttime (when there is no blue sky visible) are more serious challenges. Many birds begin to use magnetic compasses under these circumstances. Vertebrates with the ability to navigate magnetically have a small organ in the head containing magnetite; some also have a special molecular system in the retina sensitive to the Earth's field. Monarch butterflies and honey bees are also quite sensitive to the Earth's magnetic field, though currently we have no certain evidence that they use this ability in their navigation.

Long-distance travel without maps

When travelling through unfamiliar terrain, we consult a detailed map and keep track of our position relative to our goal. But for many creatures the innate requirements for moving from the breeding ground to a winter refuge seem to consist only of a compass and a brief set of instructions. The instructions for garden warblers in Europe read (approximately): 'fly southwest for 40 days or to latitude 40°, then turn southeast to latitude 10° or for another 30 days'. The latitudes are measured by using the dip angle of the Earth's magnetic field (the angle between the direction of the magnetic field and the horizon; see p. 58).

Warblers, like almost all birds, migrate at night to avoid predators and the exhausting heat of the sun. But this means neither the sun nor polarization is available. Many night migrants memorize the pattern of stars in the sky, as well as the pole point about which they rotate. A glance at a patch of clear night sky as they fly allows them to infer north, but they must update this mental picture as the seasons change (and the spring constellations they originally learnt are slowly replaced by autumn patterns). Moreover, as they move south, new constellations formerly hidden under the southern horizon become visible, while northerly ones begin to disappear, requiring more recalibration. And when the sky is overcast, they must rely on the Earth's magnetic field as a compass.

Migrating with a map

Just as bees and birds create landmark maps of their home ranges, some migrating birds formulate maps of their long-distance journeys. One dramatic use of this is seen in waterbirds such as geese and cranes. The parents use their memory of the migration route to lead their young, who then will be able to lead the next generation. To do this, the birds need to fly when there is enough light to see one another (and the landmarks). This is why the first picture that comes to mind when we think of migration is almost always a V-shaped flock of geese.

Many species of birds – perhaps most – as well as many aquatic animals seem to know where

Individuals in most species of birds migrate individually and at night. Many waterbirds, including ducks, geese, swans and cranes fly in groups and in daylight or twilight. Migrating in flocks allows older birds to lead the younger along the route, and thus gives the new birds a chance to memorize the landmarks along the way.

they are during migration, even if it is the individual's first journey – they have a map sense, one of the most mysterious abilities in nature. In theory, as with human navigators, latitude (the north/south location between pole and equator) should be relatively easy to determine – the elevation of the pole point in the sky subtracted from 90° is the latitude. The same calculation can be made with the magnetic dip angle (with less precision, since the magnetic pole is more than 1,000 km/610 miles from the geographic pole).

The problem is with longitude, the east/west direction. For most of human history, navigators depended on landmarks or dead reckoning. More recently extremely elaborate and delicate chronometers were used (now, of course, we have GPS satellites). The internal clock senses of animals are uselessly inaccurate (by about 15 minutes per day, which corresponds to an error of more than 350 km or 217 miles at the equator).

Ironically, most studies of map sense have used a species that does not migrate: homing pigeons. These birds can be taken hundreds of kilometres from their loft in the dark and into completely unfamiliar terrain, released, and after circling once or twice, will fly towards home. When birds are fitted with translucent goggles allowing them no ability to see shapes, it is possible to estimate the accuracy of return in the range 2–5 km (1.25–3 miles). This is a very precise map indeed.

Odour or magnetic fields?

One school of thought holds that the young pigeons in the loft memorize the odours wafting in on breezes from various directions. When released, birds would then sample the odours at the new locale and set off along the reverse of the bearing they associate with the smells they sensed at home. Even if it were true for the homing pigeons, this idea could not explain normal migration.

The other major hypothesis is that the map is based on magnetic information – perhaps the pattern of difference between total magnetic intensity, vertical intensity and horizontal intensity. This would explain why certain sorts of subtle magnetic disturbances disorient pigeons.

Experiments on sea turtles and other species displaced in latitude by a few tens of kilometres from home show that they can choose the correct direction based entirely on magnetic cues. The critical question, however, is whether the same creatures could solve the problem if only the magnetic longitude is altered.

There is no reason to believe that we yet know the full range of cues animals use for their compass orientation, nor the elaborate ways in which compasses are recalibrated and adjusted. Animal navigation therefore continues to be one of the most active and intriguing areas of research in animal behaviour.

JOAN ROUGHGARDEN

Sexual selection

55

Just as man can improve the breeds of his game-cocks by the selection of those birds which are victorious in the cockpit, so it appears that the strongest and most vigorous males, or those provided with the best weapons, have prevailed under nature, and have led to the improvement of the natural breed or species....

... Many female progenitors of the peacock must, during a long line of descent, have appreciated this superiority; for they have unconsciously, by the continued preference of the most beautiful males, rendered the peacock the most splendid of living birds.

CHARLES DARWIN, 1871

Originating with Darwin in 1871, the theory of sexual selection today offers a unified evolutionary narrative for just about everything concerning sex, gender and sexuality. The current grand scope of the theory belies a more humble origin, when Darwin focused on traits such as the peacock's tail and the stag's antlers. Darwin noted that gaudy tails and heavy antlers could hardly promote survival, so he theorized that these traits aided reproduction. In his view, females have the ability and opportunity to choose their mate and he postulated that they possess an innate aesthetic preference for gaudy male ornaments. He further conjectured that males compete for access to females, and that females choose those with armaments which succeed in male–male combat. By their choice of mates, females thus breed their male offspring to have both ornaments and armaments.

According to Darwin: 'Males of almost all animals have stronger passions than females', and 'the female ... with the rarest of exceptions is less eager than the male ... she is coy'. Throughout nature, then, males and females possess near-universal characteristics: passionate males fight one another and seek female favour, while females coyly choose males who bring beauty and/or weaponry to the bedroom.

There is a heated debate at present among animal behaviourists about sexual selection, and the points at issue hinge on the evidence: do we see the processes and the gender roles simply as Darwin did, or do we dig deeper and look for evidence of sexual selection in action and the heritability of supposedly sexual traits?

The Darwinian view

The Darwinian narrative of sex roles is often considered proven fact. In a review, the geneticist Jerry Coyne wrote in 2004 that 'Males, who can produce many offspring with only minimal investment, spread their genes most effectively by mating promiscuously.... Female reproductive output is far more constrained by the metabolic costs of producing eggs or offspring, and thus a female's interests are served more by mate quality than by mate quantity.' The passionate male has become the promiscuous male, the coy female has become the constrained female, and aesthetic preference has become genetic preference.

The spirit of today's narrative is identical to Darwin's of around 130 years ago. A strategic conflict of interest springing from the size difference between egg and sperm is thought to result in

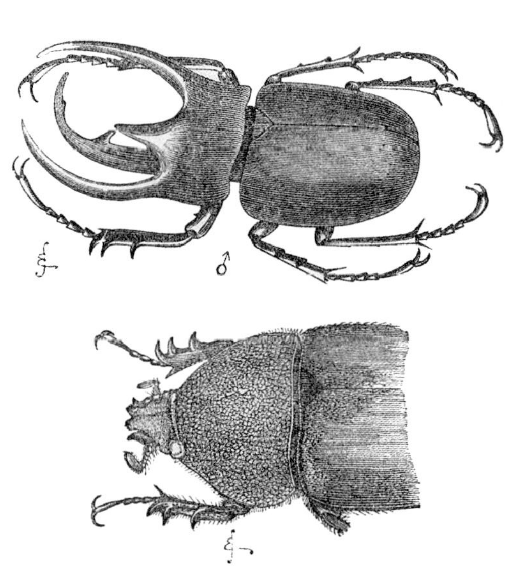

A plate from Darwin's The Descent of Man and Selection in Relation to Sex *(1888) showing* Chalcosoma atlas, *a beetle of southeast Asia: the upper one, with large 'horns' is a male (reduced size); the lower is female.*

Above *A pair of great frigatebirds on Genovesa Island, Galápagos – the male displays his brightly coloured pouch to communicate with a possible mate.*

Opposite *An example of sex-role reversal: a male Atlantic seahorse expelling fry from his pouch; male seahorses provide the parental care.*

universal sexual conflict. Although this sexual selection narrative is widely taught, its truth is increasingly under question for three reasons, leaving evolutionary biology today without any convincing account for sex, gender and sexuality in nature.

Three challenges to the Darwinian view

The first problem is the huge diversity in sex, gender and sexuality now known to exist among animals. The male/female binary taken as the base-line for sexual selection is merely a specialization. The most common body plan among plants and animals is for an individual to produce both eggs and sperm. Even among vertebrates, about 30% of the fishes on a coral reef belong to species whose individuals may change sex or be both sexes simultaneously – all beyond the scope of original sexual selection theory. And of the species where individuals do persist in one sex throughout life, many contain multiple types of males and females with entirely different sizes, shapes, colours and life histories, including age at first reproduction, longevity and propensity to maintain territories. Neither males nor females line up with a single behavioural template such as passionate or coy.

Still other species exhibit 'sex-role reversal', in which the female is gaudy and the male drab – the opposite of the peacock. In these, like the sea-horses and pipefish, or the sea-shore inhabitating phalarope or freshwater jacana, the male provides the parental care and the female is the more promiscuous. These species show that sex role cannot be derived from gamete size because male seahorse, pipefish, phalaropes and jacanas make tiny sperm (the reason they are defined as males), while the females make relatively expensive large eggs, yet their roles are reversed relative to the sexual selection 'norm'.

Finally, the commonness of same-sex sexuality among animals is becoming appreciated. Over

A pair of male flamingos (Carlos and Fernando) at Slimbridge Wetlands Centre, in central England, with their adopted chick. The couple had tried stealing eggs before, and so were given an abandoned chick to rear.

300 species of vertebrates employing same-sex sexuality under natural conditions in the wild are documented in peer-reviewed literature. According to sexual selection, this activity must be seen as mistaken, deceitful, wasteful or maladaptive. Yet this phenomenon occurs in high enough frequency that it almost surely serves some adaptive social function. Sexual selection theorists have augmented the original narrative with special-case explanations for all this diversity. While these arguments might ultimately prove satisfactory, they also invite scepticism that the entire approach is on the wrong track.

A second difficulty for sexual selection is that the species that were thought to be its exemplars turn out not to support the theory after all. A multi-year study of the European collared flycatcher shows that the 'badge', a white spot on the forehead of males, is not an ornament that females can use to increase the evolutionary prospects of their young – the variation in male fitness is not heritable, so searching for a male whose genes are better than any other male's, including genes correlated with badge size, is a waste of time. Similarly, the forehead colour patch on the British blue tit has no heritability, and so female choice of a large colour patch is not reflected in her offspring. And variations in the long tail feathers of the European barn swallow are not related to sexual selection but to manoeuvrability.

Furthermore, a review of examples of extra-pair paternity, in which eggs in a nest are sired by a male from a nearby nest and not by the male tending the nest, found that fewer than 50% of over 100 published studies claimed support for sexual selection. According to the standard view of sexual selection theory, these examples of nest-hopping were seen as evidence that females were mating with nearby males to acquire a genetic upgrade relative to her consort, an interpretation usually refuted by the data.

Another source of trouble for sexual selection theory is the logical consistency of the theory itself. If, as sexual selection claims, female choice weeds out bad genes in males, then after about 20 generations, all these genes should have disappeared, leaving a situation where all the remaining genes are equivalent with respect to evolutionary fitness, thereby rendering female choice for male genetic quality redundant. This dilemma is called the 'paradox of the lek' (leks are gatherings of males of certain species, such as black grouse, for mating displays). Papers continue to offer 'resolutions' of the paradox, but the proliferation of proposed solutions itself indicates that the paradox remains unresolved, and invites suspicion that it never will be.

The opportunity therefore now exists for evolutionary biologists to rethink the subjects presently addressed by sexual selection theory and to frame fundamentally new theoretical approaches to the subject.

DARIN A. CROFT

Why do deer have antlers: fighting or display?

Oh give me a home where the buffalo roam and the deer and the antelope play.
DR BREWSTER M. HIGLEY, 1873

In addition to having cloven hooves, buffalo (bison), deer and antelope (pronghorn) all have another conspicuous trait in common – cranial appendages in the form of horns or antlers. Although these structures are designed differently in these different animals, and may have evolved independently in each lineage, their presence in all three suggests they serve some important function. Perhaps such structures facilitate the recognition of other members of their own species on the open prairie. Or maybe they advertise the health and physique of the individual displaying them, attracting members of the opposite sex. They might also be used for battling rivals and fending off would-be predators. These various functions are not mutually exclusive, of course, and each may be of varying importance in different species.

Species recognition

The great variation in horn and antler morphology makes it relatively easy to identify an ungulate based only on their structure. Is it possible such cranial appendages serve the same function for the animals themselves? Although such a species-recognition function cannot be ruled out, it is unlikely this is primarily responsible for the widespread occurrence of antlers and horns; body shape, coloration and patterning also permit species to be distinguished visually, and having such elaborate structures would seem to be an evolutionarily expensive method of advertising identity compared to these other attributes. Moreover, deer and their relatives only bear antlers seasonally; how then would it be possible for these animals to recognize each other during the rest of the year?

Below left *Many hooved mammals (ungulates) bear either horns or antlers, such as this caribou.*

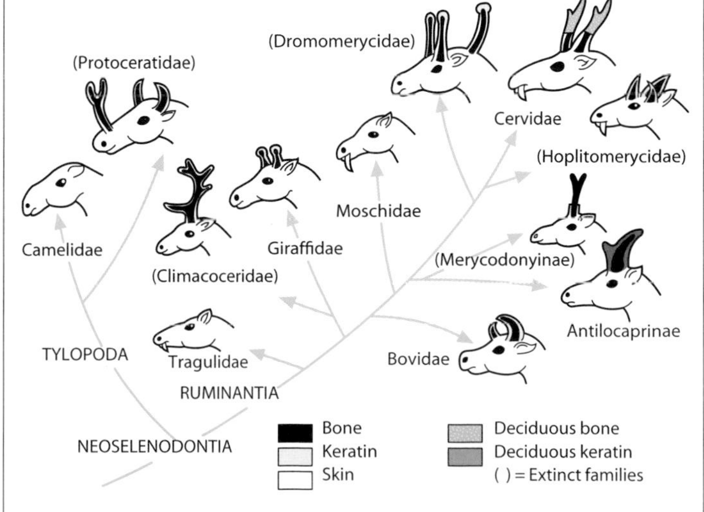

Below right *Ungulates display a great range of types of skull ornamentation and each has a different contribution of bone, skin and keratin.*

Far right *Male deer rub their antlers against trees to remove velvet, to advertise their presence to other deer and to apply scents to their antlers.*

THE 'IRISH ELK'

The most impressive antlers ever seen belonged to the 'Irish elk', *Megaloceros giganteus*, which went extinct around 8,000 years ago. A member of the deer family, its antlers spanned up to 3.5 m (over 11 ft) and weighed up to 40 kg (88 lb), making them about one-third larger than those of the modern moose (*Alces alces*). Rather than having extraordinarily large antlers for its size, mathematical relationships demonstrate that *Megaloceros* was proportioned correctly relative to other living deer; the moose, on the other hand, has antlers that are smaller than expected for its large body size. Most scholars now believe it is unlikely that the antlers of *Megaloceros* were the result of 'runaway' sexual selection.

Above *The extinct 'Irish elk' had the largest antlers of any deer.*

Attracting females

Could antlers and horns be the mammalian behavioural equivalent of the showy plumage that some male birds use to attract mates? Antlers and horns occur only in the males of many ungulate species, the expected pattern when 'female choice' is involved. The huge antlers of the extinct 'Irish elk' were once thought to be an example of 'runaway' sexual selection – a characteristic that was selected by females to the eventual detriment of the male bearing them. Among species in which both sexes have cranial appendages, those of males are often larger and/or more elaborate than those of females; this also makes sense in the context of female choice. However, most behavioural studies suggest that female deer and other such ungulates have relatively little choice with whom they mate, casting doubt on the significance of female choice in the evolution and maintenance of antlers and horns (see p. 233).

Antlers may have another role in communicating with females: scent. Like most mammals, deer and other ungulates use scents to mark their territory and to communicate with other members of their species. The velvet that covers growing antlers is known to produce pheromones (sexual olfactory signals; p. 241), and once the velvet is shed, urine is often applied to the antlers directly (by spraying) or indirectly (by rubbing on scent-marked trees or bushes) for similar effect. Branched antlers provide abundant surface area for the diffusion of odours, and their head-level position permits the odours to be distributed via air currents. Although this phenomenon appears to be common among deer, it does not occur in species that have horns instead of antlers.

Battles among males

Dominance hierarchies are common in mammals, and such social structures can be established in a variety of different ways. In many species, males battle each other to determine the social order, to defend resource patches or to win access to females. So are antlers and horns primarily duelling weapons? They would seem to be well designed for this purpose and such competitions are commonly observed in wild populations. Deer, for instance, lock their antlers together to permit them to test each other's strength, while male sheep run at each other and crash their horns together, resulting in a sound that can be heard for miles. Various antelope joust and parry with their long horns. Although all these battles can result in injury to one or both competitors, this is usually not the intended result.

The generally uneven distribution of antlers and horns between the sexes makes sense in the context of male–male combat, as does the timing of antler growth in some species. In deer, antlers grow during late spring and summer and are fully formed by autumn, in time for the mating season ('rut'). During rut, male deer use their antlers in contests to win the right to mate with one or more females; the antlers are then shed. Males with small or broken antlers generally do not win the chance to mate, though they may have another opportunity the next season when their antlers regrow. Antlers and horns as duelling weapons therefore seems a good fit with the evidence.

Defence against predators

Just as antlers and horns can be used to battle other males, they can also be used in defence; any

The main function of antlers and horns is combat between males in the mating season. Here two bison lock horns, to determine which will have access to a receptive female.

A DIVERSITY OF HORNS

The extinct Hoplitomeryx *(left), closely related to deer, had five horns.* Hexameryx *(right), also extinct, is a member of the pronghorn family.*

Only a single living ungulate has more than one pair of horns: the four-horned antelope (*Tetracerus quadricornis*). In contrast, many extinct ungulates displayed an amazing diversity of combinations of horns and hornlike structures. Among the more unusual of these was *Hoplitomeryx*, a close relative of the deer family that lived in the region of Gargano, Italy, *c.* 5–10 million years ago; it bore five relatively large horns and sported a pair of imposing upper canine teeth. A contemporary from Florida, *Hexameryx*, had three pairs of horns.

toreador can testify to the deadly effectiveness of a bull's horns, and instances of a potential prey animal wounding a would-be predator have been documented in the wild. Such encounters are not common, however, and are usually a secondary defence strategy used only when fleeing is no longer an option. Antlers would probably occur in both sexes if defence against predators were their primary function, and one would not expect such structures to be shed seasonally, as occurs in deer. Thus, although antlers and horns can be used for defence, this probably has little to do with their evolution or present distribution.

The function of antlers

The primary function of antlers in deer and their close relatives thus appears to be male–male combat; female choice probably also has some role, as does signalling (especially by scent). In other words, antlers are most likely the result of some form of sexual selection, either intrasexual (duelling) or intersexual (female choice). In caribou, the only species of deer in which females have antlers, such structures may aid in feeding during winter (by helping remove snow cover from vegetation) or help with calcium storage. Even so, antlers are larger in male caribou than in females, evidence of their importance in male–male combat. For Chinese water deer, the only deer species that lacks antlers altogether, males use their long upper canine teeth for combat instead. Male–male combat also appears to be the primary function of horns, though the importance of secondary functions such as signalling and defence can vary depending on the ecology of the particular species in question.

Invisible signalling: pheromones

Candle in hand, we entered the room. What we saw is unforgettable. With a soft flic-flac *the great night-moths were flying round the [female].... They rushed at the candle and extinguished it with a flap of the wing; they fluttered on our shoulders, clung to our clothing, grazed our faces. My study had become a cave of a necromancer, the darkness alive with creatures of the night!*

J. H. FABRÉ, 1911

We are such visual animals that we miss a method of signalling used by other species that is going on right under our noses. Pheromones (chemical communication) are the main signals employed across the animal kingdom, mediating behaviours of all kinds. Animals can detect a huge range of complex information by smell. When dogs sniff each other (or lamp-posts) they learn 'who' the other dog is, its age, sex, maturity, health – and whether it is on heat. Ant colonies are organized by smell, from the building of nests to leaving pheromone trails that lead to food.

It is therefore an irony that this most prevalent communication channel is also the least understood – and that goes for human pheromones too. The challenge is that pheromones are usually invisible and are released in vanishingly small quantities. It used to be a great mystery how male moths located females in the dark. All sorts of theories were proposed – for example that the males were detecting the sound of females or even that the male's large feathery antennae picked up infrared. At the turn of the 20th century, the French entomologist J. H. Fabré formed good ideas from simple experiments in which he eliminated every possible sense the moths could be using apart from smell, but since he couldn't smell anything he mistakenly thought that the females were not producing one. It is only relatively recently, since the 1950s, that the first pheromones have been identified.

The large feathery antennae of a male Chinese oak silkworm moth (Antheraea pernyi) *have evolved to detect minute concentrations of pheromone released by female moths into the air. The pheromone stimulates the male to fly upwind to find her.*

*A recent discovery is that frogs and other amphibians use pheromones. The magnificent tree frog (*Litoria splendida*) attracts mates with a potent soluble pheromone. The females' response from across the pond is immediate.*

New examples of pheromones are still being found across the animal kingdom. For example, the magnificent tree frog *Litoria splendida* in Australia releases a potent female sex pheromone into the water – the pheromone seems to travel very fast across the surface tension (and is being explored as a surfactant for introducing asthma medicines into the lungs).

For the majority of species we still do not know the chemicals of the pheromones they use – and some of the omissions are surprising. For instance, we haven't yet discovered – despite some false leads – what the female dog pheromone is, though every dog-owner is well aware of how powerful it can be.

How are pheromones perceived?

Pheromones are perceived with the sense of smell, which is a different kind of sense from the others: actual molecules have to travel from sender to receiver (unlike visual or sound signals). This leads to special characteristics of pheromones – a smell can 'shout' even though the animal is not there, an attribute exploited by dogs and lions marking their territories. Pheromones can go round corners, and they work at night, so many nocturnal animals use pheromones extensively. The chemical characteristics of pheromones often seem to have evolved to match the signal or environment – for example, the territorial pheromone marks left by hyenas are long-lasting, high-molecular weight compounds. By contrast, ant alarm pheromones are low-molecular weight compounds that diffuse quickly and then rapidly disperse once the alarm has passed.

What is needed in pheromone research is something as simple as a microphone for picking up smells and amplifying them – but we don't yet have this. We also have a problem identifying what the chemicals are, as the quantities (unless you are studying elephant pheromones) are so small. The machine that has made most of the analysis possible is the gas chromatograph coupled to a mass spectrophotometer. But we are often working at the limits of chemical analysis.

'Replaying' smells is the other unsolved challenge in understanding pheromones. There is no equivalent of a tape-recorder for playing back smells. And the synthesis of the chemicals used as pheromones, in all their variety, can be difficult, because the result has to be exactly the same as the animal uses itself, and pheromones are rarely single chemicals.

Finding an odour source

One of the hardest problems for flying and swimming animals is how to locate the source of a pheromone. An odour plume is unpredictable, like a smoke plume from a chimney, billowing and changing direction in the wind. We know animals can follow a pheromone plume – male moths do it with great success. It seems moths use a moment-by-moment response to 'packets' of pheromone to track the odour plume. They only know they are making progress upwind by the changing image of the ground beneath them (like looking down out of an airplane window). Some fascinating work involves transponder-tagged moths and uses radar to track them as they fly up a plume in the wild. A mystery still is how fish in mid-ocean can track odour plumes since they can't see the bottom and there would appear to be no visual cues to show that they are making progress.

Human pheromones?

How good is our sense of smell? The evidence is that it is better than we imagine. In part it is a matter of noting the difference between the number of different smells we are able to distinguish – which might be as good as a dog's ability – and the lowest threshold concentrations that we can detect – where dogs beat us paws down. Dogs may also be better at remembering or distinguishing smells. We can, however, distinguish people by smell – just as dogs can.

Do humans have pheromones? This is still the big mystery. The circumstantial evidence is good: we are mammals, and almost all other mammals seem to have pheromones. However, despite what you may see on the internet, there is no evidence for a pheromone that can make you irresistible. There is some evidence that human pheromones might be the stimulus for the menstrual synchronicity shown by women living together in close proximity, though we do not know what the pheromone(s) is.

Finally, we appear to choose as partners people who smell different from us. The smell is one that is related to our immune system, the 'major histocompatibility complex' (MHC). The details of the smell(s) remain a mystery. Humans are hard to study, and so is smell. There is a great deal still left to learn.

Many honey bee behaviours are coordinated by pheromones. One response is exploited for 'bee beards' at country fairs: bees harmlessly cover a man with a queen honey bee (or her pheromone) under his chin.

An aphid nymph secretes its alarm pheromone in response to a touch by a predator. Neighbouring aphids will drop off the leaf to escape the danger.

MARC BEKOFF

Do animals have emotions?

But man himself cannot express love and humility by external signs, so plainly as does a dog ...
CHARLES DARWIN, 1872

Do animals have emotions? Of course they do. You only have to listen to them and look at them – at their faces, tails, bodies, the way they walk, and most importantly their eyes. What we see on the outside tells us a lot about what is happening inside animals' heads and hearts. Animal emotions aren't all that mysterious. And if we could detect all the pheromones that they use as a method of signalling (p. 241), we'd know a lot more.

In the 1970s, researchers were mostly sceptics who wondered whether dogs, cats, chimpanzees and other animals felt anything. Questions such as 'What does it feel like to be a dog or a wolf?' were thought to be non-scientific and therefore ignored (see also p. 251). But there are far fewer sceptics now. Prestigious scientific journals regularly publish essays on joy in rats, grief in elephants and empathy in mice. The real question today is not so much *whether* animals have emotions, as *why* such emotions have evolved.

Emotions are evolved adaptations found in numerous species. They serve as a social glue to bond animals with one another and also catalyze

An angry male yellow baboon in the Masai Mara, Kenya, threatening when approached.

An improbable relationship: Aochan, a rat snake, and Gohan, his dwarf hamster friend, are normally predator and prey.

and regulate a wide variety of social encounters among friends and foes. They permit animals to behave adaptively and flexibly using various behaviour patterns in a wide variety of situations. Research has shown that mice are empathic rodents and also fun-loving. There are accounts of pleasure-seeking iguanas, amorous whales, angry baboons, elephants who suffer from psychological flashbacks and post-traumatic stress disorder (elephants have a huge hippocampus, a brain structure in the limbic system that's important in processing emotions), grieving elephants, otters, magpies and donkeys, sentient fish and a sighted dog who serves as a 'seeing-eye' or guide dog for his canine friend.

Anthropomorphism

Many researchers recognize that we are inevitably anthropomorphic (we attribute human traits to animals) when we discuss animal emotions, but that if we do it carefully we can give due consideration to the animals' point of view ('biocentric' anthropomorphism). Animals and humans share many traits, including emotions. Thus, we're not inserting something human into animals, but are identifying commonalities and then using human language to communicate what we observe.

Neuroscientific research by Andrea Heberlein and Ralph Adolphs shows that we anthropomorphize because we feel – the same neural systems underlie the human capacity for anthropomorphizing as well as basic emotional responses. We are even programmed to see humanlike mentality in events where it cannot possibly be involved (for example, 'angry' weather patterns).

It is not surprising to find close, enduring and endearing emotional relationships between members of the same species, but improbable relationships often occur between animals of wildly different species, even between animals who are normally predator and prey. Examples include Aochan, a rat snake, befriending a dwarf hamster named Gohan at Tokyo's Mutsugoro Okoku Zoo, and a lioness in northern Kenya adopting a baby oryx on five different occasions.

Evolutionary continuity

Scientific research in evolutionary biology, cognitive ethology (the study of animal minds) and social neuroscience support the view that numerous and diverse animals have rich and deep emotional lives. The focus here is on mammals although there are data showing that birds and perhaps fish experience various emotions.

Charles Darwin's well-accepted ideas about evolutionary continuity – that differences among species are differences in degree rather than kind (there are shades of grey among different species, not stark black and white differences) – which he

Drawings of a baboon, first in a placid condition and then 'pleased by being caressed', from Charles Darwin's The Expression of the Emotions in Man and Animals *(1872).*

Humpback whales openly display their feelings and possess spindle cells that are important in processing emotions.

set out in *On the Origin of Species*, argue strongly for the presence of animal emotions, empathy and moral behaviour. Continuity allows us to connect the 'evolutionary dots' between different species to highlight similarities in evolved traits, including individual feelings and passions. All mammals (including humans) share neuro-anatomical structures such as the amygdala and neurochemical pathways in the limbic system that are important for feelings.

Spindle cells and mirror neurons

In scientific research there are always surprises that force us both to rethink what we know and to revise our stereotypes. Thus, spindle cells, which were long thought to exist only in humans and other great apes, have been discovered in hump-back whales, fin whales, killer whales and sperm whales in the same area of their brains as spindle cells in human brains. This brain region is linked with social organization, empathy and intuition

about the feelings of others, as well as rapid instinctive reactions. Spindle cells are important in processing emotions. In late 2005 newspapers around the world published a story about a humpback whale who, after being untangled from a net, swam up to each of the rescuers and winked at them before swimming off. The rescuers all agreed she was expressing gratitude.

Mirror neurons help explain feelings such as empathy. Research on them supports the notion that individuals can feel the feelings of others. Mirror neurons allow us to understand another individual's behaviour by imagining ourselves performing the same behaviour and then mentally projecting ourselves into the other individual's place. There is compelling evidence that humans are not alone in possessing this capability. Diana monkeys and chimpanzees help one another obtain food and elephants comfort others in distress. Mirror neurons also help to explain observations of rhesus monkeys who will not accept food if another monkey suffers when they do so, and empathic mice who react more strongly to painful stimuli after they observe other mice in pain.

Baby elephants often touch other elephants with their trunks, as if to give and receive reassurance.

Animal emotions and why they matter

Another big question for which answers are revealed by studying animal passions is 'Can animals be moral beings?' Many animals know right from wrong and live according to a moral code, a phenomenon I call 'wild justice'.

There is no doubt whatsoever that when we consider the ethics of what we can and cannot do to other animals their emotions should inform our discussions and our actions. Scientific research clearly shows that animals are emotional and empathic beings, but we already knew it.

JAMES L. GOULD

59 The language of honey bees

Those hours at the observation hive when the bees revealed this secret to me remain unforgettable.
KARL VON FRISCH, 1974

Karl von Frisch won the Nobel Prize for his work on honey bee language, learning, sensory physiology and navigation.

Language is often defined as a communication system based on abstract conventions which conveys information about events distant in space and time. This elaborate formulation excludes emotional, real-time responses to food, mates and threats – grunting excitedly at a banana, for instance – while still including human sign languages and written communication. For decades, linguists and philosophers felt confident that defining language in this way would always exclude animal signalling. They were wrong.

In 1945 Karl von Frisch, Germany's pre-eminent biologist, discovered what remains easily the second-most complex language on the planet. He already knew from years of research that forager bees returning from good sources of nectar or pollen would perform dances in the darkness of the hive on the vertical sheets of comb. He also knew that recruit bees attending these dances would often leave the home cavity and subsequently arrive at the food source. Careful tests had convinced him that the recruit bees used the odour carried back on the waxy hairs of the foragers to locate the productive site they had been visiting. But now, as the Second World War ended, experiments carried out with feeding stations hundreds of metres from the hive showed that the recruits already knew the distance, as well as direction, of the target before they set out searching.

The dance

Von Frisch observed that the forager dance, in which this information had to be encoded, is shaped roughly like a figure-8. The central part of the '8' consists of two intersecting straight lines; while moving along this part of the pattern foragers 'waggle' their bodies from side to side at 13 Hz, producing a motorboat-like sound. These waggle runs, von Frisch realized, are virtually identical among foragers visiting the same food source, but different from the dances performed by bees visiting other locations. Odder still, the average angle of the waggling by a group trained to one of his stations would shift anticlockwise as the day wore on.

In fact, the mysterious language has two main parts. The direction code involves the average angle of the dance relative to vertical in the dark hive; this corresponds to the angle between the station and the sun's azimuth. Since the sun appears to move to the right across the sky (in the northern hemisphere), the angle between vertical and the waggle direction shifts slowly to the left in the hive as the minutes pass.

As he trained the foragers ever farther from the hive, von Frisch also observed that the duration of the waggle runs grew steadily longer. Distance must be encoded as the number of waggles, or the duration of waggling, or the duration of sound production. Each waggle, it turned out, corresponded to about 50 m (165 ft).

The dance-communication system is clearly a language. First, it uses arbitrary conventions: for direction, 'up' is defined as the sun's direction; 'down' would work equally well, as would north or the direction the hive faces. The essence of a language is that all members of the community accept a single set of conventions.

The arbitrariness of the dance language is underscored by the later discovery that the

Left *A returning forager bee dances on the vertical comb in the hive, surrounded by potential recruits.*

Below left *The figure-8 dance has two intersecting waggle runs during which the forager vibrates its body and produces a sound. The average direction of the runs is shown in blue.*

Left *The angle of the dance in the hive relative to vertical matches the angle between the food and the sun's azimuth in the field. In example II, for instance, the dance is 80 degrees left of vertical, and the food is 80 degrees left of the sun.*

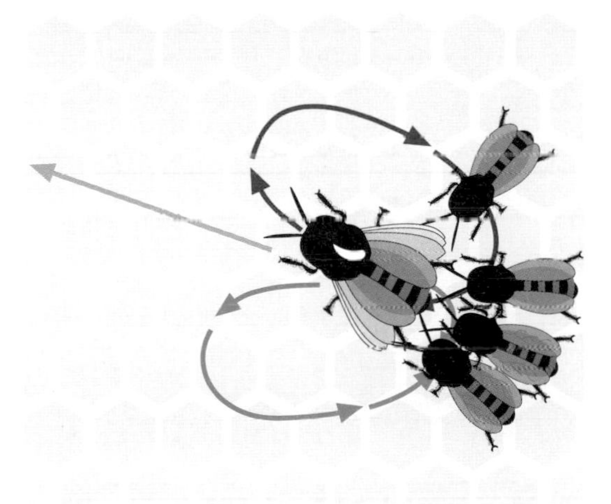

'waggle = 50 m' convention differs between subspecies: for bees from southern Europe, a waggle indicates 20 m (65 ft), while for bees in Egypt the value is about 10 m (33 ft). Unlike human language, however, these dialects are innate: Italian bees reared from the egg by German bees use the Italian dialect.

Swarming and decision making

Honey bees, von Frisch found, use the dance language to communicate the location of nectar, pollen and water (used to cool the hive). For nectar and pollen, the dance allows bees rapidly to exploit the ever-changing landscape of blooming plants, increasing the number of foragers at a good source by as much as ten-fold every hour.

Bees also use their language outside the hive when they reproduce by swarming. As the colony outgrows its hive cavity, new queens are reared and then roughly half the population leaves with the old queen and forms a cluster on a nearby branch. From here, scouts search for a suitable cavity in which to establish a new home. In the wild, the group depends on these natural hollows

A swarm of bees: about half the workers and the old queen leave the hive and cluster on a tree branch. Scouts explore the surrounding terrain and return to the swarm to dance and advertise their discoveries.

in trees to provide protection from predators and insulation from the weather.

Scouts evaluate the cavities they find on the basis of numerous factors, each of which has an optimum value. The parameters for assessment include distance from the hive, volume of the cavity, its internal shape, the size of the entrance, the direction the entrance faces, the height of the entrance above the ground, the cavity's freedom from leaks and draughts, and so on. These scout bees return to the swarm and dance on its surface, encouraging other bees to visit the cavity and judge its quality. After three or four days the dancers almost always reach a consensus and the swarm departs for the colony's chosen new home.

Maps

Researchers have used the dancing of trained foragers returning from feeding stations to decode the navigational system of bees – including how they compensate for the unseen motion of the sun, use polarized light in the sky when the sun is not visible, orient to the Earth's magnetic field and integrate their memory of landmarks with celestial cues (see also p. 229).

A variety of tests have now demonstrated that honey bee foragers build up landmark maps of their home range, and thus are able to find their way home if they are blown off course (or kidnapped by experimentors). Recruits also seem able to 'place' the location indicated by a dance before leaving the hive, apparently using a mental map to judge the 'plausibility' of a dance's information.

Numerous other tests reflect well on the cognitive ability of honey bees: they can rotate shapes in their minds and form concepts about shapes as well as the ideas of 'same' and 'different'. They can extrapolate, interpolate and coordinate. Only in the last decade have the intellectual abilities of these linguistically gifted insects been taken seriously and investigated properly. As with language, the results have consistently made our own species seem less special, and our abilities less mysterious.

COLIN ALLEN

Animal consciousness

He who understands baboon would do more towards metaphysics than Locke.
CHARLES DARWIN IN A NOTEBOOK OF 1838

It seems obvious that other animals besides humans are conscious (see p. 244). But science has a history of overturning things that seem obvious. The world is not flat, and it does move. Whales aren't fish. Still, science frequently confirms common sense too, and often leads us to a deeper appreciation of the facts. Planets are different from stars. The Moon does affect the tides. So, what about animal consciousness?

Some facts are undisputed. Many animals go through sleep and wake cycles. All animals are oblivious to some features of their environments: we cannot see in the ultraviolet range, but many insects and birds can. Consequently we are not conscious of patterns on flowers that are perfectly visible to many species of bees, butterflies and birds. Science has confirmed and deepened our understanding of these facts, for instance by showing how specialized neurons respond to light of different wavelengths.

Other facts about animals are more disputed. Experiments seem to indicate that some animals besides humans can recognize themselves in mirrors, are aware of what others can or cannot see, are capable of remembering specific events, can fashion tools and can take actions that anticipate future needs. Experiments on 'metacognition' seem to show that monkeys, dolphins (and perhaps even rats) seem to sense when they don't know how to respond to a difficult task. Thus science has pointed us towards the view that many animals have capacities that were often assumed to be uniquely human. Some of the most surprising recent discoveries are about birds. Scrub jays can remember where and when they buried food, what they buried and whether they were being watched at the time. They will spontaneously hide food in locations where they had previously been deprived of a morning meal.

From uncertainty to scepticism

Although science can help us understand non-human consciousness in relation to the mysteries of sleep and the mechanisms of sensory perception, memory and planning, there remains a mystery about a particular aspect that some

Looking into the mirror. Does the dolphin understand who is looking back?

Alex (1976–2007), the African Grey parrot whose cognitive abilities were studied for two decades by Dr Irene Pepperberg. Alex could answer a variety of questions about the size, shape, material, colour and number of objects.

people think science can never bridge. This is subjective experience. What is it like to have ultraviolet vision? What is it like to have the ability to identify objects using sound, like bats or dolphins? Indeed, is it 'like' anything at all for these animals? According to some scientists and philosophers, no matter how clever animals seem to be and no matter how much we learn about the neurological mechanisms underlying perception and cognition, there is no scientific basis for saying anything testable about their subjective conscious experience – or whether such a thing exists at all.

However, this most radical form of scepticism leaves human consciousness out of the scientific picture too. Slightly less radical, at least with regard to humans, is the idea that consciousness and language are tightly connected. Normal adult humans can vividly describe their conscious experiences from a subjective point of view.

Scientific studies with Alex (an African Grey parrot) and Kanzi (a bonobo) show that given the right developmental experiences these animals can understand dozens of English words, hundreds of sentences and can respond appropriately to new questions and requests. But Alex and Kanzi cannot describe their own subjective points of view. Alex could say what colour a novel object was, what shape it was or what it was made of. He could also tell us how many objects of a particular type were in front of him. He could not tell us what it felt like to be a parrot.

Although common sense suggests that Alex had conscious experiences of the colours he described, the radical sceptic wants to deny this. There is no experimental refutation for this view, however, only philosophical arguments about the limits of science or the nature of consciousness. This is not a wholly convincing (or wholly scientific) way of overturning common sense about animal consciousness. But it does suggest why animal consciousness remains one of the great mysteries of the natural world: neither sceptic nor proponent knows exactly what a science of animal consciousness should look like.

Towards a science of animal consciousness?

Conway Lloyd Morgan, a founding figure of evolutionary comparative psychology in the 19th century, struggled with the question of how to justify attributing subjective consciousness to non-human animals. In the end he accepted a combined approach of behavioral experiments on animals and introspection about our own consciousness as a basis for reasoning by analogy about animal minds, but he warned repeatedly against jumping too quickly to 'higher' explanations when 'lower' ones would suffice – an injunction that came to be known as Morgan's canon.

In the early 20th century, behavioristic psychologists rejected introspection and took Morgan's canon further, arguing that 'lower' explanations in terms of learned associations were always more justified than speculation about conscious mental processes. Ethologists studying the behaviour and evolution of wild animals disagreed with the behaviorists' methods, but were just as unwilling to commit themselves to anything as hard to define as consciousness.

Much has happened in psychology, ethology and the neurosciences since Morgan published his canon. The basics of neuronal signalling, learning and memory have been found to be the same in sea slugs as in humans. We now have the capacity to map our conscious experiences on to activity in the human brain (although not yet as precisely as we would like). The basic circuitry of fear, pain, pleasure, moods and feelings has been mapped out for rodent and human brains. There are differences, but there are also similarities. Meanwhile, comparative psychologists and ethologists are devising new experiments that exploit situations of natural significance to the animals themselves. These behavioral studies are revealing surprisingly sophisticated cognitive abilities in animals with rather different brains and which are relatively far from us in the evolutionary tree – insects, octopus and birds, for example.

Current neural and behavioral science offers converging lines of evidence that have made many scientists much more willing to discuss animal consciousness than their behavioristic forebears. But for every similarity there are differences. Which is more significant for animal consciousness: similarities or differences? Only time will tell. In the meantime, we are faced with the tantalizing possibility that animals may have much more to tell us about their subjective states than will ever be conveyed by words alone.

Kanzi, a bonobo, and Dr Sue Savage-Rumbaugh communicating with the lexigram keyboard that Kanzi first used spontaneously 25 years ago, after watching the researchers' attempts to teach his adoptive mother to use it.

Global Warming & the Future

No subject causes more consternation and confusion in public debate than global climate change. Has global warming occurred over the past century, and if so by how much? What will be the future of our Earth if climates change drastically? Most importantly, are humans to blame for climate change and, if so, what can we do about it?

Scientists all accept that global warming has taken place in the past two centuries of the industrial revolution. Factories and cars produce so-called greenhouse gases – primarily carbon dioxide, nitrous oxide and methane – in vast quantities, and current levels are twice as high as they were before industrialization. It is also true, however, that current levels of carbon dioxide in the atmosphere are not uniquely high, and that levels fluctuated before humans were on the Earth. These natural rhythms are linked to the orientation and position of the Earth in its orbit relative to the Sun, but the causes of rhythmical changes in the atmosphere are still mysterious.

Will warmer temperatures worldwide mean that more people will enjoy sunny summers and higher crop yields? Probably not. Predicting future climates is complex, but it seems likely that dry areas will become more arid, and temperate zones and high lands will become wetter. And extreme weather phenomena, such as the El Niño events in the Pacific, which bring unusually heavy rains to some areas and droughts to others, appear to be increasing in frequency as a result of global warming.

Futurologists have long warned about the dangers of overpopulation. Human populations

A satellite view of Hurricane Elena: as a result of global warming there is more energy in the atmosphere and so extreme weather events are likely to increase.

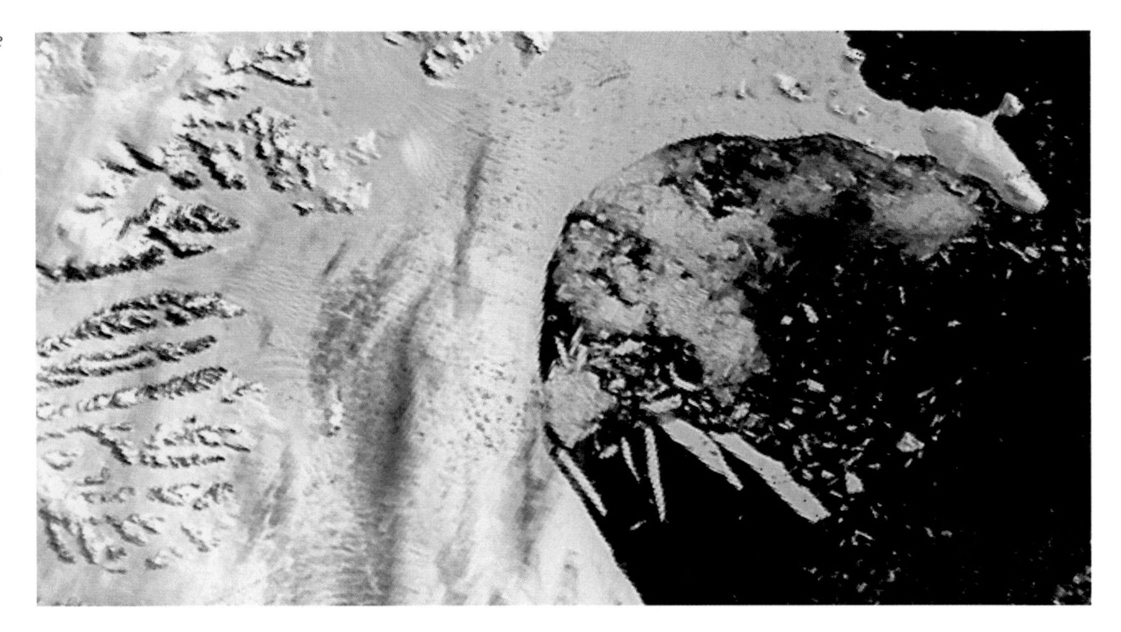

The break up of the Larsen B ice shelf, in February 2002, as recorded by NASA's MODIS satellite sensor. Ice shelves advance and retreat between winter and summer, but here great blocks of ice have been weakened by prolonged unusually warm conditions.

worldwide remained relatively low for centuries, and then the agricultural revolution allowed farmers to produce ever-increasing quantities of food and the industrial revolution led to the growth of vast cities around the world. Will human population numbers continue to rocket upwards? Current projections are that world population might rise from the current level of around 6.5 billion to over 9 billion, and then fall back to a more sustainable 2 billion or so as birth rates in wealthier nations fall. Population is limited by development, but also by other factors, including pandemics. Many such global diseases seem to originate in eastern Asia and are closely related to poultry – a repeated pattern that seems to defy sanitary regulation.

There are two reasons for concern about future population levels: wildlife and resources. Every day a forest is cut down or the natural range of a bird or mammal is reduced by agriculture. There are now many international treaties and protocols aimed at halting the deterioration of the world's remaining natural spaces. But why bother? A case can be made that every species has a value and that humans have no right wilfully to drive any other species to extinction.

The main resources that cause anxieties for the future are hydrocarbons – coal, oil and gas – which fuel our modern lifestyle. When will the oil run out? The search is on for alternative sources of power – new types of hydrocarbons, as well as biofuels and hydrogen.

Another way of tackling this problem is to make our use of current energy sources more efficient. Efficiency of energy use is measured by our ecological footprint, a term coined in 1992 by William E. Rees. At present, citizens of the United States consume more than anyone else, and so their ecological footprint is large, about twice that of the average citizen of Europe, and five or six times the size of that of an African. And yet, for sustainability, all citizens of the world should aspire to an ecological footprint at the lower end of the scale. Globally, we must reduce our impact on resources by 22%, and yet at the same time seek to ensure a fairer distribution of resources. The last decade has seen a remarkable growth in awareness of the pressing problems of global sustainability, and people seem increasingly ready to embrace some changes to their lifestyles.

Interest in the natural sciences has never been keener, and the question of what will happen to the Earth in the next century is on every thinking person's mind. The discipline of futurology has had a bad press in the past, but new tools allow ever-more sophisticated predictions, which mean that everyone can have a greater understanding of the issues which closely concern us all.

ANDY RIDGWELL

Global warming

61

If the quantity of carbonic acid increases in geometric progression,
the augmentation of the temperature will increase nearly in arithmetic progression.
Svante Arrhenius, 1896

If you were to try to count all the molecules that make up the air we breathe, you would probably reach over two thousand before you came across a molecule of the colourless and odourless gas called carbon dioxide (CO_2). So if it is present only at such a low concentration, how can it possibly affect the climate of the entire planet? We also know that CO_2 must have been naturally present in our atmosphere throughout the history of the Earth, and in addition is essential to all plants. So why should we worry about releasing more CO_2 into the atmosphere by burning carbon-containing 'fossil fuels' such as oil, gas and coal?

In the beginning

In 1896, the Swedish scientist Svante Arrhenius put forward a radical new theory to explain the ice ages that had so prominently sculpted the landscape of northern Europe and North America – he proposed that changes in the concentration of carbon dioxide in the atmosphere altered the surface temperature of the Earth and caused ice sheets to wax and wane. He went on to suggest that industrial activities and the burning of coal might drive a new warming of the Earth. Ironically, Arrhenius considered this prospect rather a good thing because of the harsh winters in Sweden, and even wondered whether coal should be burned more quickly to accelerate the warming.

At that time, most scientists were not ready to accept a critical role for CO_2 in the climate. It is only in recent years that we have accumulated enough information about historical climate changes to allow scientists to make the landmark pronouncement in 2007 that: 'Warming of the climate system is unequivocal, as is now evident from observations of increases in global average air and ocean temperatures, widespread melting of snow and ice, and rising global mean sea level.'

The second missing piece of the puzzle concerned the fate of CO_2 released from the burning of coal (and other concentrated forms of 'fossil' carbon such as oil and natural gas). Like most atmospheric gases, CO_2 is soluble in water and it was once widely assumed that the oceans would quickly mop up all industrial emissions into the atmosphere. Now, thanks to pioneering measurements made in remote parts of the world, we have unequivocal evidence that CO_2 released from fossil fuel combustion is actually accumulating in the atmosphere, year-on-year.

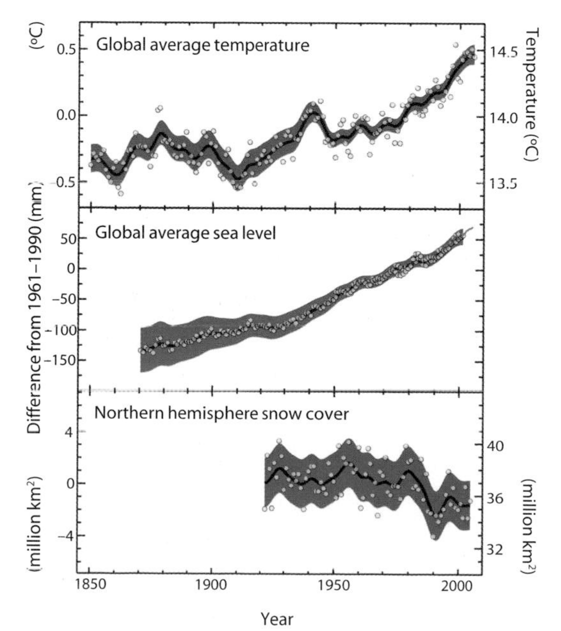

Observed changes in temperature, sea level and northern hemisphere snow cover, 1850–2000. From the 2007 report of the Intergovernmental Panel on Climate Change.

But how are these two observations – of increases in both global temperatures and atmospheric carbon dioxide concentrations – linked?

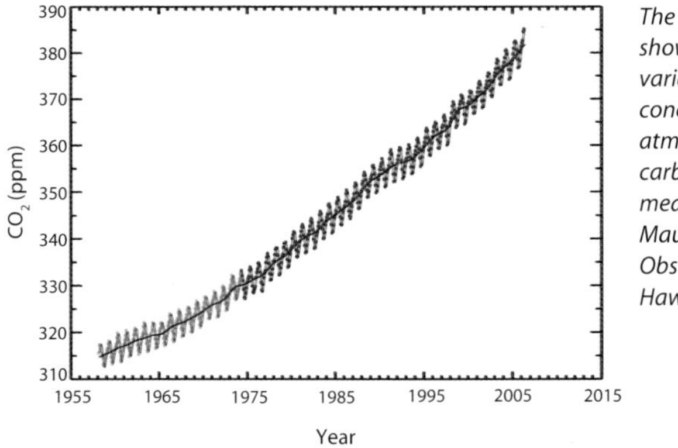

The 'Keeling Curve': showing the variation in the concentration of atmospheric carbon dioxide measured at the Mauna Loa Observatory, Hawaii.

CO_2 – the greenhouse gas

To unravel the mysterious role of CO_2 in climate we must first look at what happens to sunlight reaching the Earth. Snow and clouds, which are fairly reflective, are said to have a high 'albedo', absorbing only 20–60% of the visible light that strikes them and reflecting the rest back the way it came. In contrast, most of the surface of the planet – leaves, rocks, soils and the ocean – has a low albedo and absorbs as much as 70–95% of sunlight that reaches it.

Clearly, there has to be a loss of energy that balances the gain from visible light, otherwise the Earth would overheat. This is invisible 'infrared' radiation – the warmth you feel from a distance from a hot water radiator even though you can see nothing if you turn off the lights. The Earth balances the absorbed sunlight by continually radiating energy to space in the form of invisible infrared radiation. The emitted infrared radiation has a range of different wavelengths called a spectrum, just as sunlight has a spectrum of different wavelengths corresponding to the different colours of the rainbow. This has significance in the action of greenhouse gases such as CO_2.

The CO_2 molecule is important because it has a special property – it traps certain wavelengths of the infrared radiation spectrum that would otherwise escape into space and sends some of it back down to the surface. In fact, CO_2 is extremely good at this. Water vapour is another good absorber of infrared radiation, as is methane (CH_4), which is released from decaying vegetation in swamps and from cows as they digest grass (scientists even have a polite technical term for cow burps and farts – 'enteric fermentation'), as well as

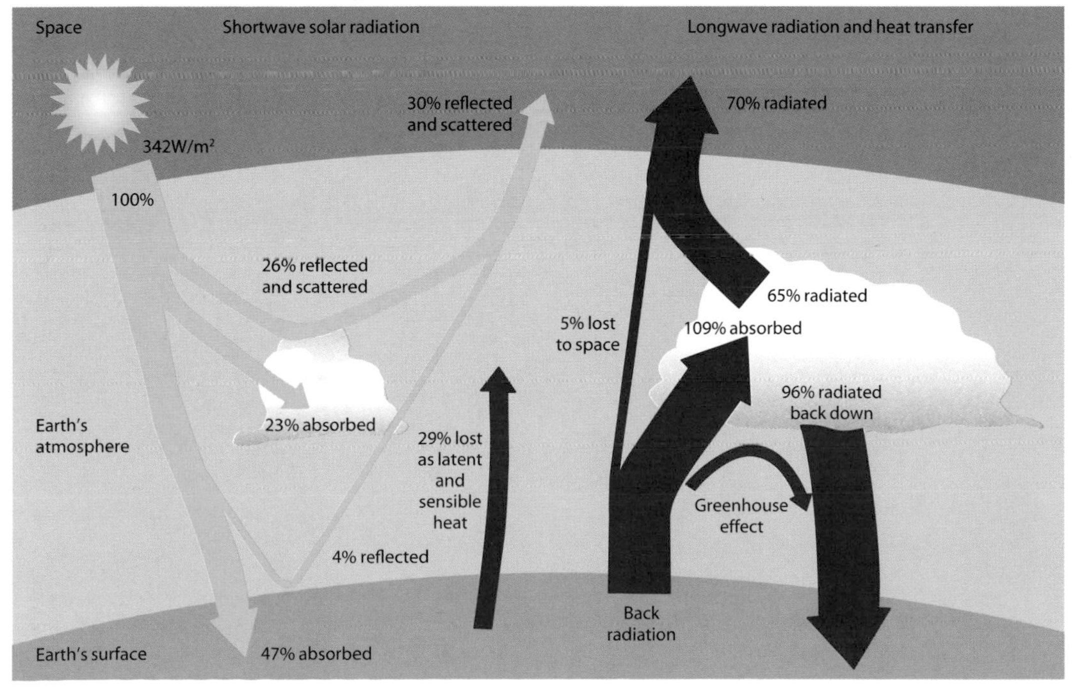

Left *Simplified representation of the flows of energy between space, the atmosphere and the Earth's surface, illustrating how heat is trapped near the surface in the Greenhouse Effect.*

Opposite *An oil refinery at Grangemouth, Scotland. Use of liquid fuels currently represents almost half of all CO_2 emissions from fossil fuel burning.*

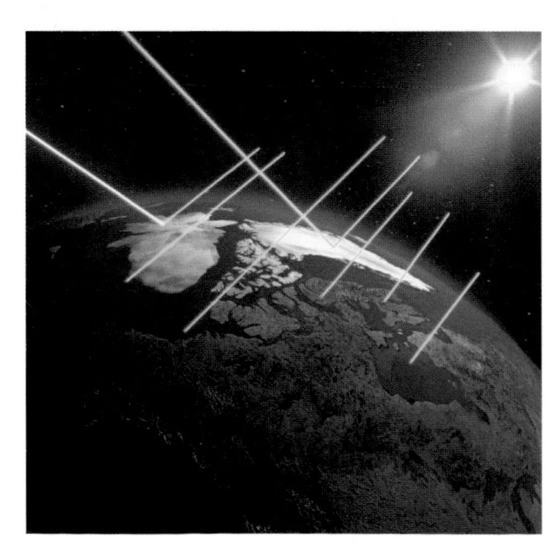

Series of images showing how polar ice reflects light from the sun. As this ice begins to melt, less sunlight gets reflected back into space. Instead, it is absorbed into the oceans and land, raising the overall temperature, and fuelling further melting. Scientists call this a 'feedback' effect.

nitrous oxide (N_2O), also partly to be blamed on cows and other livestock.

Much of the infrared spectrum is blocked by water vapour, and the atmosphere finds it difficult to lose heat at these particular frequencies. However, there is an important 'window' region through which most infrared radiation escapes to space. This is where the concentration of CO_2 in the atmosphere has a particularly important effect on climate, because one of the frequencies it absorbs most effectively at is immediately next to this 'window' – adding more CO_2 molecules to the atmosphere means that the band of the spectrum that CO_2 absorbs gets broader, and the width of the window consequently narrows.

The presence of CO_2 in the atmosphere, along with water vapour, methane and nitrous oxide, means that the Earth's surface is warmer than it would be if there were no greenhouse gases at all in the atmosphere. In fact, computer models of the Earth's climate system (see p. 262) predict that if the Earth had no atmosphere, the average annual surface temperature would be a chilly −19°C (−2.2°F), rather than the global average of around 14°C (57.2°F) that we experience today.

The past importance of CO_2 is hinted at in events some 300 million years ago when much of the coal on Earth was being formed. The CO_2 taken out of the atmosphere by plants and locked up into peat and coal deposits when they died and were buried should have resulted in colder temperatures. We see evidence for a great ice age at this time, suggesting that the Greenhouse Effect has indeed been an important controller of climate in the past (see p. 270).

Fast-forward to the future

The Greenhouse Effect, then, could be viewed as a helpful natural phenomenon. So why is there a 'problem' with rising concentrations in the atmosphere now, particularly since there have been other times in the history of the Earth when CO_2 has been present at much higher levels than at present?

One thing that Svante Arrhenius got wrong was his estimate that global warming would take 3,000 years to occur. In fact, all the ancient CO_2

Cattle grazing in front of a methane fuel plant at El Centro, California. Power is generated from the methane released by bacterial digestion of the manure of the cattle.

that was removed by plants on land and in the ocean, and which accumulated over tens of millions of years in geologic formations, is now being released by us in just a few short centuries.

Can ecosystems adapt to perhaps one of the fastest changes in climate that has ever taken place on Earth? And can society modify farming practices in response to fluctuating temperatures and rainfall, and move cities away from low-lying coastal areas quickly enough to avoid inundation by rising sea-levels? These are questions we must take decisions about. Fundamentally, however, we should ask ourselves whether it would not be preferable to act first rather than react, and ensure that the climate did not change to such a potentially disastrous degree.

ANDY RIDGWELL

What will Earth's climate be like in the future?

In tackling climate change, helping others is helping oneself.
HU JINTAO, PRESIDENT OF THE PEOPLE'S REPUBLIC OF CHINA, 2007

We have no spare copies of our planet on which we can experiment. This might sound obvious, but it highlights the problem of how we can determine what the future climate of Earth might look like in response to greenhouse gas emissions from the burning of fossil fuels and other human activities, other than by just waiting for it to happen. Clearly, it would be beneficial to know the answer well in advance, as this would give us the chance to avert the possibility of undesirable consequences. Here, we look into the computer modeller's crystal ball and consider some of the predicted effects of human activities on climate.

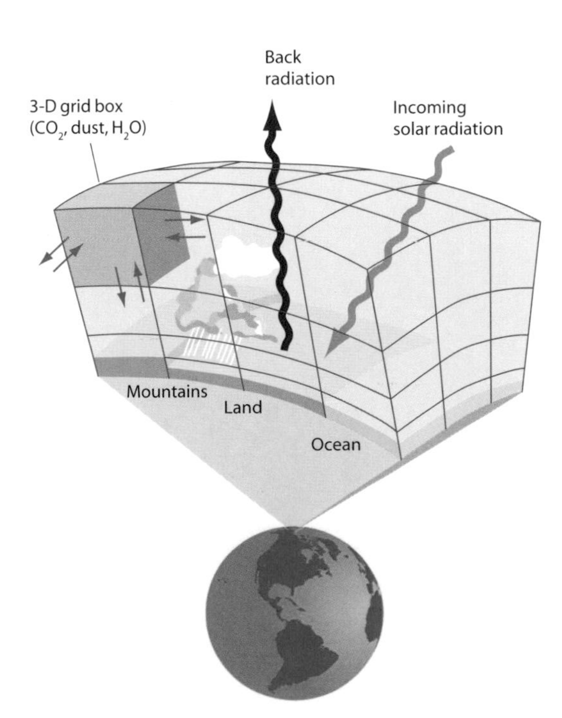

Modelling climate – and climate change – by dividing up the atmosphere into a three-dimensional grid of boxes, on which computers can then perform calculations.

Modelling climates in computers

We are used to seeing forecasts on our TV screens predicting what the local weather might be like in the next few days. These predictions are generated by feeding into a computer details of how processes in the atmosphere such as cloud formation work, together with how the atmosphere reacts when it meets the underlying land and ocean surface. Rather than attempt to keep track of every single gas molecule and rain droplet in the atmosphere – an impossible task – the atmosphere is divided up into a three-dimensional grid, often involving tens of thousands of boxes. Each box keeps track of important properties within it, such as temperature, moisture content, and direction and speed of winds, and can interact with other boxes adjoining it. By making consecutive small steps forward in time (of 30 minutes, say), heat and moisture circulate through the atmosphere, storms move across the ocean and precipitation falls, as snow during the winter and rain during the summer – all inside a computer.

In essence, the techniques required for predicting the climate of the Earth over the coming decades and centuries are much the same as this. Of course, a distinction must be made between 'weather' and 'climate' – climate being the weather averaged over time and space. So the uncertainties that begin to dominate in a weather forecast when it is extended beyond more than about a week do not mean that we cannot predict the Earth's climate more than seven days in advance. But how can we be certain that our state-of-the-art models of the Earth's future climate are credible and accurate?

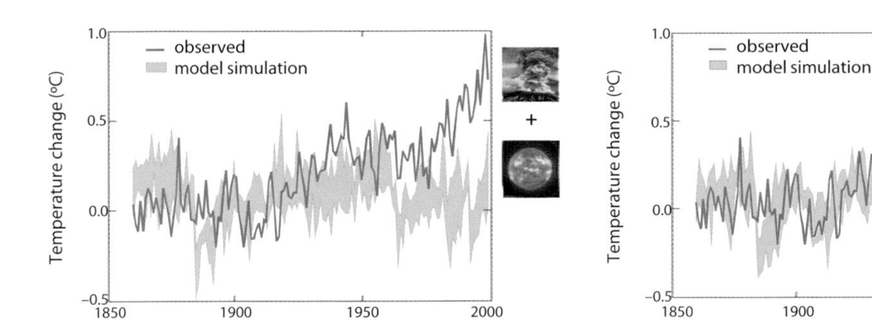

Graphs showing how computer models compare with actual recorded temperature changes. Natural causes alone cannot explain recent warming (far left). It is only when manmade factors are added (right) that the two match well.

One important check is whether our models of the climate system can reproduce the historical changes in surface temperatures that we have actual records of. Impressively, although the various models made by research groups around the world use different techniques, for instance in the number of boxes, the methods of representing cloud formation processes, as well as in other details, all agree on a vital point – that the temperature signal resulting only from natural phenomena such as volcanic eruptions and El Niño events and the variability in the energy supplied by the Sun, either in isolation or together, is a very poor fit for the records made by instrumental observations. It is only when the warming influence of the known increase in greenhouse gas concentrations in the atmosphere (and some cooling due to the industrial emissions high into the atmosphere of reflective sulphate aerosols) are taken into account that the instrumental record is reasonably reproduced.

Such is the economic and political importance attached to getting the predictions right that the United Nations established the Intergovernmental Panel on Climate Change (IPCC). This brings together the world's leading experts on climate change to scrutinize and evaluate the available observations and models. In 2007 it was awarded the Nobel Peace Prize jointly with Al Gore.

Extreme weather on the increase? (Clockwise from top left) Hurricane Linda off the west coast of North America; flooding in the aftermath of Hurricane Wilma in Florida, 2005; drought in Botswana; a tornado in South Dakota.

Where are we going?

The most recent assessment by the IPCC of projected changes in climate is of globally averaged surface air temperatures increasing by anywhere from about 1°C (about twice the observed warming since the industrial revolution) to over 6°C; most of the uncertainty is the result of not knowing how much CO_2 and other greenhouse gases we are likely to release into the atmosphere

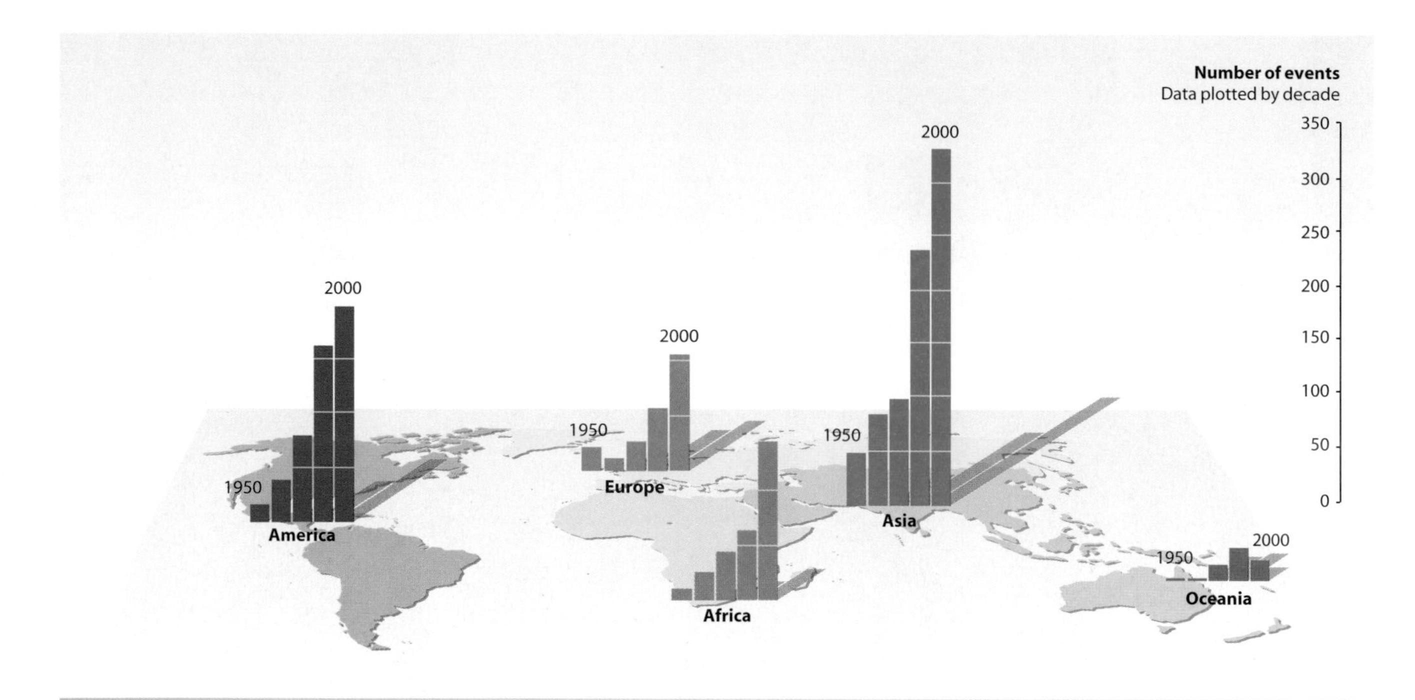

Above *Increasing flood events 1950–2000, plotted by decade.*

Opposite *Polar bears search for food at the edge of the pack ice in Churchill, Canada. In some climate projections (IPCC), Arctic late-summer sea ice will disappear almost entirely by the latter part of the 21st century.*

in the future. But thinking only in terms of a simple global average temperature when the climatic conditions on our planet vary so much from place to place is not very helpful. We therefore need to look in more detail at regional and seasonal patterns, which are of greater meaning to our personal experiences.

For instance, it is considered very likely that heat waves will become more frequent. And it is hardly very surprising that snow cover is projected to contract in response to the higher temperatures. This has important implications for adequate summer fresh water supply (from melting of the winter snow-pack) in regions such as western Canada and the Himalayas. Sea-ice is projected to shrink, with polar bears progressively losing their hunting habitat – just one of many changes to Arctic ecosystems. Another consequence of warmer surface temperatures is the release of more energy into the atmosphere, with the result that typhoons and hurricanes will probably become both more frequent and more intense in the future, with obvious economic impacts.

Another important future change concerns rainfall. The higher the temperature of the ocean, the more evaporation there will be. And more moisture in the atmosphere of course means more rainfall in total. However, the detailed picture is again not so simple, and some places on Earth are expected to get drier not wetter. For instance, high latitudes are likely to receive increases in the amount of precipitation, especially in the winter. One result of this could be an increased frequency and severity of flooding in already susceptible areas. But many of the already arid and semi-arid regions on Earth could see a decrease in rainfall, implying an increased frequency and severity of drought. There is nothing particularly 'fair' about the reaction of the climate system in response to our interfering with it.

The greatest uncertainty of all is the unpredictability of we humans in our personal and collective political and economic decisions. Because the degree of climate change increases with higher concentrations of CO_2 in the atmosphere, the more fossil fuels we burn and the faster we burn them, the more CO_2 will build up in the atmosphere. We may perhaps choose to reduce our fossil fuel use, or, conversely, we may continue our love affair with large cars. Whether we will simply run out of oil is another question we ask shortly (p. 282).

CÉSAR N. CAVIEDES

63 El Niño: extreme weather on the increase?

No other environmental perturbation, but ENSO, has such amplitude or far-reaching impact, capable of bringing hardship to a quarter of the human race on five continents.
MICHAEL DAVIS, 2001

In the spring of 1804, Commander Ivan F. Krusenstern was amazed and intrigued when, sailing from the Marquesas Islands to Hawaii, his vessel was barely making headway against a mighty current from the west. Two years earlier, during his stay in Lima, Peru, Alexander von Humboldt heard the story of a Spanish galleon sailing from Manila to Callao, Peru, that was swept across the tropical Pacific in as little as 76 days – a voyage that normally took four months. Both events took place in years during which usual westward currents in the equatorial Pacific seemed to have reversed direction.

At a gathering of local scientists convened to review the causes of the warm ocean conditions and torrential rains that had beset northern Peru early in 1891, it was agreed to name those phenomena *El Niño* (the Jesus Child) because of their onset around Christmas. Historians at the meeting remarked that such occurrences were nothing new, since chronicles and anecdotal accounts from 1877, 1844, 1828, 1791, 1728, 1720, 1687 and 1578 – to name just a few – told of heavy rains, flooding, devastation of settlements, famines and misery. The best informed among them reminded the audience that precisely in those years, catastrophic droughts or devastating deluges had been visited on distant countries such as Japan, China, India and Ethiopia.

The warm waters of El Niño cripple the coastal fisheries in Peru and worsen the chronic poverty of the fishing communities.

Other El Niños followed, and it was confirmed that climatic disasters in coastal Peru and southern Ecuador coincided with anomalous ocean warming off the west coast of South America. By the mid-1960s oceanographers and climatologists discovered that such events involved the entire tropical Pacific, from South America to Australia, affecting even the Philippines, Indonesia and southern Asia.

Strong warming of the tropical Pacific in 1972–73, 1982–83 and 1997–98 revealed that El Niño led also to rainy and snowy winters in North America, and to droughts in the Caribbean basin, northeast Brazil, sub-Saharan Africa, India and Australasia. Climate records of the 20th century indicated further that an increased frequency of El Niño episodes since 1975 was coupled to global warming. Today, El Niño and La Niña – warm and cold periods in the tropical Pacific – count among the most significant and widely studied contemporary climate variabilities.

The origin of El Niño

El Niño and La Niña episodes are the outcome of the convergence of several natural cycles (*natural forcing*) which, when peaking at the same time,

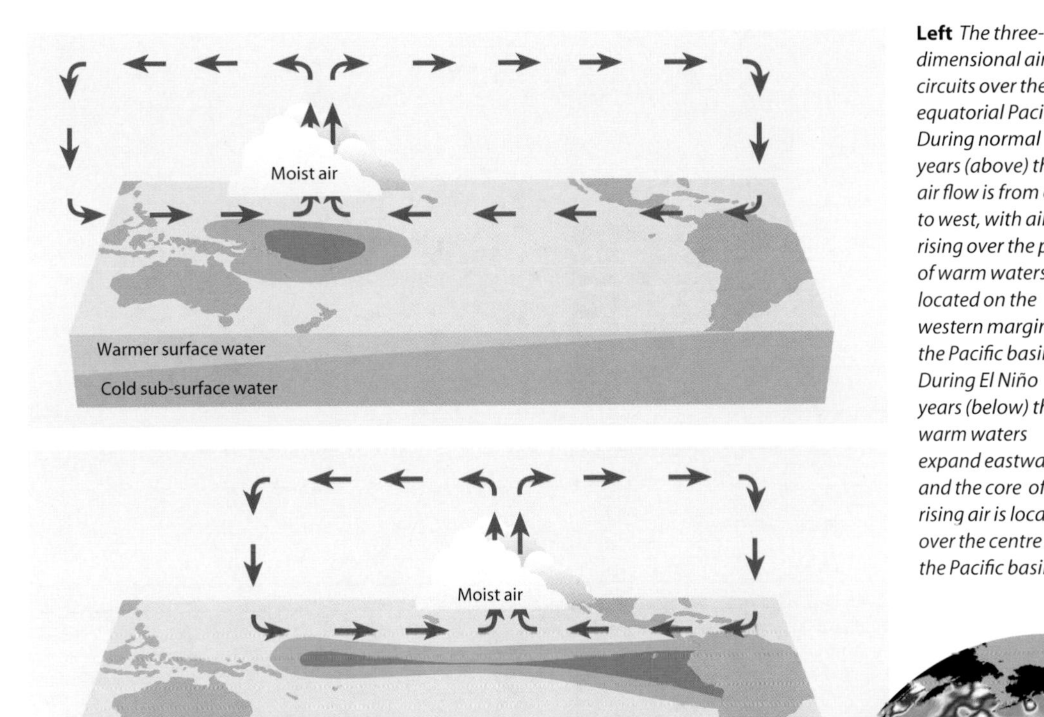

Left *The three-dimensional air circuits over the equatorial Pacific. During normal years (above) the air flow is from east to west, with air rising over the pool of warm waters located on the western margin of the Pacific basin. During El Niño years (below) the warm waters expand eastwards, and the core of rising air is located over the centre of the Pacific basin.*

cause surges in the Earth's heat budget. One of these is an 18- to 20-month cycle that affects the temperature in the upper layers of the cold Peru Current and in the eastern South Pacific, overriding the annual solar regime. This cycle is noticeably close to the 20- to 22-month cycle of the Quasi-Biennial Oscillation (the east–west/west–east shifts of lower stratospheric flows over the equator) which controls the upper winds over the tropical Pacific and Atlantic oceans. In turn, the QBO has remarkable similarities with the peaks and lows of the Southern Oscillation – the barometric seesaw between high pressure over Tahiti and low pressure over Indonesia that modulates flows between the eastern Pacific and western Pacific basins. For this reason the term *El Niño-Southern Oscillation* (*ENSO*) is used to describe this relationship between oceanic and atmospheric variabilities over the Pacific basin.

When these various naturally occurring pulses are superimposed, ENSO events can be seen to take place at return intervals of 3.6 and 6.7 years. In addition, there is the 11-year sunspot cycle which was once thought to lead to El Niño episodes. There is also a theory that ocean warming is caused by sporadic outpourings from magma vents on the Pacific floor, but this is questionable because they are too small in size to account for sizeable water temperature increases in that huge ocean. All these natural processes are interrupted by random volcanic eruptions whose ejections of ashes, chlorines and sulphides into the troposphere upset the energy budget and possibly lead to El Niño events.

Above *During an El Niño event in 1997–98 a tongue of abnormal warm waters (white-orange colours) established itself in the eastern Pacific from Peru up to the west coast of North America. By contrast, abnormal cooler waters (purple) occupied the western tropical Pacific.*

The devastating consequences of El Niño

During El Niño events warm waters in the tropical Pacific coincide with a heat and humidity surplus in the lower atmosphere. However, some low-latitude regions are affected by compensatory humidity deficits (drought). The ocean warming

Painted relief from Huaca de la Luna, Moche Valley, Peru, showing a stylized octopus surrounded by catfishes. Moche art contains frequent references to animals affected by El Niño conditions, which caused repeated droughts and floods throughout the history of Moche civilization and probably contributed to its collapse in the 1st millennium AD.

upsets the primary productivity of cool areas in the Pacific Ocean, which leads to reductions in fish stocks and consequently mortality among seabirds and sea mammals. Fishing communities in coastal Peru and Ecuador are also adversely affected. These oceanic anomalies are coupled with heavy rains, high waters and inundations, which devastate coastal and riverine settlements.

Furthermore, the high humidity and standing pools of rainwater favour the outbreak and spread of diseases such as malaria, typhus, cholera and encephalitis – even incidences of pest (*Yersinia pestis*) were reported in Andean villages during the El Niño year of 1997–98. In semi-arid environments of North and South America outbreaks of a fatal lung disease, transmitted through the Hanta virus in rodent faeces, coincide with population explosions of rodents due to the vegetation greening as a result of the El Niño rains. And European epidemiologists observe increases in the cases of influenza during the cool and humid El Niño winters in middle latitudes.

In those regions where droughts are the sequel of El Niño, crops wither and livestock starve, leading to famines, especially in sub-Saharan Africa and India. In Australia sheep are decimated and animals in the wild, such as kangaroos and koalas, perish. Dryness results in wildfires that have devastated forests in Australia, Indonesia, South Africa and Amazonia.

How old is El Niño?

Since there are no direct references to such phenomena by pre-Hispanic Peruvian cultures which had no system of writing, clues about ancient El Niños come from archaeological and geological findings. Climatic and oceanic occurrences similar in their effects to contemporary ENSOs have been pinpointed near AD 1453, from 1317 to 1464, around 1300 and around 1259, when mighty alluvial episodes destroyed settlements and interrupted cultural development in coastal Peru and highland Bolivia. Towards AD 1100 repeated and powerful oceanic El Niños upset the circulation of water and air masses and may have led to the expansion of Polynesian culture to Easter Island and possibly South America. Similar events and thick alluvia wreaked havoc in coastal Peru from AD 800 to 1300. Around AD 600–690 and also in AD 300, southern Peruvian cultures bore the brunt of such natural catastrophes.

Further back, only the traces of very extreme ENSO events have been uncovered by geological investigations. These indicate an enhancement of the phenomena during the global warming following the last glacial period. Opinions as to the onset of El Niños in the eastern Pacific and on the west coast of South America after that vary. Some posit that it started with the beginning of the last deglaciation, around 11,400 BC. Others propose that traces of ENSO are noticeable from 17,000 BC, or even 18,250 BC – *before* the end of the last glaciation. If this is true, then El Niños occurred not only during the warm periods of Earth's history, but also during the cold periods and thus may have been around through the whole Quaternary, that is, for the last 2 million years.

ENSO in the future

The heat balance of the Earth's atmosphere is affected by heat-absorbing gases such as carbon dioxide, nitrous oxide, methane, halocarbons, surface-level ozone and also by aerosols (black carbon from fossil fuel burning). Increases in the volume of these greenhouse gases and solid particles are caused by natural processes but also result from human activity (*anthropogenic forcing*). Global warming is due, to a large extent, to anthropogenic forcing (see p. 257). Consequently, with the continuation of these human activities the frequency and intensity of extreme warm events is expected to rise.

In fact, the two most powerful El Niños of the 20th century occurred in 1982–83 and 1997–98, when the contemporary warming trend – established around 1975 – was already well under way. The cold La Niña episodes have become less frequent, though their intensity has increased. As temperatures rise, so will the carbon dioxide releases from decayed or burned vegetation, contributing further to the atmospheric heating. Since La Niña in the Pacific Ocean is phased with warming in the tropical Atlantic, hurricanes will be more violent when they occur. It has also been contended that the number of tornadoes in North America grows during El Niño years, but that relationship is feeble because of strong continental controls which are not influenced by ENSOs in the Pacific. Typhoons over the western Pacific are likely to increase due to the so-called thermostat effect operating between that ocean's western and eastern hemispheres. Droughts in Australia, India, sub-Saharan Africa and South Africa will also intensify, as will flooding in the tropical lowlands of South America. In Europe, the winters have become wetter and more temperate, while the summers tend to be sunnier and drier. The 3.5- to 7-year return period of ENSO has not varied during the contemporary warming period.

In countries located at the eastern margin of the tropical Pacific, torrential rains and river swelling wreak havoc among riverine agrarian settlements.

Faced with the prospect of future severe ENSO episodes, we will have to address the conditions that have led to increases in the human-induced component of the present global warming. However, since we have no control over the natural causes of this trend, we will also have to adjust, and learn how to cope better with the devastating power of El Niño.

Many shallow lakes in Australia are transformed into clay pans during the droughts that are the sequel of El Niño warming events in the tropical Pacific.

Emissions from chimneys at a coal-fired power station, near Goole, Yorkshire, northern England. The increase in the concentration of carbon dioxide in the atmosphere has paralleled our increased fossil fuel use.

explanations have been sought deeper in atmospheric chemistry – in other, non-greenhouse gases that also react with OH. One suspect is isoprene (C_5H_8), a (non-greenhouse) gas emitted by trees in hot climates. Perhaps far less isoprene was produced in the ice-age world, leaving more hydroxyl radicals to oxidize methane? But new evidence shows that plants produce *more* isoprene when carbon dioxide is low, counteracting the effect of lower temperatures.

Another speculation involves increased production of reactive nitrogen oxides in tropical fires; these gases promote hydroxyl production. But it's doubtful whether this mechanism could explain the size of the change. One recent study has suggested that plants themselves produce methane, but this is controversial and, in any case, the amounts involved are almost certainly too small.

Nitrous oxide

Nitrous oxide (N_2O) is produced by microbial activity in moist soils and seawater. It is broken down slowly, over 100–200 years, by light and reaction with the stratospheric ozone layer. Its natural variations are largely mysterious. Nitrous oxide increases by about 30% from ice ages to warm periods. However, increases of the same size occurred during the ice ages, linked to the fast (non-periodic) rapid climate warming events in the Northern Hemisphere known as Dansgaard-Oeschger events. These abrupt temperature jumps (up to 6°C/10.8°F per century) are believed to be caused by the rapid switching-on of the thermohaline 'pump' circulating salty water in the North Atlantic. The N_2O record parallels them all.

Methane also increases somewhat 25–70 years after these warming events. Nitrous oxide, on the other hand, starts to rise hundreds of years earlier. How can a greenhouse gas increase *before* temperature begins to change? It may be that nitrous oxide is responding to changes in the Southern Hemisphere, which are out of phase with those in the North. But the mechanism is unknown.

Implications

All three gases have increased during the industrial period; all three now have atmospheric concentrations well above their natural range. The human cause for these increases is unequivocal. No natural rhythm can explain them (p. 262). The carbon dioxide increase has paralleled the increase in fossil fuel use. The methane increase is from industrial and agricultural sources, the nitrous oxide increase from intensified agriculture. Stabilizing climate will entail cutting emissions, especially of carbon dioxide. This much is straightforward. However, our ignorance about how greenhouse gases were regulated before humans intervened is worrying. The past tells us that the transitions from cold to warm periods have been accompanied by positive feedbacks: that is, warming begets greenhouse gas increases, which further warm the Earth. We don't understand much about these feedbacks, and by the same token, we have little idea what feedbacks may be lurking in the future.

PAUL R. EHRLICH & ANNE H. EHRLICH

Predicting future human population levels

As human numbers further increase, the potential for irreversible changes of far-reaching magnitude also increases.
POPULATION SUMMIT OF THE WORLD'S SCIENTIFIC ACADEMIES, 1993

The above statement on the population problem was issued by 58 of the world's academies of science in 1993. But how far is population likely to increase? That is a key question about the human future. In general, the past projections of demographers have been quite accurate. In 1974, Thomas Frejka projected that the global population in 2000 would be between 5.9 and 6.7 billion – its actual size was 6.1 billion. India's population was projected to be between 948 million and 1 billion, and was actually 1 billion.

For population trends on a global scale, migration can be ignored; the main factors are inputs (births) and outputs (deaths). Most demographers in the 1970s assumed that birthrates would top out and start to decline around 1970 and that death rates would continue to decline slowly until they were caught by dropping birthrates and growth stopped. So far, the path to the end of growth that they projected has been fairly accurate – the actual reduction in growth has just been a little slower than expected.

Future birthrates

Future projections now may present special difficulties, because humanity is moving into uncharted territory. For the first time in history, growth rates for a significant portion of humanity have levelled off and, in some cases, turned negative (populations declining). The population of eastern Europe as a whole is shrinking by about half a per cent per year and is projected to decline

A crowded New Delhi street: while birth rates in some parts of the world are increasing, in others they are declining.

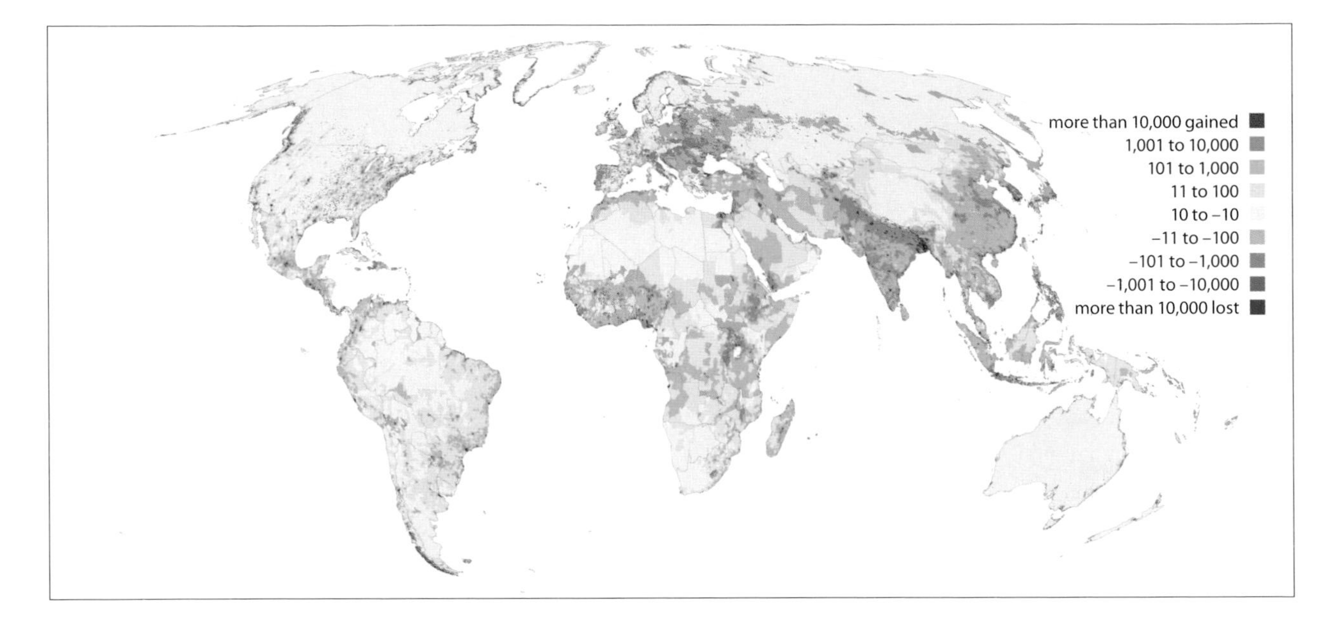

Above
Predictions for future population growth, shown as change in density per square kilometre between 1995 and 2025.

Right *A nearly empty newborn ward in Seligenstadt, Germany, one of the countries that is making efforts to reverse the downward trend in birthrates.*

from 296 million in 2006 to 230 million in 2050. The whole of Europe is projected to fall from 732 million today to 665 million in 2050, while the world population as a whole climbs from 6.5 billion to 9.2 billion. A big question is, will the very low birthrates of Europe spread to the rest of the world (where in most places birthrates *are* declining) and lead to an end to significant population growth by the end of this century? Demographers still base their projections on that assumption, but it is hardly assured. Many politicians are afraid of declining populations, even though they provide huge environmental benefits. Frantic efforts are now being instigated by some countries (e.g., Russia, Germany and Australia) to reverse the downward trend in birthrates and restore growth.

Future death rates

Even more uncertain is the situation with death rates. Several major events could produce very steep rises in mortality unanticipated by demographers. Sea level rise resulting from global warming is likely to occur slowly, yet large numbers of people could die from the disruptions it will cause. And all bets will be off if the melting of Greenland's ice sheet accelerates, or if major parts of the Antarctic ice cap slide into the sea.

Few people perhaps realize that large-scale famines may result from climate change-induced crop failures. Crops are dependent not only on appropriate temperatures, but also on precise patterns of rainfall, and both are changing. Farmers will have a hard time moving to new areas when local conditions become unsatisfactory, because they must find not only the right climate, but suitable soils and day-lengths (many crops require the right photoperiod to flower and

Weather-induced crop failure, Maryland, USA, August 1997: a farmer looks at the total loss of his sweet corn crop due to lack of rain. Such crop failures as a result of climate change may lead to large-scale famines.

produce seeds or fruit). Particularly vulnerable are likely to be grain belts such as the wheat-growing regions of India, Pakistan and Thailand – but production in the US grain belt could also be threatened. It seems likely that people in poor tropical regions, who did least to cause climate change, will suffer most. Billions could die of starvation and disease, and the global population in 2100 might consequently be smaller than today's.

Novel epidemic diseases are a third great peril. The larger the human population, the more likely it is that another new disease or flu strain will become established and spread (p. 276). And the more hungry (and so immune-compromised) people there are, the more likely it is that a global plague could be triggered. Given high-speed transport that can move disease carriers to the far corners of Earth in a day or two, we have an epidemic time bomb. Nobody knows if a novel virus, for instance, could evolve and kill everybody, but even that possibility cannot be excluded.

Toxification of the planet is another potential wild card in predicting future population levels. Civilization is inundating itself with novel toxic chemicals, each of whose individual impacts is largely unknown, let alone synergistic effects from combined exposures. Millions die from toxic exposures annually, largely of heart and lung problems traceable to air pollutants, and other toxics may induce cancers. There could be a tipping-point at which death rates escalate from the load of toxins now carried by every human. Ironically, the biggest population impact from pollution could be a drop in sperm counts, which could greatly lower the birthrate. Some scientists claim to have observed such a drop already.

Nuclear warfare is the fifth of this set, triggered by a famine or pandemic, as well as any of the other classic sources of conflict. Enough nuclear weapons still exist to cause severe climate change ('Nuclear Winter') if used. From the impacts of blast, radiation, infrastructure destruction, exposure to severe weather and famine, deaths could easily number in the billions.

Uncertainty requires action

So future human population levels are now extremely tough to predict. If humanity has simultaneous bad luck with climate change, epidemics, sperm counts and nuclear weapons, the global population could be knocked back to the 5 million people or so that lived 10,000 years ago at the time of the agricultural revolution. But with appropriate actions, the population in 2100 could be under 8 billion and gradually and purposefully declining towards a sustainable size of 1.5 to 2 billion. The human future is in the hands of this generation and those that follow us.

MICHAEL GREGER

Flu pandemics and eastern Asia

In our efforts to streamline farming practices to produce more meat for more people, we have inadvertently created conditions by which a harmless parasite of wild ducks can be converted into a lethal killer of humans.

R. H. YOLKEN & E. F. TORREY 2005

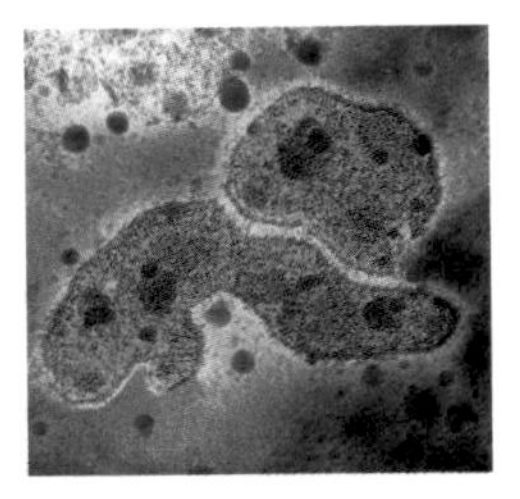

H5N1 is considered by many to be the most likely candidate for the next human pandemic flu virus.

H5N1, the strain of avian influenza currently threatening to trigger the next human flu pandemic, probably originated in South China. Likewise, the last two pandemics, the 'Asian flu' of 1957 and the 'Hong Kong flu' of 1968, arose from this same region. In fact, historical records dating back centuries link emerging influenza epidemics and pandemics to this area of the world, where the pandemic of 1889, the first for which we have cogent data, was also spawned.

Human flu is believed to have started 4,500 years ago with the domestication of waterfowl, such as ducks, the original source of all influenza viruses. Yet ducks, along with their viruses, have been domesticated all over the planet. Why is Asia thought to have been ground zero for so many outbreaks?

Waterfowl in a flooded rice field. Rice–duck farming may facilitate human exposure to intestinal waterfowl viruses such as influenza virus A.

Integrative farming

Farming practices unique to eastern Asia have inadvertently set the stage for the proliferation and species-to-species jump of influenza viruses. Farmers moved ducks from the rivers and tributaries on to flooded rice fields to be used as an adjunct to crop production, benefiting in several ways. The ducks eat the weeds and insects that might otherwise harm the crop, and thus themselves require less extra feed, reducing costs. They also fertilize the crops, but, because the influenza virus exists as an intestinal bug in waterfowl, the ducks may also release billions of virus particles into the water, creating a permanent, year-round gene pool of avian influenza viruses in close proximity to other domesticated animals and also to humans.

The so-called pig-hen-fish aquaculture system unique to Asia may have also played its part. Cages of chickens are perched directly over feeding troughs in pig pens which, in turn, are positioned above fish ponds. The pigs eat the birds' droppings and then defecate into the ponds. Depending on the species of fish, the pig excrement is then eaten directly by the fish or serves as fertilizer for aquatic plant fish food. The pond water may then be piped back up for drinking water for the pigs and hens. Human faeces may also be added to the ponds for additional enrichment. Due to the growing industrialization and pollution of migratory aquatic flyways, wild ducks are landing in increasing numbers on these farmed fish ponds, affording the naturally benign waterborne duck virus a unique opportunity to cycle through, and potentially adapt to, a mammalian species. The once harmless influenza can become dangerous in land-based birds or mammals, as viral mutants naturally selected to be better adapted to terrestrial species may be better suited for airborne spread.

These integrative farming techniques are efficient, but may have unknowingly facilitated the ability of the influenza virus to jump species. Similar to the livestock-rendering industry's recycling of slaughterhouse waste into animal feed – an ecologically sound, but, we now realize, hazardous practice (having given rise to mad cow disease) – efficiency cannot be the sole criterion for food production systems.

Globalization of risk

As new strains of influenza continue to emerge and spread, China continues to expand its extraordinary population of both farm animals and people. Since 1968, the year of the last pandemic, the number of chickens in mainland China alone has increased a thousand-fold from 13 million to 13 billion. Such industrialization of the global poultry sector may place birds in unhygienic crowded conditions conducive to further viral mutation and spread.

With such intensification around the world in recent decades, outbreaks of highly pathogenic avian influenza viruses other than H5N1 in industrial poultry operations in Australia, the Middle East, Europe, and North and South America show that eastern Asia may not be the only potential pandemic epicentre of the world.

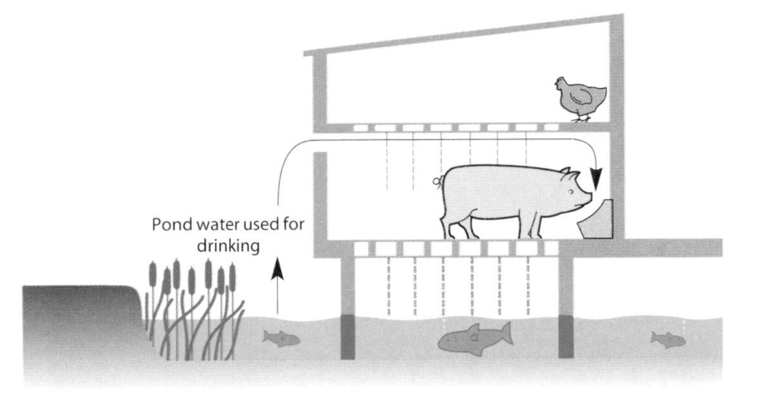

Diagram of pig-hen-fish aquaculture. The integration of swine farming with aquaculture may create conditions in which avian influenza viruses can become adapted to mammalian species.

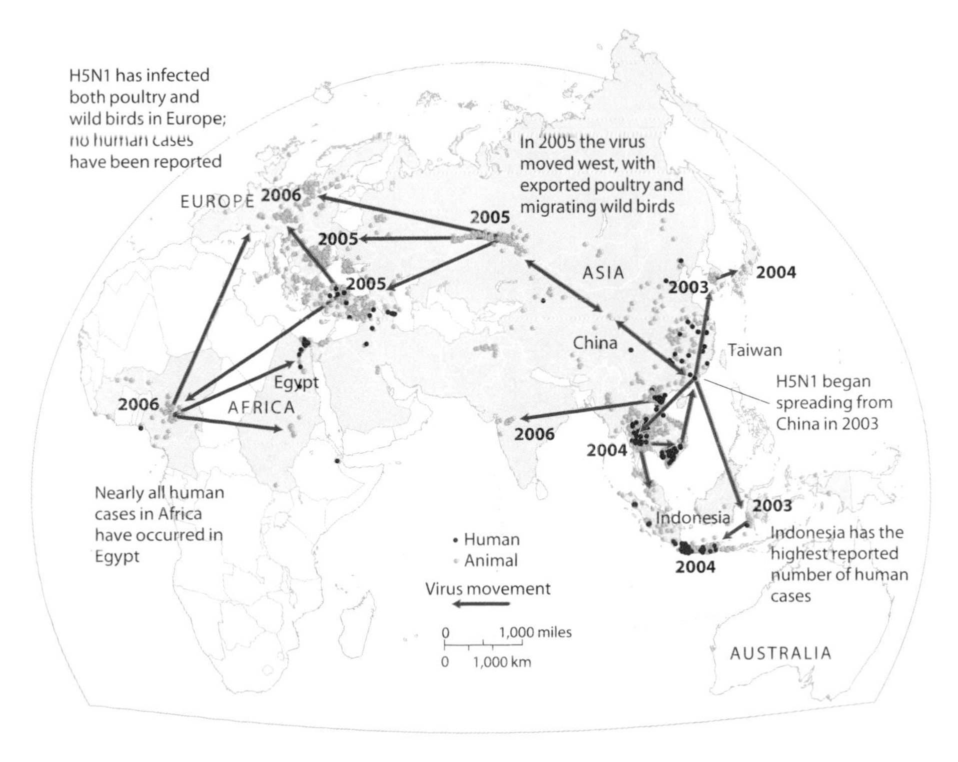

Map showing the spread of the H5N1 virus around the globe. The speed and scope of H5N1 geographic spread is unprecedented.

KATHERINE F. SMITH

Wildlife extinction and conservation

The one process now going on that will take millions of years to correct is the loss of genetic and species diversity by the destruction of natural habitats. This is the folly our descendants are least likely to forgive us.
E. O. WILSON, 1984

We live in a time when the Earth teems with life. There are more species alive today than at any time in the Earth's 4.5 billion year history. Approximately 1.7–1.8 million species have been identified by scientists, though it is believed that at least twice this number remain undescribed (p. 177). Humans have long reaped the benefits of this biological diversity, but it has come at a cost to wildlife.

Scientists believe we are in the midst of a biodiversity crisis. Ecosystems around the world are being degraded, over-exploited and destroyed. Animal and plant populations are shrinking and entire species are being driven to extinction. The primary cause of these widespread losses is human activity – species extinction is arguably the most significant outcome of our impact on the environment. Once a species is eliminated from the planet, its role in the natural environment and its unique genetic make-up are lost forever. In addition to diminishing the natural world, the current loss of biodiversity can harm humans. Humans may be causing the current biodiversity crisis, but it is also within our power to stop it.

Replica of a woolly mammoth: the Pleistocene megafauna of North and South America, including mammoths, sabre-toothed cats and giant ground sloths, is believed to have been driven to extinction by hunting and deforestation.

Species extinctions through time

The number of species on the planet has been growing since life originated some 3,500 million years ago (p. 23). This increase has not been continuous, but characterized by periods of high speciation, relatively little change and mass extinction. Scientists are aware of five mass extinctions in Earth's history, all of which were caused by natural environmental changes such as asteroid impacts and volcanic eruptions. There is growing evidence that we have entered a sixth mass extinction, driven not by nature but by human activity.

Coincident with the global spread of modern humans 30,000 years ago was a rapid decline in the number of species on the planet. Among the first species driven to extinction by humans were the Pleistocene megafauna of the Americas between around 15,000 and 8,000 years ago. Sabre toothed cats and mammoths disappeared, most likely as a result of hunting and deforestation.

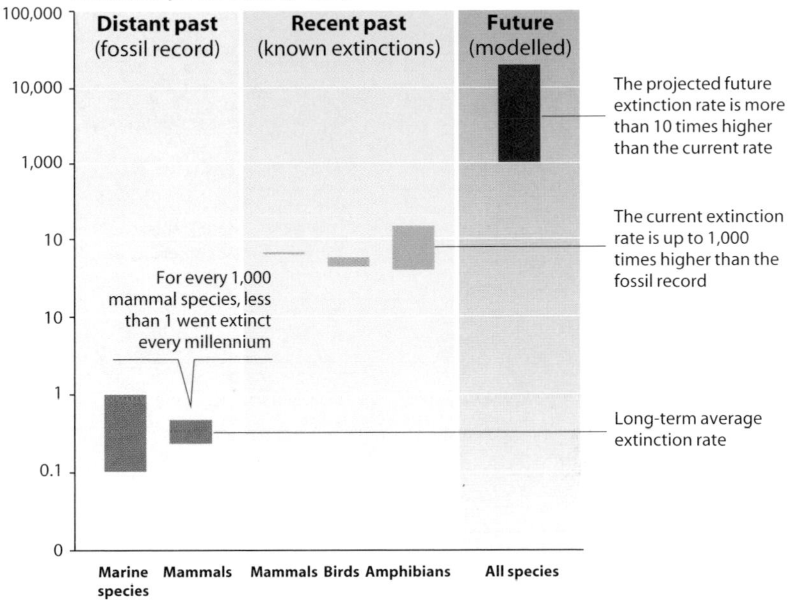

Species extinctions have increased through time for many taxonomic groups, and are expected to increase well into the future

The sixth, human-caused mass extinction

Humans sequester 20–40% of the total primary production of terrestrial ecosystems (p. 286), exploit around 80% of marine fish production and use over half of all accessible fresh surface water. There is little doubt that our domination of the planet's natural resources has caused the current biodiversity crisis.

The rate of species extinction is greater now than at any time before. More than 1,000 species are known to have gone extinct in the last 400 years, the majority in the last 150 years, and extinction is occurring at an increasingly rapid rate. Some famous extinctions in the last four centuries include the passenger pigeon of North America, the Mauritius Island dodo, the Tasmanian tiger and Steller's sea cow. There is a growing number of threatened species, suggesting that extinction rates will remain high throughout the 21st century. Some 20% of mammals, 12% of birds, 4% of reptiles, 31% of amphibians and 4% of fish are currently threatened with extinction.

Leading causes of species extinction

Conservation biologists have identified seven major threats to biological diversity that result from human activity. Nearly all species at risk of extinction face more than two of these threats.

Habitat destruction, fragmentation and degradation result from numerous processes including development, clearing land for agriculture, erosion, pollution, water diversion, logging and others. Together these represent the leading causes of species extinction. As habitat is lost, the remaining fragments shrink and become isolated, segregating wildlife and leading to population declines and ultimately extinction. Tropical moist forests occupy 7% of the Earth's land surface, but they are estimated to contain over 50% of its species. Currently, 1% of the original area of these forests is lost each year to human activity. Many believe that the most important means of protecting biological diversity is habitat preservation.

Climate change has occurred at a rapid pace over the last century. Burning of fossil fuels and large-scale forest clearance have increased levels of greenhouse gases in the atmosphere, causing an overall warming of the globe (p. 259). Species' response to global warming has already begun, with poleward shifts in the distribution of bird and plant species, as well as reproduction occurring earlier in the spring. However, the rapid rate of warming will make it difficult for many species to disperse quickly enough to move with the

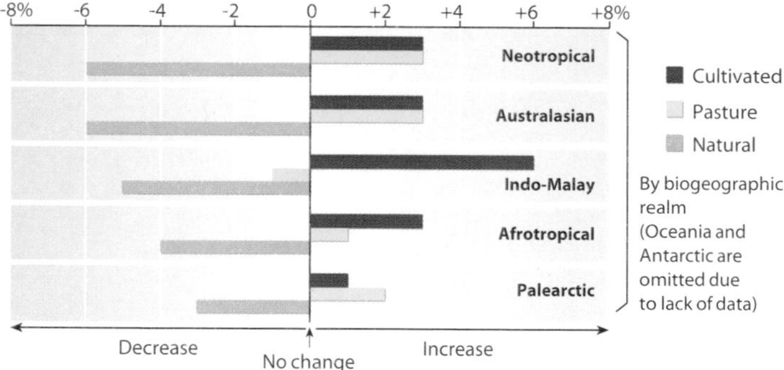

Above *Changing percentages of land use around the world reveal an increase in cultivated and pasture lands and a loss of natural habitat.*

Right *Zebra mussels, native to southeast Russia, have been introduced into the US Great Lakes where they out-compete native species for food.*

changing climate or adapt to the new conditions in their present-day range.

Overexploitation includes collecting so many individuals from a species that it can no longer reproduce sufficiently quickly to withstand the harvest. Humans have always harvested nature's bounty, but technological advances have made methods of collection so efficient that entire areas have been depleted of sought-after species. Such activities have caused the extinction of many species, including the black mamo bird of Hawaii; kings wore ceremonial cloaks that required feathers from more than 70,000 birds of the species.

Invasive species are those that people have intentionally or inadvertently moved beyond their native range. Introduced predators can eradicate native species that lack the ability to recognize them, while others threaten native species by out-competing them for limited resources. In the eastern United States, introduced zebra mussels have led to the extinction of native freshwater mussels by growing on top of them, filtering the majority of food particles out of the water and thus starving out natives.

Diseases have spread to all corners of the world with the help of humans. Infectious diseases include viruses, bacteria, parasitic worms, fungi and protozoa, all of which have the potential to drive species to extinction. When individuals in a

population have no immunity to an infectious agent, there is a greater likelihood that the resulting disease will weaken individuals, lowering survival and reproduction rates and ultimately causing extinction. One of the most significant diseases causing conservation concern is an infectious fungus, chytridiomycosis, which is causing extinctions and population declines in amphibian species around the world.

PRIMARY TOOLS USED BY CONSERVATION BIOLOGISTS

- Establishment of protected areas to guarantee intact habitat for continued species proliferation
- Habitat management to protect the supply of ecosystem services
- Enforcement of agreements such as the Convention on Biological Diversity to create a global conservation network
- Creating incentives for conservation to engage diverse segments of society, including industry and business
- Captive breeding programmes to preserve species on the brink of extinction
- Citizen science programmes to involve local communities in protecting healthy environments

Opposite above
Habitat destruction: loggers watch forest burning in the distance, eastern Madagascar.

The value of biological diversity

Biodiversity is valuable in two distinct ways. First, it has utilitarian value because it benefits humans directly. Biological diversity provides plants and animals for food, decomposing organisms that release nutrients into the soil and water, and natural resources that are used for medicines. Biological diversity also provides the support structure that maintains intact ecosystems which are necessary for performing vital functions such as air and water purification, erosion and flood prevention, and climate moderation.

Second, biodiversity has inherent value because it has worth beyond the goods and services it provides. Nature plays a prominent role in the belief systems of numerous cultures. Many people value biological diversity for the personal, religious, aesthetic and recreational rewards it provides. Others believe humans have a moral obligation to protect nature. Regardless of their personal beliefs, most people agree that it is important to prevent species extinction.

Preventing human-induced extinction

The most difficult aspect of protecting biological diversity is determining what to do when human needs conflict with those of other species. It takes significant efforts at all levels, from local to global, to develop effective conservation programmes. An added challenge is the fact that many of the drivers of biodiversity loss are projected to remain constant or increase in the near future. Nonetheless, opportunities exist to reduce the rate of biodiversity loss if ecosystems are protected, restored and managed. Ultimately the choice of protecting biodiversity will be determined by society.

Below
Conservation in action: a canopy feeding platform at Nyaru Menteng Orangutan Rehabilitation Centre, on Palas Island, Central Kalimantan, Borneo.

DAVID L. GREENE

68 What will replace liquid hydrocarbon fuels?

The fossil energy era will be a tiny blip in human history.
NEBOJSA NAKICENOVIC, 2006

Modern humans use enormous quantities of fossil petroleum: more than 30 billion barrels of oil each year. It took nature on the order of 200 million years to create the fossil energy we began using just over 200 years ago. How long can human beings continue to rely on the Earth's fossil resources for liquid fuels?

About a quarter of the fossil energy we use is petroleum, the liquid form of fossil hydrocarbons (see p. 94). Petroleum fuels have a high energy content per unit of both volume and weight – this, as well as relative ease of handling and storage, makes them ideally suited for mobile energy uses. Compared to the value they provide, liquid hydrocarbon fuels are inexpensive. Even though prices are rising, motorists would probably pay much more to avoid giving up the mobility that liquid hydrocarbon fuels enable. The enormous gap between the cost of producing petroleum fuels and their value in use is why petroleum has been called 'the prize', a coveted source of both wealth and conflict.

Human beings now burn petroleum at a rate in excess of 150,000 litres per second, a rate that has been increasing exponentially since petroleum became a significant energy source early in the last century. The world's transport systems are all but totally dependent on petroleum. How long can such rates of liquid hydrocarbon consumption be sustained?

World oil consumption by region, 1965 to 2005.The rapid growth of oil demand was interrupted as a result of the oil crises of 1973 and 1979, partly brought on by the peaking of US oil supply in 1970.

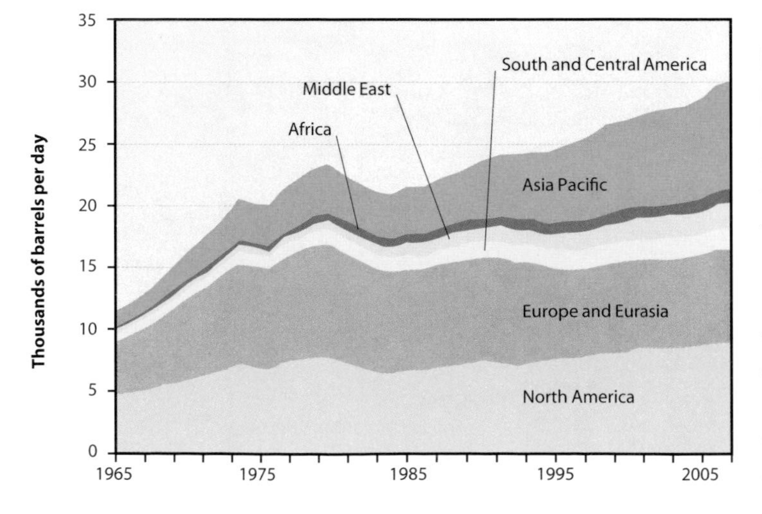

How much is left?

In part, the answer depends on how much petroleum there is in the world. But it also depends on the rates at which we produce and consume it. The emphasis on rates is critical since it is the rate of production rather than the quantity remaining that will determine the availability of petroleum in the future. It is simply not possible to continue increasing the rate of production of a finite resource until the last drop is gone.

As of 2005, we had used up nearly a trillion barrels of the Earth's petroleum resources. So how much is left? That depends on three things: first, how one defines petroleum; secondly, how optimistic one is that technological advances will increase the amount of oil that can be extracted from existing reservoirs; and thirdly which of the range of estimates one accepts.

Conventional petroleum is defined by its chemical composition, relatively low density and ability to flow in underground reservoirs. Examples of unconventional hydrocarbons include the extra heavy oil in Venezuela, Canada's oil sands and the United States' shale oil deposits. While

Traffic gridlock in Guangzhou, Guandong Province, China, 2005. The world's transportation systems now consume liquid hydrocarbons at the rate of 150,000 litres per second.

these are not petroleum, they can be refined, with extra effort, into conventional motor fuels.

Historically, only about one-third of the petroleum in a reservoir was actually extracted, but technological advances in detection and recovery have allowed a steadily increasing percentage of the oil in underground reservoirs to be produced. Future technological advances will no doubt further expand the boundaries.

Numerous attempts have been made to measure the world's ultimate resources of petroleum. The most recent estimates, and perhaps the most comprehensive, by the United States Geological Survey imply that in 1995 between 1.7 trillion and 3.7 trillion barrels of oil resources remained to be used. While it is tempting to take the current rate of world oil consumption and divide it into the range of estimated ultimately recoverable resources to get an estimate of how many years worth of oil is left, this would be a serious mistake. World oil use has increased dramatically over the past century and is expected to continue to increase as motor vehicle ownership skyrockets in growing economies like China and India. Also, as already noted, it is not possible to continue increasing the rate of oil production until it is all consumed – at some point the rate of production must slow and eventually decline.

What are the alternatives?

Substitutes for conventional petroleum must therefore be found. In fact, it can reasonably be argued that the transition is already underway. Canada now produces more than a million barrels per day of synthetic crude oil from its oil sands, and the International Energy Agency (IEA) predicts that liquid fuels produced from coal, a very minor factor today, will amount to nearly 1 million barrels per day by 2030.

Gasoline and distillate fuels can be made from coal or natural gas by a process known as Fischer-Tropsch synthesis. But not only is the process just 60–70% energy efficient (the figure for conventional petroleum refining is 80–90%), but it produces more of the climate-changing greenhouse gas, carbon dioxide. On the positive side, conversion of oil sands, heavy oil, coal and natural gas into transportation fuels can be accomplished with existing technology and at costs that are no higher than the cost of petroleum today.

A harvested field of sugar cane, near the Cerradinho Ethanol and Sugar Mill, Cantaduva, Brazil. Ethanol from Brazilian sugar cane is one of the few liquid fuels capable of competing with petroleum.

Fortunately, liquid hydrocarbons can be made from any hydrocarbon or any source of carbon. Plants, coal, natural gas, even garbage can be made into perfectly suitable fuels and have been for decades. Brazil began its Proalcool program, making ethanol for motor fuel from sugar cane, in the 1970s in response to the oil crises of that decade. Since that time, manufacturers have developed vehicles capable of adjusting automatically to any mixture from 100% gasoline to 85% ethanol with complete transparency to the motorist. At the same time, ethanol production costs have decreased to the point where it is now economically very competitive with petroleum-based fuels. Other countries have produced ethanol from corn, sugar beet and many sources of starch or sugar. In the future, scientists hope to develop economical methods of producing ethanol from the woody parts of plants in which most of the solar energy they capture is stored.

But there is a limit to how much of the world's demand for hydrocarbons could be satisfied by biofuels. By 2020, the IEA estimates that about 10% of gasoline and 3% of diesel fuel could be replaced by biofuels. In the much longer term, 2050–2100, perhaps a third or more of world gasoline and diesel could be replaced by biofuels given advances in technology and crop productivity. These figures, however, are conditional on minimal conflicts with other agricultural land uses, which is currently an increasing concern.

Liquid hydrocarbons will almost certainly one day be replaced by electricity or hydrogen. Already, hybrid vehicles combine internal combustion engines with electric motors to enhance the efficiency of fuel consumption. With further breakthroughs in battery technology, all-electric vehicles might become practical. Then, any form of energy that can produce electricity, from solar to nuclear, could be used.

Hydrogen can be produced using almost any source of energy and is likely to become a feasible replacement for liquid hydrocarbon fuels. The fuel cells that provide clean energy to spacecraft by combining oxygen and hydrogen to produce electricity have become cheaper, more durable and more powerful over the past two decades. Yet they

Bags of algae hang outside the Redhawk power plant near Phoenix, Arizona. The hope is that the fast-growing green algae could soak up carbon dioxide while at the same time producing biodiesel.

have a long way to go before they can compete economically with the internal combustion engine. Current costs must come down by nearly an order of magnitude, and much better ways found to store adequate amounts of hydrogen, the lightest element in the Universe, in vehicles.

Unfortunately, the sources of energy that can produce hydrogen without any greenhouse gas emissions (e.g., solar, wind or nuclear electrolysis) are generally much more expensive than those that do (e.g., hydrogen synthesis from natural gas or coal). But the carbon dioxide generated when hydrogen is produced from fossil fuels could be captured and stored. If strong controls are in place, vehicles powered by hydrogen can be almost entirely pollution free, regardless of the source of energy used to produce the hydrogen.

Both hydrogen and electricity will require entirely new energy delivery infrastructures, as well as vehicles with entirely new propulsion systems. Synthetic fuels from unconventional fossil sources, on the other hand, are consistent with existing fuel delivery systems and with the vehicles themselves. They are undoubtedly the path of least resistance.

Energy efficiency

Perhaps the most important change in liquid hydrocarbon fuel use is not an energy resource at all, but energy efficiency. Professor John Heywood of the Massachusetts Institute of Technology has pointed out that the typical internal combustion engine vehicle on the road today converts only about 16% of the available energy in its fuel into useful work propelling the vehicle. While other transport modes are more efficient than the passenger car, clearly enormous potential for efficiency improvement still exists. Over the next decade or two, it should be possible to reduce the fuel consumption per kilometre of cars and light trucks by a third. In three or four decades, energy consumption rates might be cut in half without affecting the usefulness of vehicles.

Improving energy efficiency will not only stretch existing fossil hydrocarbon resources, but also make alternatives more economical. By using energy efficiency to substitute for petroleum fuels we can reduce environmental impacts, extend the life of conventional resources and buy time to develop sustainable alternatives to liquid hydrocarbon fuels.

WILLIAM E. REES

Humanity's ecological footprint

Humanity's [eco]footprint first grew larger than global biocapacity in the 1980s; this overshoot has been increasing every year since, with demand exceeding supply by about 25 per cent in 2003. This means it took approximately a year and three months for the Earth to produce the ecological resources we used in that year.
WWF 2006

How much of the Earth's surface is dedicated to supporting just me at my present standard of living? This should be the first question of human ecology, but, because the Earth seems automatically to provide those of us privileged to live in rich countries with everything we need, we rarely think to ask. Things may be changing. Ominous headlines about climate chaos, acidifying oceans, fisheries collapse, tropical deforestation, the toxic contamination of food supplies etc., warn us daily that, with 6.5 billion people on the planet, the size of our footprint matters.

Growing flowers in Ethiopia for western markets. Our eco-footprints 'wander' all over the planet.

The logic of eco-footprint analysis

The eco-footprint has arguably become the world's best-known indicator of human (un)sustainability. Eco-footprint analysis (EFA) was explicitly developed to reopen the debate on human carrying capacity. The method asks 'how large a productive area does a particular population need to support its current level of consumption indefinitely?' Answering this question enables any population to compare its demand on nature with its domestic supply of productive ecosystems. This in turn reveals the extent to which that population is living beyond its local ecological means and is dependent on distant 'elsewheres' for survival.

Eco-footprinting is based on a set of simple premises, which can be summed up as follows. People are part of the ecosystems that support them. Most human impacts on ecosystems are associated with energy and material extraction and subsequent production/consumption. Furthermore, most of the energy and material resources consumed, and waste discharges produced, can be converted to corresponding productive or assimilative ecosystem areas. And there is a measurable, finite area of productive land/water on Earth.

Eco-footprint analysis has been the subject of extensive debate in the scientific literature,

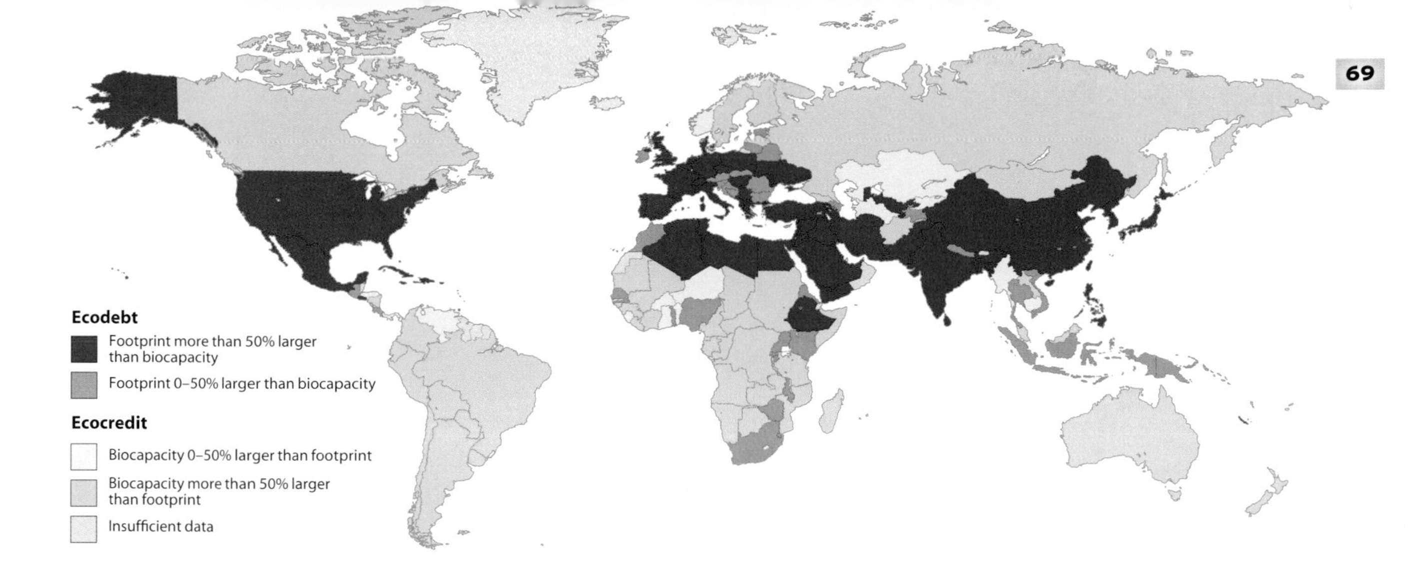

Ecological creditor and debtor countries, 2003, showing national ecological footprint relative to nationally available biocapacity, based on the WWF Living Planet Report, 2006.

contributing to the improvement of inherent methodological strengths that are both scientifically well founded and reflect thinking people's intuitive sense of reality.

Implications for sustainability

A population's ecological footprint is the area of productive land and water (ecosystems) required on a continuous basis both to produce the resources that the population consumes and to assimilate its wastes. Eco-footprints increase with income – the more money people earn, the more of nature's products they consume. People in rich countries need an average of 4 to 10 global average hectares (10 to 25 acres) to support their lifestyles, while the poor get by on the product of less than half a hectare (1 acre).

Densely populated rich countries such as the Netherlands, the United Kingdom and Japan have eco-footprints several times larger than their domestic productive areas – these countries are running massive ecological deficits with the rest of the world. In today's trade-oriented world, our eco-footprints 'wander' all over the planet but, because of globalization, most people are blind to the fact that their survival depends on the sustainable management of land- and waterscapes half a world away.

What does all this mean for sustainability? The average global citizen has an eco-footprint of about 2.2 ha (5.4 acres). This is worrisome because there are only 1.8 ha (4.4 acres) of bio-productive land and water per person on Earth. In other words, the total human eco-footprint is already 22% larger than the productive area of the planet. This means that we are living, in part, not on sustainable production but by depleting critical ecosystems.

There is a potentially even bigger problem. Eco-footprints constitute *mutually exclusive* appropriations. This means that we compete with each other for available biocapacity. As implied above, an equitable 'Earth-share' would be about 1.8 global average ha. However, wealthy North America's per capita eco-footprint is over 9 ha (22.3 acres) per capita and the 'sustainability gap' is still increasing. It would take four additional Earth-like planets to support just the present world population at North American material standards.

Sustainability with equity would require that rich nations use every means at their disposal to reduce their eco footprints by a factor of three to five. Other studies concur that the world's wealthy countries must achieve per capita reductions in energy and material consumption of up to 80% or more to create the ecological space for needed growth in the developing world. The alternative may be ecological collapse and geopolitical chaos.

Fortunately, we actually have the knowledge and technology to make such massive cuts in resource use while improving quality of life. Regrettably, however, we have not yet mustered the public pressure and political will to implement the tax and regulatory incentives required to do the job (p. 288). The growing human eco-footprint may therefore yet trample any prospect that techno-industrial civilization will survive the 21st century.

WILLIAM E. REES

Human behaviour and saving the planet

[For survival] we must rely on highly evolved genetically based biological mechanisms, as well as on suprainstinctual survival strategies that have developed in society, are transmitted by culture, and require, for their application, consciousness, reasoned deliberation and willpower.

A. Damasio, 1994

Humanity is a conflicted species. Science shows that human over-consumption of energy and material resources is threatening to deplete stocks of vital natural capital and has already filled waste sinks to over-flowing. Both these trends undermine global life-support systems and threaten human civilization, if not the rest of life on Earth. We know what must be done, but to date neither ordinary people nor their governments seem capable of the kind of 'deep' behavioural change necessary to reverse the trends that threaten our survival.

Opposite *Wasted wealth: the consumer society will have to curtail its ongoing romance with the private car. 'Oxford Tire Pile No. 1, Westley, California 1999': photograph by Edward Burtynsky.*

Left *Heat loss from an apartment block in Frankfurt-am-Main. The 1950s block was refurbished in 2006 using Passivhaus measures. The dark colours on the thermogram on the right show the reduction in the amount of heat escaping.*

Nature and nurture conspire against us

Both our genes (natural endowment) and our memes (cultural endowment of beliefs, values and assumptions) work against the global community's becoming truly sustainable. Remember, our behavioural biology has been fashioned by natural selection. Like all other organisms, therefore, humans have evolved a natural predisposition to expand to fill all accessible habitats and to consume all available resources.

However, bolstered by our superior intelligence, capacity for language and cumulative technology, we have left competing species wallowing in our wake. Work by Charles Fowler and Larry Hobbs has shown that *Homo sapiens'* geographic range, population size, energy use, carbon dioxide emissions and biomass consumption are larger than those of non-human species by orders of magnitude.

For example, the consumption of biomass by humans is 100 times above the upper 95% confidence limits for biomass consumption by 96 other mammalian species. Humans have become the dominant macro-consumer in all major terrestrial and accessible marine ecosystems on the planet, and may well be the most voraciously successful predatory and herbivorous vertebrate ever to walk the Earth. Ironically, it is precisely this extraordinary evolutionary success that may be killing us.

It doesn't help that humanity's natural tendency to overexploit resources is being reinforced by powerful cultural norms and meme complexes. For example, after the Second World War, industry in the Western developed countries set about deliberately to create what we now call 'the consumer society'. Retail analyst Victor Lebow set out the mission in 1955: 'Our enormously productive economy demands that we make consumption our way of life, that we convert the buying and use of goods into rituals, that we seek our spiritual satisfaction and our ego satisfaction in consumption.' This socially constructed consumer ethic is, in turn, reinforced by national governments and official international development agencies, virtually all of whom subscribe to a common myth of global 'development' rooted in economic integration, unlimited growth and liberalized trade.

Yet another human quality confounds sustainability. Peoples' behaviour, particularly group behaviour, is innately conservative. This may be a legacy of evolutionary times when 'the environment' changed relatively slowly and it was adaptive for people to remain in seemingly successful behavioural ruts. The consequence today is extreme difficulty in inducing mass social change when the payoff is uncertain. This reality is the more daunting in our multi-national, multi-cultural, multi-religious global village, where social inequity and eco-apartheid are facts of life and internecine conflict overwhelms efforts to focus attention on the human family's mutual interest in saving the planet.

Turning things around

For all these reasons, mere awareness of environmental risks or even imminent catastrophe is generally insufficient to induce intelligently responsive changes to personal or collective

Opposite
Turbines of a wind farm on Euboea, Greece. Wind power is one of the sustainable new technologies that will have to be developed to reduce our dependence on burning fossil fuels.

behaviour. But all is not lost – in the right circumstances, societies can evolve quickly. For example, short of actual catastrophe, strong economic incentives can be an effective way to force significant socio-behavioural change.

Sharply rising energy prices in the late 1970s and early 1980s had a dramatic impact on consumer preferences, producing a well-documented, first-ever decline in per capita (and therefore global) energy use in modern history. Moreover, the world did not implode. Instead, economies responded by becoming more efficient and beginning the *serious* development of potentially sustainable new technologies, some of which (e.g. wind power) we benefit from today. Regrettably, the subsequent return to low energy prices heralded the rebirth of consumerism.

ESSENTIAL STEPS TO SUSTAINABILITY

New cultural 'memes' must be developed to over-ride expansionist instincts and replace today's destructive global development paradigm. The world must recognize that:

- The human enterprise is a fully contained, totally dependent subsystem of a finite ecosphere.
- There are biophysical limits to the growth of the aggregate human economy.
- The world has already exceeded long-term carrying capacity by about 22% and must reduce its total ecological footprint by at least that amount (p. 286).
- Sustainability requires greater social equity (under prevailing conditions, the richest 20% of the human family receive 75% of global income; the poorest 20% barely subsist on 1.5% of income).
- High-income countries must reduce energy and material consumption by 80% to open up the ecological space for necessary consumption growth in developing countries.

Despite this setback, the lesson is there to learn. The basic solution requires, first, that the global community unite in recognizing that in today's ecologically stressed and geopolitically unstable world, blind obeisance to individual or tribal instincts that served so well in our evolutionary past has become maladaptive. Instead, there must be universal recognition that our individual interests may now be best served by honouring our common interests. Sustainability is a collective enterprise (see Box).

The (un)sustainability conundrum

As demanding as these measures seem, they are to everyone's advantage and the transition to sustainability could be achieved with minimum pain. Despite the world's addiction to growth, neither objective indicators of population health nor subjective indicators of well-being show further improvement with income growth in wealthy countries. Moreover, we have the technology to make the needed deep cuts in resource consumption with only modest effects on lifestyles.

In short, we have nothing to lose and literally everything to gain. Determined government policies are needed to make it happen, together with new transnational legal instruments and institutions to ensure that they are implemented and enforced. High on the list of priorities is 'ecological fiscal reform' – adjustments to national tax systems to correct for market failures and ensure that the prices paid for energy and material-intense goods and services 'tell the truth' about their ecological and social impacts. There is no more effective way to change peoples' consuming behaviour.

The (un)sustainability conundrum clearly poses the ultimate challenge to human intelligence and self-awareness. Rising fully to this challenge would finally separate humankind from species that remain slaves to crude survival instincts. Indeed, the triumph of new wilfully constructed global cultural memes over genetically scripted determinism would mark the next leap forward in human evolution. I leave it to the reader to ponder both the probability of success and the consequences of failure to make the leap.

Further reading

Origins

1 How did the Earth form?

Chambers, J. E. & Halliday, A. N., 'The origin of the solar system', in McFadden, L., Weissman, P. & Johnson, T. (eds), *Encyclopedia of the Solar System*, 2nd ed. (Amsterdam & London, 2007), 29–52

Wood, B. J., Walter, M. J. & Wade, J., 'Accretion of the Earth and segregation of its core', *Nature* 441 (2006), 825–33

2 The origins of life

Cairns-Smith, A. G., *Seven Clues to the Origin of Life* (Cambridge & New York, 1985)

Davies, P., *The Origin of Life* (London, 2003)

Knoll, A. H., *Life on a Young Planet* (Princeton, 2003)

Wolfson, A., *Life Without Genes* (London, 2000)

3 The origins of multicellular life

Bonner, J. T., *The Evolution of Complexity, By Means of Natural Selection* (Princeton, 1988)

Buss, L. W., *The Evolution of Individuality* (Princeton, 1987)

Butterfield, N. J., '*Bangiomorpha pubescens* n. gen., n. sp.: implications for the evolution of sex, multicellularity and the Mesoproterozoic-Neoproterozoic radiation of eukaryotes', *Paleobiology* 26 (2000), 386–404

Butterfield, N. J., 'Macroecovolution and macroecology through deep time', *Palaeontology* 50 (2007), 41–55

Maynard Smith, J. & Szathmáry, E., *The Major Transitions in Evolution* (Oxford & New York, 1995)

Michod, R. E., 'The group covariance effect and fitness trade-offs during evolutionary transitions in individuality', *Proceedings of the National Academy of Sciences* 103 (2006), 9113–17

4 The Cambrian evolutionary 'explosion'

Briggs, D. E. G., Collier, F. J. & Erwin, D. H., *Fossils of the Burgess Shale* (Washington, 1994)

Hou, X. & others, *The Cambrian Fossils of Chengjiang, China: The Flowering of Early Animal Life* (Oxford, 2004)

Morris, S. C., *The Crucible of Creation* (Oxford & New York, 1998)

Parker, A. R., *In the Blink of an Eye* (London & Cambridge, MA, 2003)

Valentine, J. W., *On the Origin of Phyla* (Chicago, 2004)

5 The biggest mass extinction of all time

Benton, M. J., *When Life Nearly Died* (London & New York, 2003; pb 2008)

Benton, M. J. & Twitchett, R. J., 'How to kill (almost) all life: the end-Permian extinction event', *Trends in Ecology and Evolution* 18 (2003), 358–65

Erwin, D. H., *Extinction: How Life on Earth Nearly Ended 250 Million Years Ago* (Princeton, 2006)

Jin, Y. G. & others, 'Pattern of marine mass extinction near the Permian-Triassic boundary in South China', *Science* 289 (2000), 432–36

Wignall, P. B. & Twitchett, R. J., 'Oceanic anoxia and the end Permian mass extinction', *Science* 272 (1996), 1155–58

6 Were the dinosaurs warm-blooded?

Chinsamy-Turan, A., *The Microstructure of Dinosaur Bone: Deciphering Biology with Fine Scale Techniques* (Baltimore & London, 2005)

Chinsamy-Turan, A. & Hillenius, W., 'Dinosaur physiology' in Weishampel, D. B., Dodson, P. & Osmólska, H. (eds), *The Dinosauria* (Berkeley, 2004)

Farlow, J. O., 'Dinosaur energetics and thermal biology' in Weishampel, D. B., Dodson, P. & Osmólska, H. (eds), *The Dinosauria* (Berkeley, 2004)

Reid, R. E. H., 'Dinosaurian physiology: the case for "intermediate" dinosaurs', in Farlow, J. O. & Brett-Surman, M. K. (eds), *The Complete Dinosaur* (Bloomington, IN, 1997)

7 Giant dinosaurs: how did they get so big?

Alexander, R. McNeill, 'All-time giants: the largest animals and their problems', *Palaeontology* 41 (1998), 1231–45

Burness, G. P., Diamond, J. & Flannery, T., 'Dinosaurs, dragons, and dwarfs: the evolution of maximal body size', *Proceedings of the National Academy of Sciences* 98 (2001), 14518–23

Curry-Rogers, K. & Wilson, J. A., *The Sauropods: Evolution and Paleobiology* (Berkeley, 2005)

Fastovsky, D. E. & Weishampel, D. B., *The Evolution and Extinction of the Dinosaurs*, 2nd ed. (Cambridge & New York, 2005)

Hone, D. W. E. & Benton, M. J., 'The evolution of large size: how does Cope's Rule work?', *Trends in Ecology and Evolution* 20 (2005), 4–6

Sander, P. M. & others, 'Adaptive radiation in sauropod dinosaurs: bone histology indicates rapid evolution of giant body size through acceleration', *Organisms, Diversity & Evolution* 4 (2004), 65–73

8 Why did the dinosaurs die out?

Alvarez, L. W. & others, 'Extraterrestrial cause for the Cretaceous-Tertiary extinction', *Science* 208 (1980), 1095–108

Alvarez, W., *T. Rex and the Crater of Doom* (Princeton & London, 1997)

Hildebrand, A. R. & others, 'Chicxulub Crater: A possible Cretaceous-Tertiary boundary impact crater on the Yucatan peninsula', *Geology* 19 (1991), 867–71

Smit, J., 'The global stratigraphy of the Cretaceous-Tertiary boundary impact ejecta', *Annual Review of Earth and Planetary Sciences* 27 (1999), 75–113

Smit, J. & van der Kaars, S., 'Terminal Cretaceous extinctions in the Hell Creek area, Montana: compatible with catastrophic extinction', *Science* 223 (1984), 1177–79

9 Why do mammals rule the world?

Bininda-Emonds, O. R. P. & others, 'The delayed rise of present-day mammals', *Nature* 446 (2007), 507–12

Kemp, T. S., *The Origin and Evolution of Mammals* (Oxford & New York, 2005)

Qiang, J. & others, 'The earliest known eutherian mammal', *Nature* 416 (2002), 816–22

Radinsky, L., 'Evolution of brain sizes in carnivores and ungulates', *The American Naturalist* 112 (1978), 815–31

Zhe-Xi, L. & others, 'An Early Cretaceous tribosphenic mammal and metatherian evolution', *Science* 302 (2003), 1934–40

10 The hunt for the earliest human ancestor

Alemseged, Z. & others, 'A juvenile early hominin skeleton from Dikika, Ethiopia', *Nature* 443 (2006), 296–301

Brunet, M. & others, 'A new hominid from the Upper Miocene of Chad, Central Africa', *Nature* 418 (2002), 145–51

Collard, M. & Wood, B., 'Defining the genus Homo', in Henke, W. & Tattersall, I. (eds), *Handbook of Paleoanthropology. Volume 3. Phylogeny of Hominids* (Berlin, 2007), 1575–611

Stringer, C. & Andrews, P., *The Complete World of Human Evolution* (London & New York, 2005)

Wood, B. A., *Human Evolution: A Very Short Introduction* (Oxford & New York, 2005)

Wood, B. & Constantino, P., 'Human origins: life at the top of the tree', in Cracraft, J. & Donoghue, M. J. (eds), *Assembling the Tree of Life* (Oxford & New York, 2004), 517–35

The Earth

11 Why does the compass point north?

Bloxham, J. & Gubbins, D., 'The secular variation of the Earth's magnetic field', *Nature* 317 (1985), 777–81

Gubbins, D. & Herrero-Bervera, E. (eds), *Encyclopedia of Geomagnetism and Paleomagnetism* (Heidelberg, 2007)
Jacobs, J. A., *Reversals of the Earth's Magnetic Field* (Cambridge & New York, 1995)
Jonkers, A. R. T., *Earth's Magnetism in the Age of Sail* (Baltimore, 2003)
Laj, C. A. & others, 'Geomagnetic reversal paths', *Nature*, 351(1991), 447
Merrill, R. T., McElhinny, M. W. & McFadden, P. L., *The Magnetic Field of the Earth: Paleomagnetism, the Core, and the Deep Mantle* (London, 1996)

12 How continents and oceans form
Davidson, J. P., Reed, W. E. & Davis, P. M., *Exploring Earth – An introduction to Physical Geology* (New Jersey, 1997)
Hawkesworth, C. J. & Kemp, A. I. S., 'Evolution of the continental crust', *Nature* 443 (2006), 811–17
Kemp, A. I. S. & Hawkesworth, C. J., 'Granitic perspectives on the generation and secular evolution of the continental crust', in Rudnick, R. L. (ed.), *The Crust*, 349–410, vol. 3, Holldand, H. D. & Turekian, K. K. (eds), *Treatise on Geochemistry* (Oxford, 2003)
Patchett, P. J. & Samson, S. D., 'Ages and growth of the continental crust from radiogenic isotopes', in Rudnick, R. L. (ed.), *The Crust*, 321–48, vol. 3, Holldand, H. D. & Turekian, K. K. (eds), *Treatise on Geochemistry* (Oxford, 2003)
Valley, J. W., 'A cool Early Earth?' *Scientific American* (October, 2005), 58–65

13 Is Mount Everest getting higher?
Chen, J. & others, 'Progress in technology for the 2005 height determination of Qomolungma Feng (Mt Everest)', *Science in China: Series D, Earth Sciences* 49 (2006), 531–38
Dickey, P. A., 'Who Discovered Mount Everest?', *EOS Transactions American Geophysical Union*, 66 (1985), 54–59
Gulatee, B. L., 'The Height of Mount Everest: A new determination (1952–54)', *Survey of India Technical Paper no. 8* (1954)
Poretti, G., Marchesini, C. & Beinat, A., 'GPS surveys Mount Everest', *GPS World* (October 1994), 32–44
Poretti, G., Mandler, R. & Lipizer, R., 'The height of mountains', *Bollettino di Geofisica Teorica ed Applicata* 47 (2006), 557–75
Poretti, G., 'The GPS station at the Pyramid Geodetic Laboratory', *Bollettino di Geofisica Teorica ed Applicata* 47 (2006), 25–32

14 Where did the oxygen in the atmosphere come from?
Holland, H. D., *The Chemical Evolution of the Atmosphere and Oceans* (Princeton, 1984)
Lane, N., *Oxygen: The Molecule That Made the World* (New York, 2003)
Morton, O., *Eating the Sun: How Plants Power the Planet* (London, 2007)
Walker, J. C. G., *Earth History: The Several Ages of the Earth* (Boston, 1986)

15 What makes volcanoes explode?
Schmincke, H. U., *Volcanism* (Berlin, 2004)
Sigurdsson, H., Houghton, B., Rymer, H. & Stix, J., *Encyclopedia of Volcanoes* (San Diego & London, 1999)
Sparks, R. S. J. & others, *Volcanic Plumes*. (Chichester & New York, 1997)

16 The formation of diamonds
Bari, H., *Diamonds of the World: In the Heart of the Earth, in the Heart of Stars, at the Heart of Power* (Paris, 2001)
Fortey, R., *The Earth: An Intimate History* (London & New York, 2004)
Grotzinger, J. & others, *Understanding Earth*, 5th ed. (New York, 2006)

17 Which was the largest volcanic eruption ever?
Francis, P. & Oppenheimer, C., *Volcanoes* (Oxford & New York, 2004)
Sparks, S. & others, 'Super-eruptions: global effects and future threats', *Report of a Geological Society of London Working Group*, 2nd (print) ed., (2005) www.geolsoc.org.uk

18 The mystery of tsunamis
Bondevik, S. & others, 'Record-breaking height for 8000-year-old tsunami in the North Atlantic', *EOS Transactions American Geophysical Union* 84, 31 (2003), 289
Geist, E. L., Titov, V. V. & Synolakis, C. E., 'Tsunami: wave of change', *Scientific American* (January, 2006), 56–63
Gonzalez, F. I., 'Tsunami!', *Scientific American* (May, 1999), 44–55
Gregg, C. E. & others, 'Tsunami warnings: understanding in Hawaii', *Natural Hazards* 40, 1 (2007), 71–87
Kanamori, H., 'Lessons from the 2004 Sumatra-Andaman earthquake', *Philosophical transactions – Royal Society. Mathematical, Physical and Engineering Sciences* 364,1845 (2006),1927–45
Keller, G. & others, 'Age, stratigraphy and deposit of near –K/T siliciclastic deposits in Mexico: relation to bolide impact?', *Geological Society of America Bulletin* 4 (1997), 410–28
Lay, T. & others, 'The great Sumatra-Andaman earthquake of 26 December 2004', *Science* 308, 5725 (2005),1127–33
Sieh, K., 'Sumatran megathrust earthquakes: from science to saving lives', *Philosophical Transactions – Royal Society. Mathematical, Physical and Engineering Sciences* 364,1845 (2005), 1947–63

19 Asteroid and comet impacts on Earth
Claeys, P., 'Impact events and evolution of the Earth' in Gargaud, M., Martin, H. & Claeys, P. (eds), *Lectures in Astrobiology, Volume II* (Berlin, 2007)
French, B. M., *Traces of Catastrophe: A Handbook of Shock-Metamorphic Effects in Terrestrial Meteorite Impact Structures*, Lunar and Planetary Institute, Houston, contribution 954, (1998)
Melosh, H. J., *Impact Cratering: A Geologic Process* (NewYork, 1989)
Montanari, A. & Koeberl, C., *Impact Stratigraphy* (Berlin, 2000)

20 Where does oil come from?
Peters, K. E., Walters, C. M., & Moldowan, J. M., *The Biomarker Guide*, 2nd ed. (Cambridge, 2005)
Mello, M. R., *Chemical and molecular studies of the depositional environments of source rocks and their derived oils from the Brazilian marginal basins* (PhD thesis, Bristol University, 1988)
Mello, M. R. & others, 'Petroleum geochemistry applied to petroleum system investigation', in Mello, M. R. & Katz, B. J. (eds), *Petroleum Systems of South Atlantic Margins*, AAPG Memoir 73 (2000), 41–52

Evolution

21 The evidence for evolution
Bowler, P., *Evolution: The History of an Idea*, 3rd ed. (Berkeley & London, 2003)
Freeman, S. & Herron, J. C., *Evolutionary Analysis*, 4th ed. (San Francisco, 2006)
Mayr, E., *What Evolution Is* (New York & London, 2002)
Zimmer, C., *Evolution: The Triumph of an Idea* (New York, 2006)

22 How did the eye evolve?
Fernald, R. D., 'Casting a genetic light on the evolution of eyes', *Science* 313 (2006), 1914–18
Land, M. F. & Nilsson, D.-E., *Animal Eyes* (New York & Oxford, 2002)
Parker, A. R., *In the Blink of an Eye* (London & Cambridge, MA, 2003)
Parker, A. R., *Seven Deadly Colours* (London, 2005)

23 Why do so many people not accept evolution?
Davis, P. & Kenyon, D. H., *Of Pandas and People: The Central Question of Biological Origins* (Dallas, 1993)
Forrest, B. & Gross, P. R., *Creationism's Trojan Horse: The Wedge of Intelligent Design* (Oxford & New York, 2004)
Gould, S. J., *The Panda's Thumb: More Reflections in Natural History* (New York, 1980)
Isaak, M., *The Counter-Creationism Handbook* (Berkeley, 2006)
Jones, John J., III, Tammy Kitzmiller et al. v. Dover Area School District et al., 2005 WL 578974 (MD Pa. 2005), 20 December 2005 (available as www.pamd.uscourts.gov/kitzmiller/kitzmiller_342.pdf)
Miller, J. D., Scott, E. C., & Okamoto, S., 'Public acceptance of evolution', *Science* 313 (2006), 765–66
Morris, H. M., *Scientific Creationism* (San Diego, 1974)

Numbers, R. L., *The Creationists: From Scientific Creationism to Intelligent Design*, expanded ed. (Chicago, 2006)
Padian, K. & Angielczyk, K. D., 'Are there transitional forms in the fossil record?', in Petto, A. J. & Godfrey, L. (eds), *Scientists Confront Creationists*, 2nd ed. (New York, 2007)
Scott, E. C., *Evolution vs. Creationism* (Berkeley, 2006)

24 Disentangling the genetic code
Freeland S. J. & Hurst L. D., 'Evolution encoded', *Scientific American*, 290(4) (2004), 84–91
Hayes, B., 'Ode to the Code', *American Scientist*, 92(6) (2004), 494–98
Hunter, L., Molecular Biology for Computer Scientists,
http://compbio.uchsc.edu/Hunter/01-Hunter.pdf
Knight R. D., Freeland S. J. & Landweber, L. F., 'Selection, history and chemistry: the three faces of the genetic code', *Trends in Biochemical Sciences*, 24(6) (1999), 241–47
Knight R. D., Freeland S. J. & Landweber L. F., 'Rewiring the keyboard: evolvability of the genetic code', *Nature Reviews Genetics*, 2(1), (January, 2001) 49–58

25 Selfish-gene theory
Burt, A. & Trivers, R. L., *Genes in Conflict: The Biology of Selfish Genetic Elements* (Cambridge, MA, 2006)
Darwin, C. R., *The Variation of Animals and Plants under Domestication*, 2nd ed. (London,1875)
Dawkins, R., *The Selfish Gene*, 3rd ed. (Oxford & New York, 2006)
Dawkins, R., *The Extended Phenotype: The Gene as the Unit of Selection* (Oxford & San Francisco, 1982)
Dawkins, R., *The Ancestor's Tale* (Phoenix, 2005)
Maynard Smith, J. & Szathmáry, E., *The Origins of Life: From the Birth of Life to the Origin of Language* (Oxford & New York, 1999)
Ridley, M., *The Origins of Virtue* (London & New York, 1997)

26 Drawing the tree of life
Bininda-Emonds, O. R. P. & others, 'The delayed rise of present-day mammals', *Nature* 446 (2007), 507–12
Darwin, C., *On the Origin of Species* (London, 1859)
Forey, P. L. & others, *Cladistics*, 2nd ed. (Oxford & New York, 1998)
Futuyma, D., *Evolutionary Biology*, 3rd ed. (Sunderland, MA, 1998)
Mayr, E., *What Evolution Is* (New York & London, 2002)
Ridley, M., *Evolution*, 3rd ed. (Oxford & New York, 2004)

27 Human genetic variation
Demuth, J. P. & others, 'The evolution of mammalian gene families', PLoSONE 1(1), e85 (2006)
Edwards, A. W., 'Human genetic diversity: Lewontin's fallacy', *Bioessays* 25:8 (2003), 798–801
Estivill, X. & Armengol, L., 'Copy number variants and common disorders: filling the gaps and exploring complexity in genome-wide association studies,' PLoS Genetics 3, e190 (2007)
Pennisi, E., 'Breakthrough of the year: human genetic variation', *Science* 318 (2007), 1842–43
Stranger, B. E. & others, 'Population genomics of human gene expression', *Nature Genetics* 39, (2007), 1217
http://genome.wellcome.ac.uk/
http://science.education.nih.gov/supplements/nih1/genetic/default.htm
http://www.genome.gov/10005107

28 How do new species form?
Coyne, J. A. & Orr, H. A., *Speciation* (Sunderland, MA, 2004)
Howard, D. J. & Berlocher, S. H., *Endless Forms: Species and Speciation* (New York, 1998)
Magurran, A. E. & May, R. M., *Evolution of Biological Diversity* (Oxford & New York, 1999)
Schilthuizen, M., *Frogs, Flies and Dandelions: The Making of Species* (Oxford & New York, 2001)

29 Exploring the links between evolution and development
Carroll, S. B., Grenier, J. K. & Weatherbee, S. D., *From DNA to Diversity: Molecular Genetics and the Evolution of Animal Design* (Malden, MA, 2001)
Carroll, S. B., *Endless Forms Most Beautiful: The New Science of Evo Devo and the Making of the Animal Kingdom* (New York & London, 2005)
Carroll, S. B., *The Making of the Fittest: DNA and the Ultimate Forensic Record of Evolution* (New York & London, 2006)
Davidson, E. H. & Erwin, D. H., 'Gene regulatory networks and the evolution of animal body plans', *Science* 311 (2006), 796–800
Raff, R. A., *The Shape of Life: Genes, Development, and the Evolution of Animal Form* (London & Chicago, 1996)
Nielsen, C., *Animal Evolution: Interrelationships of the Living Phyla* (Oxford & New York, 2001)

30 Are five fingers essential?
Coates, M. I., & Ruta M., 'Skeletal changes in the transition from fins to limbs', in Hall, B. K. (ed.), *Fins into Limbs* (Chicago, 2007)
Gould, S. J., *The Panda's Thumb: More Reflections in Natural History* (New York, 1980)
Mortlock, D. P. & Innis, J., 'Mutation of HOXA13 in hand-foot genital syndrome', *Nature Genetics* 15 (1997), 179–80
Zakany, J. & others, 'Regulation of number and size of digits by posterior Hox genes: A dose-dependent mechanism with potential evolutionary implications', *Proceedings of the National Academy of Sciences, USA* 94 (1997), 13695–700

Biogeography & Environments

31 Numbers of species in the tropics and at the poles
Clarke, A. & Gaston, K. J., 'Climate, energy and diversity', *Proceedings of the Royal Society B*, 273 (2006), 2257–66
Gaston, K. J. & Spicer, J. I., *Biodiversity. An Introduction*, 2nd ed. (Oxford & Malden, MA, 2004)
MacArthur, R. H., 'Patterns of species diversity', *Biological Reviews* 40 (1965), 510–35
Terborgh, J., *Diversity and the Tropical Rain Forest* (New York, 1992)
Willig, M. R., Kaufman, D. M. & Stevens, R. D., 'Latitudinal gradients of biodiversity: pattern, process, scale and synthesis', *Annual Reviews of Ecology, Evolution and Systematics* 34 (203), 273–309
Wilson, E. O., *The Diversity of Life* (Cambridge, MA, 1992)

32 The evolution of deserts
Goudie, A. S., *Great Warm Deserts of the World: Landscapes and Evolution* (Oxford & New York, 2002)
Manguiet, M. M., *Aridity: Droughts and Human Development* (Berlin, 1999)
Matthews, J. (ed.), *The Encyclopaedic Dictionary of Environmental Change* (London & New York, 2001)
Warner, T. T., *Desert Meteorology* (Cambridge & New York, 2004)

33 How do plants and animals live in the desert?
Philips, S. J. & Comus, P. W. (eds), *A Natural History of the Sonoran Desert* (Tucson, 2000)
Sowell, J., *Desert Ecology: An Introduction to Life in the Arid Southwest* (Salt Lake City, 2001)
Whitford, W. G., *Ecology of Desert Systems* (London & San Diego, 2002)

34 Has there always been ice at the poles?
Beerling, D. J., *The Emerald Planet: How Plants Changed Earth's History* (Oxford & New York, 2007)
Christie, R. L. & McMillan, N. J., 'Tertiary fossil forests of the Geodetic Hills, Axel Heiberg Island, Arctic Archipelago', *Geological Society of Canada Bulletin* 403 (1991)
Ruddiman, W. F., *Earth's Climate Past and Future* (New York, 2000)
White, M. E., *The Greening of Gondwana* (Frenchs Forest, New South Wales, 1986)

35 How deep can life live under ice and rock?

Amy, P. S. & Halderman, D. L., *Microbiology of the Deep Subsurface* (Boca Raton, 1997)

Chapelle, F. H. & others, 'A hydrogen-based subsurface microbial community dominated by methanogens', *Nature* 415 (2002), 312–15

Lin, L. H. & others, 'Long-term sustainability of a high-energy, low-diversity crustal biome', *Science* 314 (2006), 479–82

Parkes, R. J. & others, 'Deep bacterial biosphere in Pacific Ocean sediments', *Nature* 371 (1994), 410–13

Whitman, W. B., Coleman, D. C. & Wiebe, W. J., 'Prokaryotes: the unseen majority', *Procs of the National Academy of Sciences* 95 (1998), 6578–83

Wolfe, D. W., *Tales from Underground. A Natural History of Subterranean Life* (Jackson, TN, 2002)

36 Why are islands special?

Carlquist, S., *Island Biology* (New York, 1974)

Grant, P. R., *Evolution on Islands* (Oxford & New York, 1998)

Whittaker, R. J. & Fernandez-Palacios, J. M., *Island Biogeography: Ecology, Evolution and Conservation*, 2nd ed. (Oxford & New York, 2006)

Williamson, M., *Island Populations* (Oxford & New York, 1981)

37 What do we know about Darwin's finches?

Darwin, C. R., *Journal of Researches into the Geology and Natural History of the Various Countries Visited during the Voyage of H.M.S. Beagle*, 2nd. ed. (London, 1845)

Darwin, C., *On the Origin of Species* (London, 1859)

Grant, P. R., *Ecology and Evolution of Darwin's Finches* (Princeton, 1999)

Grant, P. R. & Grant, B. R., *How and Why Species Multiply* (Princeton, 2007)

Lack, D., *Darwin's Finches* (Cambridge, 1947)

38 The origins of Australia's special wildlife

Flannery, T. F., *The Future Eaters – an Ecological History of Australian Lands and People* (Sydney, 1994)

Long, J. A. & others, *Prehistoric Mammals of Australia and New Guinea – 100 Million Years of Evolution* (Baltimore, 2002)

Murray, P. F. & Vickers-Rich, P., *Magnificent Mihirungs. The Colossal Flightless Birds of the Australian Dreamtime* (Bloomington, IN, 2004)

39 Breathing sulphur: life on the abyssal black smokers

Childress, J. J. & Fisher, C. R., 'The biology of hydrothermal vent animals: physiology, biochemistry, and autotrophic symbioses', *Oceanography Marine Biology Annual Reviews* 30 (1992), 337–441

Van Dover, C. L., *The Ecology of Deep-Sea Hydrothermal Vents* (Princeton, 2000)

Van Dover, C. L. & others, 'Biogeography and ecological setting of Indian Ocean hydrothermal vents', *Science* 294 (2001), 818–23

40 Extremophiles and life on other planets

Boston, P. J., 'Life Below and Life "Out There"', *Geotimes* 45, 8 (2000),14–17

Boston, P. J. & others, 'Cave biosignature suites: microbes, minerals and Mars', *Astrobiology Journal* 1,1 (2001), 25–55

Brock, T. D., *Life at High Temperatures* (Yellowstone National Park, 1994)

Dick, S. J., *Life on Other Worlds: The 20th Century's Extraterrestrial Life Debate* (Cambridge, 1998)

Ekers, R. D., Cullers, D. K. & Billingham, J., *SETI 2020: A Roadmap for the Search for Extraterrestrial Intelligence* (Mt View, CA, 2003)

Gross, M., *Life on the Edge: Amazing Creatures Thriving in Extreme Environments* (New York, 1998)

Plants & Animals

41 Estimating present-day global biodiversity

Erwin, T. L., 'How many species are there?: Revisited', *Conservation Biology* 5 (1991), 330–33

May, R. M., 'How many species?', *Philosophical Transactions of the Royal Society, Series B* 330 (1990), 292–304

Wilson, E. O., *The Diversity of Life* (Cambridge, MA, 1992)

42 Why are insects so diverse?

Farrell, B. D., '"Inordinate fondness" explained: why are there so many beetles?', *Science* 281 (1998), 555–59

Grimaldi, D. & Engel, M. S., *Evolution of the Insects* (New York, 2005)

Gullan, P. J. & Cranston, P. S., *The Insects: An Outline of Entomology*, 3rd ed. (Oxford & Malden, MA, 2005)

Labandeira, C. C. & Sepkoski, J. J., Jr., 'Insect diversity in the fossil record', *Science* 261 (1993), 310–15

43 Why are some organisms small and some large?

Haldane, J. B. S., *On Being the Right Size* (Oxford & New York, 1985)

Nee, S., 'More than meets the eye: Earth's real biodiversity is invisible', *Nature* 429 (2004), 804–5

44 What is the largest living organism?

Ferguson, B. A. & others, 'Coarse-scale population structure of pathogenic Armillaria species in a mixed-conifer forest in the Blue Mountains of northeastern Oregon', *Canadian Journal of Forest Research* 33 (2003), 612–23

Gould, S. J., 'A humongous fungus among us', *Natural History* 101 (1992), 10–14

Grant, M. C., Mitton, J. B. & Linhart, Y. B., 'Even larger organisms', *Nature* 360 (1992), 216

Mihail, J. D. & others, 'Fractal geometry of diffuse mycelia and rhizomorphs of Armillaria species', *Mycological Research* 99 (1995), 81–88

Mihail, J. D. & Bruhn, J. N., 'Foraging behaviour of Armillaria rhizomorph systems', *Mycological Research* 109 (2005), 1195–207

Smith, M. L., Bruhn, J. N. & Anderson, J. B., 'The fungus *Armillaria bulbosa* is among the largest and oldest living organisms', *Nature* 356 (1992), 428–31

45 Engineering limits to the body size of land animals

Alexander, R. McNeill, *Dynamics of Dinosaurs and Other Extinct Giants* (New York, 1989)

Alexander, R. McNeill, 'All-time giants: the largest animals and their problems', *Palaeontology* 41 (1998), 1231–45

Bonner, J.T., *Why Size Matters: From Bacteria to Blue Whales* (Princeton, 2006)

Vogel, S., *Comparative Biomechanics: Life's Physical World* (Princeton, 2003)

46 Running, hopping, skipping: animal locomotion

Biewener, A. A., *Animal Locomotion* (Oxford, 2003)

Cavagna, G. A., Heglund, N. C. & Taylor, C. R., 'Mechanical work in terrestrial locomotion: two basic mechanisms for minimizing energy expenditure', *American Journal of Physiology* 233 (1977), 243–61

Dawson, T. J. & Taylor, C. R., 'Energetic cost of locomotion in kangaroos', *Nature*, 246 (1973), 313–14

Kram, R. K. & Dawson, T. J., 'Energetics and biomechanics of locomotion by red kangaroos (*Macropus rufus*)', *Comparative Biochemistry and Physiology* B 120 (1998), 41–49

Minetti, A. E., 'The biomechanics of skipping gaits: a third locomotion paradigm?', *Proceedings of the Royal Society of London* B, 265 (1988), 1227–35

Srinivasan, M. & Ruina, A., 'Computer optimization of a minimal biped model discovers walking and running', *Nature* 439 (2005), 72–75

Wunderlich, R. E. & Schaum, J. C., 'Kinematics of bipedalism in *Propithecus verreauxi*', *Journal of Zoology* 272, 2 (2007), 165–75

47 Flying and walking: learning from nature

Collins, S. H. & Ruina, A., 'A bipedal walking robot with efficient and human-like gait', in *Proceedings IEEE International Conference on Robotics & Automation, Barcelona, Spain, 1983–1988* (2005)

Collins, S., Ruina, A., Tedrake, R. & Wisse, M., 'Efficient bipedal robots based on passive-dynamic walkers', *Science* 307 (2005), 1082–85

Stafford, N., 'Spy in the sky', *Nature* 445 (2007), 808–9

48 How do dogs see the world?

Miller, P. E. & Murphy, C. J., 'Vision in dogs', *Journal of the American Veterinary Medical Association* 207 (1995), 1623–34

49 Why are animals coloured?

Alcock, J., *Animal Behavior: An Evolutionary Approach,* 8th ed. (Sunderland, MA, 2005)

Andersson, M., *Sexual Selection* (Princeton, 1994)

Bennett, A. T. D. & Théry, M., 'Avian color vision and coloration: multidisciplinary evolutionary biology', *American Naturalist* 169 (2007), S1–S6

Mappes, J., Marples, N. & Endler, J. A., 'The complex business of survival by aposematism', *Trends in Ecology and Evolution* 20 (2005), 598–603

Ruxton, G. D., Sherratt, T. N. & Speed, M. P., *Avoiding Attack: The Evolutionary Ecology of Crypsis, Warning Signals & Mimicry* (Oxford & New York, 2004)

50 How do animals adapt to the deep?

Brubakk, A. O. & Neuman, T. S. Bennett, *Elliott's Physiology and Medicine of Diving*, 5th ed. (Edinburgh, 2003)

Hooker, S. K. & others, 'Ascent exhalations of Antarctic fur seals: a behavioural adaptation for breath-hold diving?', *Proceedings of the Royal Society, London*, B 272 (2005), 355–63

Kooyman, G. L., *Diverse Divers* (Berlin, 1989)

Kooyman, G. L. & Ponganis, P. J., 'The challenges of diving to depth', *American Scientist* 85 (1997), 530–39

Nouvian, C., *The Deep: The Extraordinary Creatures of the Abyss* (Chicago, 2007)

Animal Behaviour

51 Instinct and learning in animal behaviour

Gould, J. L. & Gould, C. G., *The Animal Mind*, 2nd ed. (New York,1999)

Gould, J. L. & Marler, P., 'Learning by instinct', *Scientific American* 256:1 (1987), 74–85

Heinrich, B., 'An experimental investigation of insight in common ravens', *The Auk* 112 (1995), 994–1003

Kavanau, J. L., 'Behaviour of captive white-footed mice', in Willems, E. R. & Raush, H. L. (eds), *Naturalistic Viewpoints in Psychology*, (New York, 1969), 221–70

Köhler, W., *The Mentality of Apes* (New York, 1927)

Lorenz, K. Z. & Tinbergen, N., 'Taxis and instinctive action in the egg-retrieving behaviour of the greylag goose', trans. of 1938 article in Schiller, C. H. (ed.), *Instinctive Behaviour* (New York, 1957), 176–208

Marler, P. & Tamura, M.,'Culturally transmitted patterns of vocal behaviour in sparrows', *Science* 146 (1964), 1483–86

Weir, A. S., Chappell, J. & Kacelnik, A., 'Shaping of hooks in New Caledonian Crows', *Science* 297 (2002), 981

52 Is an ant colony a superorganism?

Darwin, C., *On the Origin of Species* (London, 1859)

Gordon, D. M., *Ants at Work* (New York, 1999)

Hamilton, W. D., *Narrow Roads of Gene Land*, Vol. 1 (New York, 1997)

Wheeler, W. M., 'The ant colony as an organism', *Journal of Morphology* 22 (1911), 307–25

53 Why are animals nice to each other?

Dugatkin, L. A., *The Altruism Equation – Seven Scientists Search for the Origins of Goodness* (Princeton & Oxford, 2006)

Gadagkar, R., *Survival Strategies – Cooperation and Conflict in Animal Societies* (London & Cambridge, MA, 1997)

Gadagkar, R., *The Social Biology of* Ropalidia marginata*: Toward Understanding the Evolution of Eusociality* (Cambridge, MA, 2001)

Ratnieks, F. L. W., Foster, K. R. & Wenseleers, T., 'Conflict resolution in insect societies', *Annual Review of Entomology* 51 (2006), 581–608

Zahavi, A. & Zahavi, A., *The Handicap Principle – A Missing Piece of Darwin's Puzzle* (Oxford & New York, 1997)

54 How do animals navigate?

Emlen, S., 'The stellar orientation system of migratory birds', *Scientific American* 233 (2) (1975), 102–11

Gould, J. L., 'The case for magnetic sensitivity in birds and bees', *American Scientist* 68 (1980), 256–67

Gould, J. L., 'Animal navigation', *Current Biology* 14 (2004), 221–24

Keeton, W. T., 'The mystery of pigeon homing', *Scientific American* 231 (6) (1974), 96–107

Walcott, C., 'The homing of pigeons', *American Scientist* 62 (1974), 542–52

Williams, T. C. & Williams, J. M., 'Oceanic mass migration of land birds', *Scientific American* 239 (4) (1978), 166–76

55 Sexual selection

Akçay, E. & Roughgarden, J., 'Extra-pair paternity in birds: review of the genetic benefits', *Evolutionary Ecology Research* 9 (2007) 1–14

Darwin, C., *The Descent of Man, and Selection in Relation to Sex* (London, 1871)

Coyne, J., 'Charm Schools', *Times Literary Supplement*, 30 July, 2004

Miller, C. W. & Moore, A. J., 'A potential resolution to the lek paradox through indirect genetic effects', *Proceedings of the Royal Society B 274* (2007), 1279–86

Roughgarden, J., *Evolution's Rainbow: Diversity, Gender, and Sexuality in Nature and People* (Berkeley, 2004)

Roughgarden, J., Oishi, M. & Akçay, E., 'Reproductive social behavior: cooperative games to replace sexual selection', *Science*, 311 (2006) 965–69

Roughgarden, J., 'Challenging Darwin's theory of sexual selection', *Daedalus* 136(2) (2007), 1–14

56 Why do deer have antlers: fighting or display?

Bubenik, G. A. & Bubenik, B. B. (eds), *Horns, Pronghorns, and Antlers. Evolution, Morphology, Physiology, and Social Significance* (New York, 1990)

Janis, C. M., 'Tertiary mammal evolution in the context of changing climates, vegetation, and tectonic events', *Annual Reviews in Ecology and Systematics* 24 (1993), 467–500

Nowak, R. M., *Walker's Mammals of the World* (Baltimore, 1999)

Prothero, D. R., *Horns, Tusks, and Flippers: The Evolution of Hoofed Mammals* (Baltimore, 2002)

Savage, R. J. G. & Long, M. R., *Mammal Evolution* (Oxford, 1986)

57 Invisible signalling: pheromones

Agosta, W. C., *Chemical Communication: The Language of Pheromones* (San Francisco, 1992)

Agosta, W. C., *Bombardier Beetles and Fever Trees: A Close-Up Look at Chemical Warfare and Signals in Animals and Plants* (Reading, MA, 1996)

Müller-Schwarze, D., *Chemical Ecology of Vertebrates* (Cambridge, 2006)

Shepherd, G. M., 'The human sense of smell: are we better than we think?' Plos Biology (2004) 2(5): e146

Wyatt, T. D., *Pheromones and Animal Behaviour* (Cambridge, 2003)

Wysocki, C. J. & Preti, G., 'Facts, fallacies, fears, and frustrations with human pheromones', *Anatomical Record* 281A (2004), 1201–11

58 Do animals have emotions?

Balcombe, J., *Pleasurable Kingdom: Animals and the Nature of Feeling Good* (London & New York, 2006)

Bekoff, M., 'Wild justice and fair play: cooperation, forgiveness, and morality in animals', *Biology & Philosophy* 19 (2004), 489–520

Bekoff, M., *Animal Passions and Beastly Virtues: Reflections on Redecorating Nature* (Philadelphia, 2006)

Bekoff, M., *The Emotional Lives of Animals: A Leading Scientist Explores Animal Joy, Sorrow, and Empathy – and Why they Matter* (Novato, 2007)

Darwin, C., *The Expression of the Emotions in Man and Animals*, 3rd ed., with introduction, afterword, and commentaries by Paul Ekman (New York, 1872/1998)

Panksepp, J., 'Affective consciousness: core emotional feelings in animals and humans' *Consciousness and Cognition* 14 (2005), 30–80

59 The language of honey bees

Frisch, K. von, *The Dance Language and Orientation of Bees* (Cambridge, MA, 1976)

Gould, J. L. & Gould, C. G., *The Honey Bee* (New York, 1995).
Seeley, T. D., Visscher, P. K. & Passino, K. M., 'How do bees choose a new nesting site?', *American Scientist* 94 (2006), 220–29

60 Animal consciousness
Balcombe, J., *Pleasurable Kingdom: Animals and the Nature of Feeling Good* (London & New York, 2006)
Emery, N. J. & Clayton, N. S., 'The mentality of crows: convergent evolution of intelligence in corvids and apes', *Science* 306 (2004), 1903–7
Panksepp, J., 'Affective consciousness: core emotional feelings in animals and humans', *Consciousness and Cognition* 14 (2005), 30–80
Pepperberg, I. M., *The Alex Studies: Cognitive and Communicative Abilities of Grey Parrots* (Cambridge, MA, 2002)
Savage-Rumbaugh, S. & Lewin, R., *Kanzi: The Ape at the Brink of the Human Mind* (New York, 1996)
Smith, J. D., Shields, W. E. & Washburn, D. A., 'The comparative psychology of uncertainty monitoring and metacognition', *Behavioral and Brain Sciences* 26 (2003), 317–73

Global Warming & the Future

61 Global warming
62 What will Earth's climate be like in the future?
Archer, D., *Global Warming: Understanding the Forecast* (Malden, MA, 2006)
Evans, K., *Funny Weather* (Brighton, 2006)
Flannery, T., *The Weather Makers* (London, 2006)
Intergovernmental Panel on Climate Change (IPCC), Climate Change 2007: The Physical Science Basis, Summary for Policymakers (Geneva, 2007), http://www.ipcc.ch
Maslin, M., *Global Warming: A Very Short Introduction* (Oxford & New York, 2004)

63 El Niño: extreme weather on the increase?
Caviedes, C., *El Niño in History* (Gainesville, FL, 2001)
Davis, M., *Late Victorian Holocausts. El Niño Famines and the Making of the Third World* (London & New York, 2001)
Fagan, B. M., *Floods, Famines and Emperors: El Niño and the Fate of Civilizations* (New York, 1999)
Intergovernmental Panel on Climate Change (IPCC), Climate Change 2007: Impacts, Adaptation and Vulnerability (Geneva, 2007), http://www.ipcc.ch
Nash, J. M., *El Niño: Unlocking the Secrets of the Master Weather-maker* (New York, 2002)

64 Greenhouse gases and Earth's natural rhythms
Alverson, K. D, Bradley, R. S. & Pedersen, T. F. (eds), *Palaeoclimate, Global Change and the Future* (Berlin, 2003)
EPICA Community Members, 'Eight glacial cycles from an Antarctic ice core', *Nature* 429 (2004), 623–28
IPCC Fourth Assessment Report: Working Group 1 Report – The Physical Science Basis, http://www.ipcc.ch/ipccreports/ar4-wg1.htm
Kohfeld, K. E. & others, 'Role of marine biology in glacial-interglacial CO_2 cycles', *Science* 308 (2005), 74–78
Schulze, E.-D. & others (eds), *Global Biogeochemical Cycles in the Climate System* (San Diego, 2001)
Spahni, R. & others, 'Atmospheric methane and nitrous oxide of the late Pleistocene from Antarctic ice cores', *Science* 310 (2005), 1317–21

65 Predicting future human population levels
Daily, G. C., Ehrlich, A.H. & Ehrlich, P. R., 'Optimum human population size', *Population and Environment* 15 (1994), 469–75
Ehrlich, P. R. & Ehrlich, A. H., *The Population Explosion* (New York, 1990)
Ehrlich, P. R. & Ehrlich, A. H., *One with Nineveh: Politics, Consumption, and the Human Future* (Washington, DC, 2005)
National Academy of Sciences USA. 'A Joint Statement by Fifty-eight of the World's Scientific Academies', Population Summit of the World's Scientific Academies. New Delhi, India (1993)

66 Flu pandemics and eastern Asia
Shortridge, K. F., 'Avian influenza A viruses of southern China and Hong Kong: ecological aspects and implications for man', *Bulletin of the World Health Organization* 60 (1982), 129–35
Scholtissek, C. & Naylor, E., 'Fish farming and influenza pandemics', *Nature* 331 (1988), 215
Webster, R. G. & others, 'Evolution and ecology of influenza A viruses', *Microbiological Reviews* 56 (1992), 152–79

67 Wildlife extinction and conservation
IUCN (World Conservation Union). Red list of threatened species (Gland, 2006) http://www.iucn.org/themes/ssc/redlist.htm
Lomolino, M. V., Riddle, B. R. & Brown, J. H., *Biogeography* (Sunderland, 2006)
Primack, R. B. *Essentials of Conservation Biology*, 4th ed. (Sunderland, MA, 2006)
Raup, D. M. & Sepkoski, J. J., Jr, 'Mass extinctions in the marine fossil record', *Science* 215 (1982), 1501–03
Wilson, E. O., *The Future of Life* (New York, 2002)

68 What will replace liquid hydrocarbon fuels?
Ahlbrandt, T. S. & others, *Global Resource Estimates from Total Petroleum Systems*, AAPG Memoir 86, Association of American Petroleum Geologists (Tulsa, 2005)
Deffeyes, K. S., *Hubbert's Peak: The Impending World Oil Shortage* (Princeton, 2001)
Fulton, L. & Howes, T., *Biofuels for Transport*, International Energy Agency, OECD (Paris, 2004)
International Energy Agency, *World Energy Outlook 2006* (Paris, 2006)
International Energy Agency, *World Energy Outlook 2007 – China and India Insights* (Paris, 2007)
World Business Council for Sustainable Development, 'Mobility 2030: Meeting the challenges to sustainability', Full Report of the Sustainable Mobility Project, World Business Council for Sustainable Development, 2004, www.wbcsd.org .
Yergin, D., *The Prize* (New York, 1991)

69 Humanity's ecological footprint
Business Council for Sustainable Development, *Getting Eco-Efficient. Report of the BCSD First Antwerp Eco-Efficiency Workshop* (Geneva, 1993)
Rees, W. E., 'Revisiting carrying capacity: area-based indicators of sustainability', *Population and Environment* 17 (1996), 195–215
Rees, W. E., 'Ecological footprints and bio-capacity: essential elements in sustainability assessment', in Dewulf, L. & Van Langenhove, H. (eds), *Renewables-Based Technology: Sustainability Assessment* (Chichester & Hoboken, 2006)
Wackernagel, M. & Rees, W. E., *Our Ecological Footprint: Reducing Human Impact on the Earth* (Gabriola Island, 1996)
WWF Living Planet Report 2006 (Gland, 2006)
www.panda.org/news_facts/publications/living_planet_report/index.cfm

70 Human behaviour and saving the planet
Damasio, A., *Descartes' Error: Emotion, Reason and the Human Brain* (New York, 1994)
Diamond, J., *Collapse: How Societies Choose to Fail or Succeed* (New York, 2005)
Fowler, C. W. & Hobbs, L., 'Is humanity sustainable?', *Proceedings of the Royal Society of London, Series B: Biological Sciences* 270 (2003), 2579–83
Kollmuss, A. & Agyeman, J., 'Mind the Gap: why do people act environmentally and what are the barriers to pro-environmental behavior?', *Environmental Education Research* 8 (2002), 239–60
Lebow, V., *The Journal of Retailing* (Spring 1955), 7
Rees, W. E., 'Ecological footprints and bio-capacity: essential elements in sustainability assessment', in Dewulf, L. & Van Langenhove, H. (eds.), *Renewables-Based Technology: Sustainability Assessment* (Chichester & Hoboken, 2006)

Sources of illustrations

a = above, b = below, c = centre
1 R. Stockli, A. Nelson, F. Hasler, NASA/GSFC/NOAA/USGS; 2–3 Peter Lilja/Taxi/Getty Images; 4 NASA Jet Propulsion Laboratory (NASA-JPL); 5l J. D. Griggs/U.S. Geological Survey; 5r Heidi & Hans-Jurgen Koch/Minden Pictures/FLPA; 6l © Jack Dykinga; 6r Photo M. Cleopatra Pimienta, Cali, Colombia. Courtesy Terry Erwin; 7l Simon Litten/FLPA; 7r Randy Olson/National Geographic/Getty Images; 12 © Jane Francis; 13 © Peter Batson/DeepSeaPhotography.com; 14 Louie Psihoyos/Science Faction/Getty Images; 15 © W. Perry Conway/Corbis; 16–17 © Frans Lanting/Corbis; 18 © Karen Bass/naturepl.com; 19 NASA; 20 NASA Jet Propulsion Laboratory (NASA-JPL); 21 ML Design; 22a NASA Headquarters - GReatest Images of NASA (NASA, HQ, GRIN); 22b ML Design; 23 © Yann Arthus-Bertrand/Corbis; 24 © Roger Ressmeyer/Corbis; 25a ML Design after Dr Nicola McLoughlin; 25b © Ralph White/Corbis; 26l Dr David Wacey; 26r © Visuals Unlimited/Corbis; 27 Stephen Sharnoff & Sylvia Duran/National Geographic/Getty Images; 28a © Robert Pickett/Corbis; 29al, 29ar N.J. Butterfield; 29b © Steve Bloom/stevebloom.com; 30 © DK Ltd/Corbis; 31l © Chip Clark, Museum of Natural History, Smithsonian Institution, Washington, DC; 31r, 32l, 32r John Sibbick; 33 © Paul Wignall; 34 © Jonathan Blair/Corbis; 35 Christopher Perkins, after Anusuya Chinsamy; 36bl, 36bc, 36br Courtesy Anusuya Chinsamy; 37 © Grant Delin/Corbis; 38 ML Design after Wedel et al. (2000) and Dal Sasso, C., Maganuco, S., Buffetaut, E. & Mendez, M. A., 'New information on the skull of the enigmatic theropod Spinosaurus, with remarks on its size and affinities', *Journal of Vertebrate Paleontology*, 2005, 25(4):888–96; 39 Museum für Naturkunde, Berlin; 40 Amblin/Universal/The Kobal Collection; 41 The Field Museum, Chicago, GN89671_53c; 42 Courtesy David A. Kring ; 43 John Sibbick; 44a Christopher Perkins, based on Alvarez et al., 1980 and other sources; 44bl Courtesy David A. Kring; 44br Christopher Perkins; 45 Don Davis, NASA; 47 Elliott Neep/FLPA; 48a © Peter Trusler; 48b ML Design after Cifelli, R.L, & Gordon, C. L., 'Evolutionary biology: Re-crowning mammals', *Nature* 447, 918–20 (21 June 2007); 49a ML Design after Edinger, T., and Radinsky, L.; 49b ML Design after Kardong, K. V., *Vertebrates: Comparative Anatomy, Function, Evolution*, 3rd ed. (New York, 2002), p. 274; 51 © Kenneth Garrett; 52 Kathy Schick & Nicholas Toth; 53al © Kenneth Garrett; 53ar John Sibbick; 53b © Patrick Robert/Corbis; 54–55 Photo Tom Pfeiffer/www.decadevolcano.net; 56 © Yann Arthus-Bertrand/Corbis; 57 Royal Geographical Society; 58a ML Design after U.S. Geological Survey; 58b ML Design; 59 David Gubbins; 60a, 60b From Physical Earth Planet Int., vol. 162, Gubbins, D., Willis, A. P. & Sreenivasan, B., *Correlation of Earth's magnetic field with lower mantle thermal and seismic structure*, 256–60, © Elsevier 2007; 61 Reto Stockli, NASA Earth Observatory; 62a Christopher Perkins; 62b Courtesy Chris Hawkesworth; 63 © E.B. Watson; 64 Christopher Perkins, after Condie, K.C., 1998, 'Episodic continental growth and supercontinents: a mantle avalanche connection?', *Earth and Planetary Science Letters*, 163, 97–108; 65a © José Jácome/epa/Corbis; 65b J.D. Griggs/U.S. Geological Survey; 67 © Stephen Venables/Royal Geographical Society; 68a Courtesy Giorgio Poretti; 68b Courtesy Giorgio Poretti; 69a Courtesy Claire Ting, Department of Biology, Williams College, Williamstown; 69b ML Design after David Catling; 70 Phil Stoffer/U.S. Geological Survey; 71a, 71b ML Design after David Catling; 72 ML Design; 73 Robert Krimmel/U.S. Geological Survey; 74l ML Design after Steve Sparks; 74r Dennis Kunkel/Phototake Science/Photolibrary; 75 J.D. Griggs/U.S. Geological Survey; 76 Christopher Perkins, after Steve Sparks; 77 ML Design; 78 Courtesy Stan Celestian and Glendale Community College, Arizona; 79 Richard J. Brown; 80 ML Design after Eckstrand, O.R. et al. (1995), Geology of Canada No. 8, Geological Survey of Canada; 81 © Hubert Stadler/Corbis; 83a NASA/GSFC/METI/ERSDAC/JAROS, and U.S./Japan ASTER Science Team; 83b Christopher Perkins, after Sparks, S., Self, S., et al., 2005, *Super-eruptions: global effects and future threats*, Report of a Geological Society of London Working Group (2nd print ed.), p. 25; 84 Hulton Archive/Getty Images; 85b Dave Harlow/U.S. Geological Survey; 86l © Anders Grawin; 86r ML Design after Geosciences, Australia; 87a AFP/Getty Images; 87b ML Design after O. Grabe, GFZ Potsdam; 88 Kozak Collection, Earthquake Engineering Research Center (EERC), University of California, Berkeley; 89 A. Babeyko (University of Frankfurt/M. Geophysics) and S. Sobolev (GFZ Potsdam); 90 ML Design, after French, B.,*Traces of Catastrophes*, Lunar and Planetary Institute, Houston, Texas, Contribution 954 (1998); 91 Jonathan S. Blair/National Geographic/Getty Images; 92a ML Design after Lunar & Planetary Institute, University of Arizona; 92b NASA, Clementine Mission; 93b Philippe Claeys; 94 © Bettmann/Corbis; 95 Courtesy Marcio R. Mello, modified from Kansas Geological Survey, Education Online, 2001; 96–97 Peter Artymiuk, Wellcome Images; 98 Mitsuhiko Imamori/Minden Pictures/FLPA; 99 The Hornet, 22 March, 1871; 100 Reproduced with the permission of the Cambridge University Library, DAR.121; 101 Wellcome Library, London; 102 © Lester V. Bergman/Corbis; 103a Simon Maina/AFP/Getty Images; 103b © Kalliopi Monoyios; 104 Sunset/FLPA; 105a Medical Art Service, Munich/Wellcome Images; 105b ML Design after Land, M. F. & Nilsson, D.-E., *Animal Eyes* (Oxford University Press, 2002); 106a © Andrew Parker; 106b ML Design after Parker, A., *In the Blink of an Eye* (Simon & Schuster, 2003); 107 © Hulton-Deutsch Collection/Corbis; 108 Christopher Perkins, after Miller, J.D., Scott, E.C., Okamoto, S., 'Public Acceptance of Evolution', *Science Magazine*, AAAS, vol. 313, 11 Aug 11, 2006; 109 Bibliothèque Municipale de Bourges, France, Lauros/Giraudon/The Bridgeman Art Library; 110l Wellcome Library, London; 110r A. Barrington Brown/Science Photo Library; 111 ML Design after U.S. Department of Energy Human Genome Program; 112 MRC Lab of Molecular Biology, Wellcome Images; 113 Courtesy Desmond Morris; 114 © Theo Allofs/zefa/Corbis; 115 Neil Broomhall/Oxford Scientific (OSF)/Photolibrary; 116 Carolus Linneaus, *Systema Naturae*, 10th edition, 1758; 117al Wendy Dennis/FLPA; 117acl Richard Du Toit/Minden Pictures/FLPA; 117acr Jurgen & Christine Sohns/FLPA; 117ar Foto Natura Catalogue/FLPA; 117b Christopher Perkins, after Mike Benton; 118 Wellcome Library, London; 119 © Steve Schapiro/Corbis; 120 The Sanger Institute, Wellcome Images; 120 inset Peter Artymiuk, Wellcome Images; 121 EM Unit/Royal Free Medical School, Wellcome Images; 122 © Jon Hrusa/epa/Corbis; 123l Christopher Perkins, after Peter Mayhew; 123r Christopher Perkins, after Peter Mayhew; 124a Photo J. Rick. Courtesy Loren Rieseberg; 124b Jim Brandenburg/Minden Pictures/FLPA; 125a Jesús Mavárez, Lina González & Marcos Guerra; 125c, 125b Christopher Perkins, after Baraclough, T.G. & Vogler, A.P., 'Detecting the geographic pattern of speciation from species level phylogenies', *American Naturalist* 155 (2000), 419–34, Chicago University Press; 126al, 126ar Luiz A. Rocha; 127 Wellcome Library, London; 128 Pascal Goetgheluck/Science Photo Library; 129a Paul Martin, Wellcome Images; 129b Abigail Tucker, Wellcome Images; 130l Courtesy Paul Brakefield, Julie Gates, David Keys & Sean Carroll, in Brakefield, P.M., Gates, J., Keys, D., Kesbeke, F., Wijngaarden, P. J., Monteiro, A., French, V., & Carroll, S.B., 'Development, plasticity and evolution of butterfly eyespot patterns', *Nature*, vol. 384 (1996), pp. 236–42; 130r Mike Shapiro and David Kingsley; 131a, 131b, 132 Courtesy Mike Coates; 133 © Karen Gowlett-Holmes; 134–35 Tui De Roy/Minden Pictures/Getty Images; 136 © Jack Dykinga; 137 ML Design after W. Barthlott, 1997; 138–39 © Frans Lanting/Corbis; 139a ML Design; 139b ML Design after *Trends in Ecology and Evolution*, vol. 19, Wiens, J.J., Donoghue, M.J., 'Historical biogeography, ecology and species richness', 639–44, © Elsevier 2004; 140a Reinhard Dirscherl/FLPA; 140b altrendo nature/Getty; 141 Joe Carnegie; 142 Courtesy Michèle Clarke; 143 © Marli Bryant Miller www.marlimillerphoto.com; 144 © Merlin D. Tuttle, Bat Conservation International, www.batcon.org; 145 © Jack Dykinga; 146 Gerard Lacz/FLPA; 147 Patricio Robles Gil/Sierra Madre/Minden Pictures/Getty Images; 148 © Jane Francis; 149a © Bryan and Cherry Alexander Photography; 149b, 150 © Jane Francis; 151 Christopher Perkins, after Whitman et al., 1998; 152 Heribert Cypionka, www.microbiological-garden.net; 153a Photo Aaron Gronstal; 153b John Priscu; 154 ML Design after Ricklefs, R.E. & Lovette, I.J., 'The roles of island area per se and habitat diversity in the species-area relationships of four Lesser Antillean faunal groups', *Journal of Animal Ecology* 68, 1142–60 (Blackwell, 1999); 155 Frans Lanting/FLPA; 156a Pete Oxford/Minden Pictures/FLPA; 156b Frans Lanting/FLPA; 157 © Mataparda; 158 ML Design; 159l Greg W. Lasley; 159r pl. 44, hand coloured lithograph by John and Elizabeth Gould from Charles Darwin *Zoology of the voyage of H.M.S. Beagle*, 1838–41, Vol. 3, *Birds* (London, 1841); 160 ML Design after Grant, P.R., *Ecology and Evolution of Darwin's Finches* (Princeton, 1999); 161 David Parer & Elizabeth Parer-Cook/Ausscape; 162 ML Design; 163 Natural History Museum, London; 164a © Peter Trusler; 164b Kakadu National Park; 165 Courtesy John Long; 166 © Steven David Miller/ naturepl.com; 167l Woods Hole Oceanographic Institution; 167r © Peter Batson/DeepSeaPhotography.com; 168 ML Design after Van Dover et al., 'Biogeography and ecological setting of Indian Ocean hydrothermal vents', *Science* 294 (2001), 818-823; 169, 170a © Peter Batson/ DeepSeaPhotography.com; 170b Woods Hole Oceanographic Institution; 171 Michael N. Spilde; 172 © Peter Essick/Aurora Photos; 173a © Penny Boston; 173b © Gus Frederick; 174–75 © Steve Bloom/stevebloom.com; 176 Paul Sutherland/National Geographic/Getty Images; 177 © Alain Compost; 178 Courtesy Michael Benton; 179a Photo Christy Jo Geraci, Smithsonian Institution, Washington, DC. Courtesy Terry Erwin; 179b Photo M. Cleopatra Pimienta, Cali, Colombia. Courtesy Terry Erwin; 180 Chris Newbert/Minden Pictures/FLPA; 181 © Gary Braasch/Corbis; 182a Natural History Museum, London; 182b ML Design after Barraclough, T.G., Barclay, M.V.L. & Vogler, A.P., 'Species Richness: Does flower power explain beetle-mania?' *Current Biology* 8 (1998), R843-R845; 183a, 183b © James C. Lowen (www.pbase.com/james_lowen); 185 Hiroya Minakuchi/Minden Pictures/FLPA; 186 © Steve Bloom/stevebloom.com; 187a Thomas Mangelsen/Minden Pictures/FLPA; 187b Reproduced with the permission of Peter Schouten/National Geographic/University of Wollongong; 188l, 188r Courtesy Greg Filip, USDA Forest Service, Portland, Oregon; 189 Courtesy Jeffry B. Mitton; 190 Buyenlarge/Time Life Pictures/Getty Images; 191a ML Design; 191b Louie Psihoyos/Science Faction/Getty Images; 192 Bobby Haas/National Geographic/Getty Images; 193l Plate 63, from Vol. 1 Eadweard Muybridge *Animal Locomotion*, 1887; 193r © Arne Dedert/epa/Corbis; 194–95 Frans Lanting/FLPA; 195b Christopher Perkins, after Biewener, A.A., *Animal Locomotion* (Oxford University Press, 2003); 196

Jurgen & Christine Sohns/FLPA; 197a Jonathan Wood/Getty Images; 197b NASA Headquarters – GReatest Images of NASA (NASA, HQ, GRIN); 198 Biblioteca Reale, Turin; 199a Philippe Psaila/Science Photo Library; 199b Michael Gore/FLPA; 200–1 Courtesy Honda; 200b Dr Jeremy Burgess/Science Photo Library; 201 Christopher Perkins; 201 (left inset) Courtesy Honda; 201 (centre inset) Courtesy Andy Ruina and Steve Collins; 201 (right inset) Drazen Tomic; 202a Jim Brandenburg/Minden Pictures/ FLPA; 202b Angela Hampton/FLPA; 203 Jim & Jamie Dutcher/National Geographic/ Getty Images; 204al, 204ar Courtesy Julian Partridge; 204b © James Burley; 205 © P.J. DeVries; 206a © W. Perry Conway/Corbis; 206b © Staffan Widstrand/naturepl.com; 207a © A. T. D. Bennett; 207c Michael & Patricia Fogden/Minden Pictures/FLPA; 207b Michael & Patricia Fogden/Minden Pictures/FLPA; 208–9 Piotr Naskrecki/Minden Pictures/FLPA; 209a, 209b © A. T. D. Bennett; 210 Emory Kristof/National Geographic Image Collection; 211 © David Shale/naturepl.com; 212l ML Design; 212r Courtesy Sascha Hooker; 213 Cousteau Society/The Image Bank/Getty Images; 214–15 © Janet Horton; 216 © Evelyne Bauer/Kalahari Meerkat Project; 217a From K. Lorenz and N. Tinbergen, *Taxis und Instinkhandlung in der Eirollbewegung*, Zeitschrift fur vergleichende Physiologie 21, 699–716, 1935; 217b ML Design after Gould, J.L. & Gould, C.G., *The Animal Mind* (Scientific American Library/W. H. Freeman, 1994); 218 Nina Leen/Time & Life Pictures/Getty Images; 219 ML Design after Gould, J.L. & Marler, P., 'The Instinct to Learn', *Scientific American* 255 (1) (1987), 74–85; 220a Courtesy Behavioural Ecology Research Group, University of Oxford; 220b Mauro Fermariello/ Science Photo Library; 221 Mark Moffett/Minden Pictures/FLPA; 222 © Daniel Kronauer; 223l © Stephen Pratt; 223r Christopher Perkins; 224 Mitsuaki Iwago/Minden Pictures/FLPA; 225 © Joe McDonald/Corbis; 226 © Clive Temple; 227 Drawing Sudha Premnath. Courtesy Raghavendra Gadagkar; 228 Courtesy Thresiamma Varghese; 229 ML Design after Dyer, F.C. & Gould, J.L., 'Honey bee navigation', *American Scientist* 71, 587–97 (1983); 230l, 230r ML Design after Gould, J.L., *Ethology: The Mechanisms and Evolution of Behavior* (New York, W.W. Norton) 1982; 231 Michio Hoshino/Minden Pictures/FLPA; 232 © James C. Lowen (www.pbase.com/james_lowen); 233 Fig. 16 from Vol. 1, Charles Darwin, *The Descent of Man and Selection in Relation to Sex* (London, 1888); 234 Tui De Roy/Minden Pictures/FLPA; 235 © Doug Perrine/ naturepl.com; 236 © South West News; 237l Helen Rhode/FLPA; 237r ML Design after Janis, from Bubenik, G.A., and Bubenik, A B. (eds), *Horns, Pronghorns, and Antlers: Evolution, Morphology, Physiology, and Social Significance* (Springer-Verlag, 1990); 238l The Field Museum, Chicago, GEO84564c; 238r L. Lee Rue/FLPA; 239 Sumio Harada/Minden Pictures/FLPA; 240 Natural History Museum, London; 241 Nigel Cattlin/FLPA; 242 Michael & Patricia Fogden/Minden Pictures/FLPA; 243a © Mark Brownlow/naturepl.com; 243b Nigel Cattlin/FLPA; 244 Jan Nystrom; 245a Koichi Kamoshida/Getty Images; 245b figs 16 & 17 from Charles Darwin, *The Expression of the Emotions in Man and Animals* (London, 1904); 246–47 Kevin Schafer/Photographer's Choice/Getty Images; 247 Jan Nystrom; 248 Nina Leen/Time & Life Pictures/Getty Images; 249a © Ken Lorenzen; 249bl ML Design after von Frisch, K., *Dance Language and Orientation of Bees* (Springer-Verlag, 1965); 249br ML Design after Gould, J.L., *Ethology: The Mechanisms and Evolution of Behavior* (New York, W.W. Norton) 1982; 250 © Ken Lorenzen; 251 Joe Raedle/Newsmakers/Getty Images; 252 © Mike Lovett; 253 Frans Lanting/FLPA; 254–55 © Corbis; 256 NASA/Goddard Space Flight Center, Scientific Visualization Studio; 257 Climate Change 2007: The Physical Science Basis, Summary for Policymakers, Intergovernmental Panel on Climate Change (IPCC); 258 © Steve Morgan; 259a National Oceanic and Atmospheric Administration (NOAA); 259b ML Design; 260 NASA/Goddard Space Flight Center, Conceptual Image Lab; 261 © Paulo Fridman/Corbis; 262 ML Design after Ruddiman, W.F., *Earth's Climate: Past and Future* (W.H.Freeman & Co Ltd, 2001); 263al, 263ar © Crown Copyright 2007, The Met Office; 263ctl R. Stockli, A. Nelson, F. Hasler, NASA/GSFC/NOAA/USGS; 263ctr © Carlos Barria/Reuters/Corbis; 263cbl Jim Reed/FLPA; 263cbr Winfried Wisniewski/ FLPA; 264 From *Ecosystems and Human Well-being: Synthesis*, Millennium Ecosystem Assessment/World Resources Institute, Washington, D.C, 2005. Courtesy Millennium Ecosystem Assessment; 265 Michio Hoshino/Minden Pictures/FLPA; 266 César Caviedes; 267a ML Design, based on information supplied by David Adamec, NASA Goddard Space Flight Center, and METEO France; 267b NASA Jet Propulsion Laboratory (NASA-JPL); 268 Steve Bourget; 269a César Caviedes; 269b Ian Waldie/ Getty Images; 270 W. Berner, Climate and Environmental Physics, Physics Institute, University of Bern, 1978; 271l Laurent Augustin, CNRS, IPEV, EPICA, 271r ML Design after after Renato Spahni; 272 © Steve Morgan; 273 © Dave Bartruff/Corbis; 274a Population Action International; 274b Frank May/DPA/PA Photos; 275 Joanna B. Pinneo/Aurora/Getty Images, 276a NIBSC/Science Photo Library; 276b © Chaiwat Subprasom/Reuters/Corbis; 277a ML Design; 277b NGM Maps/National Geographic Image Collection; 278 © Jonathan Blair/Corbis; 279 From *Ecosystems and Human Well-being: Synthesis*, Millennium Ecosystem Assessment/World Resources Institute, Washington, D. C., 2005. Courtesy Millennium Ecosystem Assessment; 280a Frans Lanting/FLPA; 280c From *Ecosystems and Human Well-being: Synthesis*, Millennium Ecosystem Assessment/World Resources Institute, Washington, D. C., 2005. Courtesy Millennium Ecosystem Assessment; 280b U.S. Fish & Wildlife Service; 281 © Alain Compost; 282 ML Design after *BP Statistical Review of World Energy* 2006, Historical data series; 283 China Photos/Getty Images; 284 © Paulo Fridman/Corbis; 285 Robert Clark/National Geographic Image Collection; 286 © Roel Veyt; 287 ML Design after *World Wildlife Fund Living Planet Report*, 2006; 288 Image courtesy the artist and Flowers East, London; 289l, 289r Passivhaus Institut Darmstadt; 291 Bob Gibbons/FLPA.

Sources of quotations

p. 12 R. Buckminster Fuller, *Operating Manual for Spaceship Earth* (New York, 1963); p. 19 Immanuel Kant, *Allgemeine Naturgeschichte und Theorie des Himmels* (1755); p. 23 Erasmus Darwin, *Zoonomia* (London, 1794); p. 27 Julian S. Huxley & Thomas H. Huxley, *Evolution and Ethics: 1893–1943* (London, 1947); p. 30 Stphen J. Gould, 'Of Tongue Worms, Velvet Worms, and Water Bears', *Natural History* 104 (1995); p. 32 Doug Erwin, *The Great Paleozoic Crisis* (Columbia, 1993); p. 35 J. O. Farlow, 'Dinosaur energetics and thermal biology', in Weishampel, D. B. & others, *The Dinosauria* (Berkeley, 1992); p. 38 G. P. Burness, J. Diamond and T. Flannery, 'Dinosaurs, dragons, and dwarfs: the evolution of maximal body size', *Proceedings of the National Academy of Sciences USA*, 98 (2001); p. 42 E. H. Colbert, *The Age of Reptiles* (New York, 1965); p. 46 Winston Churchill, attr., in M. Gilbert, *Never Despair: Winston S. Churchill, 1945–1965* (London, 1988); p. 50 Charles Darwin, *The Descent of Man, and Selection in Relation to Sex* (London, 1871); p. 57 Alexander von Humboldt, *Cosmos, A Sketch of a Physical Description of the Universe* (London, 1849); p. 61 Jules Verne, *Journey to the Centre of the Earth* (1864 ; English edition 1871); p. 69 Alfred Lord Tennyson, *Harold: A Drama* (London, 1877); p. 72 Pliny the Younger, Letters, LXVI, trans. William Melmoth (London, 1746); p. 77 Rachel Carson, *The Sense of Wonder* (New York, 1965; published posthumously); p. 81 Marquis de Sade, *Philosophy in the Bedroom* (1795); p. 86 Frank I. González, 'Tsunami!', *Scientific American*, 280 (1999); p. 90 Réné Goscinny, *Asterix le Gaulois/Asterix the Gaul* (Brussels, 1961); p. 94 David Curry, pers. comm. 2008 ; p. 99 Theodosius Dobzhansky, 'Nothing in biology makes sense except in the light of evolution', *The American Biology Teacher*, 35 (1973); p. 104 Charles Darwin, *On the Origin of Species* (London, 1859); p. 107 Transcript of the Scopes Trial (1925); p. 113 Richard Dawkins, *The Extended Phenotype* (Oxford, 1982); p. 116 Charles Darwin, *On the Origin of Species* (London, 1859); p. 123 Menno Schilthuizen, *Frogs, Flies and Dandelions: The Making of Species* (Oxford & New York, 2001); p. 127 T. J. Horder, 'Syllabus for an embryological synthesis', in *Complex Organismal Functions: Integration and Evolution in Vertebrates*, D. B. Wake and G. Roth (eds) (Chichester, 1989); p. 131 Charles Darwin, *On the Origin of Species* (London, 1859); p. 137 A. R. Wallace, *Tropical Nature and Other Essays* (London, 1878); p. 141 Mildred Cable with Francesca French, *The Gobi Desert* (London, 1942); p. 144 Edward Abbey, *Desert Solitaire* (Tucson, 1988); p. 148 Roald Amundsen, *The South Pole* (London, 1912); p. 151 W. B. Whitman, D. C. Coleman, W. J. Wiebe, 'Prokaryotes: the unseen majority', *Proceedings of the National Academy of Sciences* 95 (1998); p. 154, 158 Charles Darwin, *Journal of researches into the natural history and geology of the countries visited during the voyage of H.M.S. Beagle round the world, under the Command of Capt. Fitz Roy, R N* 2nd ed. (London, 1845); p. 162 A. R. Wallace, *Australasia* (London, 1893); p. 167 Jules Verne, *Twenty Thousand Leagues under the Sea* (London, 1870); p. 171 Carl Sagan, *Cosmos* (London, 1980); p. 177 Charles Darwin, *On the Origin of Species* (london, 1859); p. 181 G. G. Simpson, *The Meaning of Evolution* (New Haven, 1949); p. 184 J. B. S. Haldane, *On Being the Right Size*, written in 1927 (Oxford, 1985); p. 188 Walt Whitman, 'Song of Myself', 1891; p. 190 Galileo Galilei, 1638, trans. in *Dialogues Concerning Two New Sciences* (London, 1914); p. 193 Aristotle, *On the Gait of Animals*; p. 202 Corey Ford, *Every Dog Should Have a Man: The Care and Feeding of Dog's Best Friend* (New York, 1952); p. 205 Charles Darwin, *On the Origin of Species*, 5th ed. (London, 1869); p. 210 Heathcote Williams, *Whale Nation* (London, 1988); p. 217, 225 Charles Darwin, *On the Origin of Species* (London, 1859); p. 233 Charles Darwin, *The Descent of Man and Selection in Relation to Sex* (London, 1871), pp. 218 and 449; p. 237 Dr Brewster M. Higley, 'My Western Home' (Kansas, 1873); p. 241 J. H. Fabré, *Social Life in the Insect World* (London, 1911); p. 244 Charles Darwin, *The Expression of the Emotions in Man and Animals* (London, 1872); p. 248 Karl von Frisch, 'Decoding the language of the bee', *Science* 185 (1974), 663–68; p. 257 Svante Arrhenius, 'On the influence of carbonic acid in the air upon the temperature of the ground', *Philosophical Magazine and Journal of Science* 41(1896), 237–76; p 262 President Hu Jintao, Asia-Pacific Economic Co-operation Summit, September 2007; p. 266 Michael Davis, *Late Victorian Holocausts. El Niño Famines and the Making of the Third World* (London & New York, 2001); p. 270 Svante Arrhenius, 'On the influence of carbonic acid in the air upon the temperature of the ground', *Philosophical Magazine and Journal of Science* 41(1896), 237–76; R. H. Yolken & E. F. Torrey, *Beasts of the Earth: Animals Humans and Disease* (New Brunswick, 2005); p. 278 E. O. Wilson, Biophilia (Cambridge, MA, 1984); p. 282 Nebojsa Nakicenovic, 'Dynamics of Global Energy Transitions' in Modeling the Oil Transition: A Summary of the Proceedings of the DOE/EPA Workshop on the Economic and Environmental Implications of Global Energy Transitions, 2007; p. 286 WWF 2006 Living Planet Report (www.panda.org); p. 288 A. Damasio, *Descartes' Error: Emotion, Reason and the Human Brain* (New York, 1994).

Index

Page numbers in *italics* refer to illustrations